Springer Proceedings in Physics

Volume 165

More information about this series at http://www.springer.com/series/361

Joachim Peinke · Gerrit Kampers
Martin Oberlack · Marta Wacławczyk
Alessandro Talamelli
Editors

Progress in Turbulence VI

Proceedings of the iTi Conference on Turbulence 2014

Springer

Editors
Joachim Peinke
Institute of Physics and ForWind
University of Oldenburg
Oldenburg
Germany

Gerrit Kampers
Institute of Physics and ForWind
University of Oldenburg
Oldenburg
Germany

Martin Oberlack
Department of Mechanical Engineering
TU Darmstadt
Darmstadt
Germany

Marta Wacławczyk
Department of Mechanical Engineering
TU Darmstadt
Darmstadt
Germany

Alessandro Talamelli
Department of Industrial Engineering, Alma Mater Studiorum
Università di Bologna
Forlì
Italy

ISSN 0930-8989 ISSN 1867-4941 (electronic)
Springer Proceedings in Physics
ISBN 978-3-319-29129-1 ISBN 978-3-319-29130-7 (eBook)
DOI 10.1007/978-3-319-29130-7

Library of Congress Control Number: 2016930819

Printed on acid-free paper

This Springer imprint is published by SpringerNature
The registered company is Springer International Publishing AG Switzerland

Final Farewell to Konrad Bajer

Konrad graduated from the Physics Faculty at the University of Warsaw in 1980. His Master's thesis was concerned with exact solutions of the Einstein equations. He then became interested in fluid dynamics and started to work at the Institute of Geophysics, which was then run by Prof. Krzysztof Haman.

Konrad's personality and his scientific attitude were fully formed during his stay at Trinity College, Cambridge, where he worked on his Ph.D. thesis in the years 1984–1988 under the supervision of the distinguished mathematician Prof. Keith Moffatt. Their collaboration turned into a lifelong friendship, which also resulted in a number of joint papers of high scientific quality in magnetohydrodynamics and the fundamentals of fluid dynamics. One of these, concerned with Stokes flows, has been widely cited and recognized and is considered by the co-author Prof. Moffatt as one of his best papers.

Konrad Bajer, May 2011. Photo by Victor L'vov

In both environments, at Trinity College, Cambridge, and at the Physics Faculty UW, Warsaw, Konrad created a friendly group of collaborators and students, with whom he developed the theory of classical and quantum vortices and of turbulence. He always listened attentively to young colleagues, who quickly became his friends, and his interactions were characterized by the great care and effort that he took to help them in their scientific and social development. He used without hesitation his scientific contacts all over the world to promote the career development of those he considered talented and promising.

His continuous efforts over 30 years, developing collaborations, teaching younger colleagues, helping them in their development, keeping active connections with numerous scientific institutions, and finally in conference organization, were recognized by the international community through the election of Konrad to the Congress Committee of the International Union of Theoretical and Applied Mechanics (IUTAM). This was a recognition not only of his scientific achievements (with over 40 papers in renowned scientific journals), but also of the great success of the 13th European Turbulence Conference, which he organized in 2011 in the wonderful setting of the Old Library of the University of Warsaw, which attracted over 400 participants, and which was followed by a series of historical lectures by distinguished scientists in the field of turbulence from all over the world.

Recently, Konrad has once again demonstrated his openness to new scientific areas, through his work on the application of physics to the theory and practice of combustion and gasification of biomass. His inter-institutional seminar on this subject has been very popular over recent years and has gathered together students of physics, mathematics, and several faculties of the Warsaw University of Technology, working on different aspects of combustion and gasification. His work in this field, which was new for him and the entire scientific community in Warsaw, was also recognized at international level.

Konrad always attached great importance to popularization of science both through very carefully prepared lectures for students and through elucidating scientific problems for non-specialists. He often took part in programs popularizing science in radio and television. He actively promoted the idea of the Copernicus Science Centre from the moment of its foundation, and until the end he worked on its Programme Council. His death has cast a shadow on the development of many new ideas and projects, one of these being the proposed Programme on Physics of Growth and Form in 2017 at the Isaac Newton Institute in Cambridge, for which he was a member of the International Advisory Committee.

Konrad has always been a very friendly person, helpful, warm, and open in the way he treated his colleagues at work and all the people he had discussions with. We will keep him like that in our memories forever.

Marek Dudyński
Marek Trippenbach

Preface

The iTi has become an established biannual conference taking place in the years between the ETC—(European Turbulence Conference) and TSFP—(Turbulence and Shear Flow Phenomena) conferences. With 50–100 participants, the iTi conference places value on the discussions and personal contacts in the location of the beautiful town of Bertinoro in nothern Italy. The size of the conference allows to have no parallel sessions and to give time to focus on special topics. The content-related focus areas of the conference are the interdisciplinary aspects of turbulence, defining our abbreviation iTi—interdisciplinary turbulence initiative.

In the 6th iTi conference 96 scientists from 19 different countries came together. In total there were 86 contributions, from which 53 were presented as talks, with 5 invited talks, covering a wide range of aspects of current turbulence research. Advances in the basics of understanding and modeling turbulence as well as practical implications like the control of turbulence were addressed.

Furthermore, it is a tradition of the iTi to organize a one day workshop before the conference. The special topic of this edition is turbulent flows generated by multi-scale excitations.

The content of the 6th iTi conference is documented in this volume by 54 contributions of participants. All contributions were thoroughly reviewed by external reviewers, to whom we want to express our thanks for their valuable and important work.

The present volume is dedicated to the memory of Prof. Konrad Bajer who passed away in Warsaw on August 29, 2014.

Oldenburg, Germany	Joachim Peinke
Oldenburg, Germany	Gerrit Kampers
Darmstadt, Germany	Martin Oberlack
Darmstadt, Germany	Marta Wacławczyk
Forlì, Italy	Alessandro Talamelli

Contents

Part II Wall Bounded Flows

Part I
Theory

Uncertainties in Turbulence

Arkady Tsinober

Abstract We argue that in order to reduce the inherent uncertainties caused by low-dimensional approaches (LES etc.) to treat turbulent flows at high Re, not accessible to DNS of NSE, it is necessary to study the inherent nonlocal (Tsinober, An informal conceptual introduction to turbulence, 2009 [6], Tsinober, The essence of turbulence, with emphasis on issues of paradigmatic nature, 2013 [7]) properties and relations, manifested in rich direct and bidirectional coupling of large/resolved and small/unresolved scales via non-trivial experiments at high Re. We show this by evidence from those with pointwise access to the full tensor of velocity derivatives.

1 Introduction

Uncertainty and its quantificationn is a generic issue in science, technology and applications and is a central concern in modeling of all complex, highly-dimensional systems [2], including systems governed by deterministic laws, as classical NSE turbulence. The apparently random behavior of turbulence in incompressible media is a manifestation of deterministic law of nature, adequately described by NSE. Nevertheless, turbulence is considered traditionally as containing inherent uncertainty, being chaotic. An important point is that complex (*chaotic*) behavior makes the law neither probabilistic nor indeterminate. The lack of predictability (and uncertainty) in chaotic deterministic systems is related to what we, human beings, are able to observe, analyze and compute, i.e., it is in a sense subjective, independently of whether this lack of predictability is due to sensitive dependence on initial conditions, or not. In other words, the common blame of these to be uncertain because they are chaotic [5] is a misinterpretation of chaotic behavior—"chaotic" and "uncertain" are far from being synonymous.

As concerns modeling, the origin of uncertainty is not in the behavior of real systems, such as described by NSE, but in their simplified, low dimensional models, due to finite computing resources, and underestimation of the role of unresolved

A. Tsinober (✉)
School of Mechanical Engineering, Tel Aviv University, Tel Aviv, Israel
e-mail: tsinober@eng.tau.ac.il

J. Peinke et al. (eds.), *Progress in Turbulence VI*,
Springer Proceedings in Physics 165, DOI 10.1007/978-3-319-29130-7_1

scales: the belief that the unresolved scales and processes associated with them can be adequately represented by relatively simple (e.g., diffusive-like) deterministic formulae [4] or other "parameterizations", representing key processes in climate models without resolving them [3]. In such an approach the small/unresolved scales are essentially abused assuming that they are 'slaved' to the explicitly treated part of the flow and serving mostly as a passive sink of energy. This is a major misconception and oversimplification due to neglecting the inherently-nonlocal nature of turbulence. This is the main issue in the sequel. We bring several examples, concerning nontrivial impact of dissipative processes on the conventionally-defined inertial range (CDIR), based on field experiments at large Re in atmospheric surface layer.

2 On *What Is Nonlocality* and a Few Words on *Locality*

Nonlocality is a broad, but much neglected issue. It is a generic property of turbulent flows independently of mean shear or other external factors.

In the context of modeling the nonlocality is manifested in a rich direct and bidirectional coupling between large/resolved and small/unresolved scales and comprises an essential part of the complex interaction between the multitude of the degrees of freedom in turbulent flows. On other aspects see [6, 7]. There is some conflict between the common view on dominance of local effects, though there are no rigorous grounds for it, and the fact that turbulence is more than suspect of being inherently nonlocal, and continuing attempts to single out "locality" and related "simpler" properties by some hypotheses, etc. The main rationale for such attempts becomes clear because nonlocality is among the main reasons of the absence of a sound theory of turbulence, based on first principles, see [6, 7] for references.

There are other reasons why non-locality is bad. With nonlocality it is far from trivial, if not impossible, to use the experimental data—all limited in space and time—for "validation" of theoretical developments for, e.g., homogeneous flows, i.e., in "infinite" domains. Also, locality is necessary for the "physical foundation" of LES modeling of turbulence, etc. Though even just looking at the equations for the small/unresolved scales it is straightforward to realize that these scales depend on the large/resolved scales via nonlinear space and history-dependent functionals, i.e., essentially non-local both spatially and temporally and also bidirectionally, which makes low-dimensional description pretty problematic. So it is unlikely—and there is accumulating evidence for this—that relations between them (such as "energy flux") would be approximately local in contradiction to K41 hypotheses and surprisingly numerous (but futile) efforts to support their validity.

3 Is the Kolomogorov 4/5 Law a Pure Inertial Relation and Is the Inertial Range a Well-Defined Concept

We start with the above as an outstanding example of interpretational abuse. One of the popular arguments for the existence of the CDIR at large Re is the Kolmogorov 4/5 law, which is the consequence of the NSE, for a review of theoretical results see [6]:

$$S_3 = -(4/5)\langle\varepsilon\rangle r + 6\nu dS_2/dr; \;\; S_n = \langle(\Delta u)^n\rangle, \tag{1}$$

and in which the last term is negligible at large Re. There is an experimental confirmation that at Re up to $\mathrm{Re}_\lambda \approx 10^4$ the term $6\nu dS_2/dr$ is negligible, and the relation $S_3 = -(4/5)\langle\varepsilon\rangle r$ does hold in a broad range of scales exceeding three decades at $\mathrm{Re}_\lambda \approx 10^4$, [6]. Thus one gets an impression that the 4/5 is a purely inertial relation at large Re. However, this would be true if the S_3, or more precisely, just velocity increments $\Delta u \equiv [\mathbf{u}(\mathbf{x}+\mathbf{r}) - \mathbf{u}(\mathbf{x})] \cdot \mathbf{r}/r$, would not contain non-negligible contributions from dissipative events, i.e., in reality what is called IR is **assumed to be purely inertial and having no contribution from viscous events**. Indeed, computing Δu one encounters also large and even very large instantaneous dissipation at the ends $(\mathbf{x}, \mathbf{x}+\mathbf{r})$. In other words, the second Kolmogorov hypothesis involves a strong assumption that the dissipative events, such that at least at one of the ends $(\mathbf{x}, \mathbf{x}+\mathbf{r})$ the instantaneous dissipation $\varepsilon > q\langle\varepsilon\rangle$ with $q > 1$ do not matter for the statistics of velocity increments. To (dis)prove this one needs access to instantaneous dissipation at large Re. Indeed, looking at Fig. 1 it is seen that there exists a substantial number

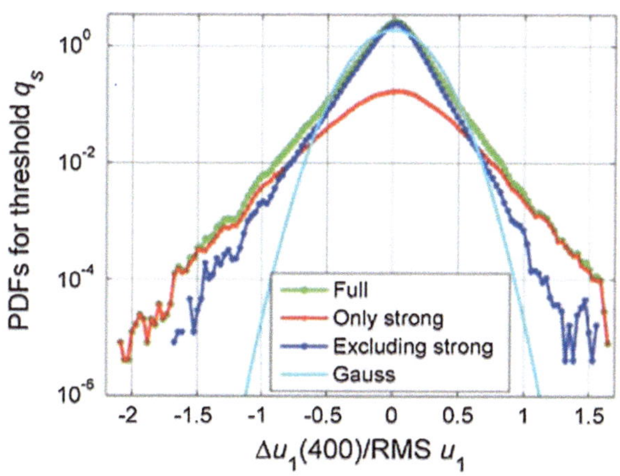

Fig. 1 Histograms of the increments of the longitudinal velocity component for the full data and the same data, in which the strong dissipative events (see text) were removed. $r/\eta = 400$ and threshold $q = 3$; field experiment, $\mathrm{Re}_\lambda \approx 10^4$; the lower edge of the inertial range is about $r/\eta = 40$. Note that the PDFs with removed strong dissipative events (*dark blue curve*) are not close to the Gaussian curve. An important feature is that dissipative events literally live within the CDIR in turbulence at high Reynolds numbers, i.e., nonlocality is a broader issue and does not necessarily involve scale separation, such as in long-range interactions, see [6, 7] for references

of dissipative (!) events (DE), living in the conventionally-defined dissipative range (CDDR) with essential contribution to the PDF of velocity increments in the CDIR at high Reynolds numbers, $\mathrm{Re}_\lambda \approx 10^4$. The key feature is that this contribution is largest to the tails of the PDF of velocity increments, i.e., removal of the strong dissipative events results in narrowing of the tails in the PDFs of Δu. Thus the CDIR is an ill-defined concept. In particular, this means that the neglected viscous term in the Kolmogorov 4/5 law does not contain all the viscous contributions. Those, present in the structure function S_3 itself, remain and keep the 4/5 law precise: without the dissipative events, just mentioned, the 4/5 law does not hold! In this sense the 4/5 law is not a pure inertial law even at $\mathrm{Re}_\lambda \approx 10^4$.

Indeed, strong dissipative events do contribute to the 4/5 law, and removing them leads to an increase of the scaling exponent above unity. It is noteworthy that the contribution of the dissipative events in the 4/5 law at large Reynolds numbers is not small in spite of considerable cancellation between negative and positive events.

Thus—contrary to common view—the 4/5 law is not a pure inertial relation at large Re. The important implication of more general nature is that nonlinear interactions (NI) are not synonymous to purely inertial ones. NI consist of purely inertial ones with an essential contribution from the viscous and cross-interactions. In case of the third order functions and energy fluxes these interactions are constrained by the 4/5 law. The consequence is that the purely inertial and dissipative events are adjusting to keep some quantities, as the total dissipation or the energy flux, approximately constant at large Re. The constraint, responsible for this adjustment, is just the 4/5 law or, more generally, the NSE. For more on this issue see Sect. 10.2 in [6].

As concerns structure functions of higher order, $p > 3$, there is no simple constraint as the 4/5 law for S_3. The consequence is that the mentioned above dissipative events are responsible for what is called anomalous scaling for S_p, $p > 3$. This is clearly seen from Fig. 2. Thus the anomalous scaling is not an attribute of the CDIR, and the latter is not a well-defined concept, just like "cascade" and the CDDR.

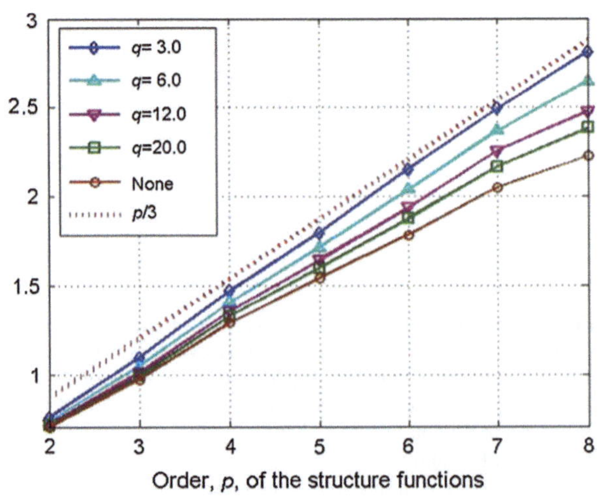

Fig. 2 Scaling exponents of structure functions at $\mathrm{Re}_\lambda \approx 10^4$ for the longitudinal velocity component corresponding to the full data and the same data, in which the strong dissipative events with various thresholds q were removed. With $q = 3$ the higher order structure functions ($p > 3$) exhibit Kolmogorov scaling $p/3$. See [6, 7] for references

It is noteworthy that handling the issue with strong dissipative events makes sense not only due to their dissipative nature, but also because these events represent strong strain and all observed nonlinearities appear to be much stronger in the strain-dominated regions, rather than in regions with concentrated vorticity [6, 7]. Finally we would like to mention that the data used were somewhat spatially underresolved. For quantitative improvement one needs a probe with sub-Kolmogorov resolution. Such a probe was recently successfully tested in laboratory, [1].

4 Concluding Remarks

Among more general contexts nonlocality leads to nontrivial paradigmatic consequences, such as illposedness of the concepts of cascade and local equilibrium, conventional inertial and dissipative ranges, the anomalous scaling being not the attribute of the (nonexistent) conventional inertial range, the 4/5 law as not a purely inertial relation, and importance of the nature of dissipation on the processes in the CDIR, [6, 7]. In the context of interest it is manifested in direct and bidirectional coupling/interaction of large/resolved and small/unresolved scales.

Consequently, low-dimensional approaches are inherently uncertain and generically deficient as, e.g., abusing and missing an essential part of physics and dynamics resided mostly with small/unresolved scales associated with, e.g., such fundamental properties of turbulence as being an essentially rotational and dissipative rather than purely inertial phenomenon.

References

1. Borisenkov, Y.: Micro-machined super miniature hot film multi-array probe for measurements in turbulent flows with sub-kolmogorov resolution. Ph.D. thesis, Tel-Aviv University (2014)
2. Boumans, M., Hon, G., Petersen, A.: Error and Uncertainty in Scientific Practice. Pickering & Chatto, London (2013)
3. McFarlane, N.: Wiley Interdisciplinary Reviews—Climate Change, pp. 482–497 (2011)
4. Palmer, T.N.: Europhys. News **36**, 42–46 (2005)
5. Palmer, T.N., Hardaker, P.J.: Philos. Trans. R. Soc. **A369**, 4681–4684 (2011)
6. Tsinober, A.: An Informal Conceptual Introduction to Turbulence. Springer, Dordrecht (2009)
7. Tsinober, A.: The Essence of Turbulence, with Emphasis on Issues of Paradigmatic Nature. Springer, Dordrecht (2013)

Complete Description of Turbulence in Terms of Hopf Functional and LMN Hierarchy: New Symmetries and Invariant Solutions

Marta Wacławczyk

Abstract This paper deals with two methods for the full statistical description of turbulent field, namely the Lundgren–Monin–Novikov hierarchy (Lundgren, Phys Fluids, 10:969–975 1967, [5]) for the multipoint probability density functions (PDFs) of velocity and Hopf functional equation for turbulence (Hopf, J Ration Mech Anal, 1:87–122 1952, [2]). These equations are invariant under certain transformations of dependent and independent variables, so called symmetry transformation. The importance of these symmetries for the turbulence theory and modelling is discussed.

1 Introduction

Although the phenomenon of turbulence is described by deterministic Navier–Stokes equations, due to its sensitivity to small variations in the initial and boundary condition the turbulent field may be treated as a stochastic field. For its full description, all multipoint statistics of arbitrary order should be known. With respect to turbulence research three complete descriptions of turbulence are known, namely the infinite hierarchy of the multi-point correlation equations (so-called Friedmann–Keller (FK) hierarchy, [3]), the infinite hierarchy of the multipoint PDF equations (Lundgren–Monin–Novikov (LMN) equations, [5]) and finally the Hopf functional approach, [2]. The two latter approaches will be discussed below.

The n-point velocity PDF $f_n = f_n(\mathbf{v}_{(1)}, \ldots, \mathbf{v}_{(n)}; \mathbf{x}_{(1)}, \ldots, \mathbf{x}_{(n)}, t)$ carries information about all statistics up to n-point statistics of infinite order which can be calculated from the PDF by integration over the sample space variables $\mathbf{v}_{(1)}, \ldots, \mathbf{v}_{(n)}$, for example

$$\langle U_{i_{(1)}}(\mathbf{x}_{(1)}, t) \cdots U_{i_{(n)}}(\mathbf{x}_{(n)}, t)\rangle = \int v_{i_{(1)}} \ldots v_{i_{(n)}} f_n \mathrm{d}\mathbf{v}_{(1)} \ldots \mathrm{d}\mathbf{v}_{(n)}. \tag{1}$$

M. Wacławczyk (✉)
Chair of Fluid Dynamics, Department of Mechanical Engineering, TU Darmstadt, Otto-Berndt Str. 2, 64287 Darmstadt, Germany
e-mail: marta.waclawczyk@igf.fuw.edu.pl

J. Peinke et al. (eds.), *Progress in Turbulence VI*,
Springer Proceedings in Physics 165, DOI 10.1007/978-3-319-29130-7_2

E. Hopf introduced another very general approach to the description of turbulence. He considered the case where the number of points in PDF goes to infinity, so that the probability density function becomes a probability density functional $F([\mathbf{v}(\mathbf{x})]; t)$ where instead of the vector of sample space variables $\mathbf{v}_{(1)}, \ldots, \mathbf{v}_{(n)}$ at points $\mathbf{x}_{(1)}, \ldots, \mathbf{x}_{(n)}$ one deals with a continuous set of sample space variables $\mathbf{v}(\mathbf{x})$. It is more convenient to consider a functional Fourier transform of the probability density functional, called the characteristic functional than $F([\mathbf{v}(\mathbf{x})]; t)$ itself

$$\Phi([\mathbf{y}(\mathbf{x})]; t) = \int \mathrm{e}^{i(\mathbf{y},\mathbf{v})} F([\mathbf{v}(\mathbf{x})]; t) D\mathbf{v}(\mathbf{x}) = \langle \mathrm{e}^{i(\mathbf{y},\mathbf{U})} \rangle \tag{2}$$

where the integration is performed with respect to the probability measure $F([\mathbf{v}(\mathbf{x})];t)$ $D\mathbf{v}(\mathbf{x})$ and $(\mathbf{y}, \mathbf{v}) = \int_G y_i v_i \mathrm{d}\mathbf{x}$ is a scalar product of two vector fields. With this definition moments of the velocity can be calculated as the functional derivatives of the characteristic functional at the origin [2]

$$\left. \frac{\delta^n \Phi([\mathbf{y}(\mathbf{x})], t)}{\delta y_{i_{(1)}}(\mathbf{x}_{(1)}) \cdots \delta y_{i_{(n)}}(\mathbf{x}_{(n)})} \right|_{\mathbf{y}=0} = i^n \langle U_{i_{(1)}}(\mathbf{x}_{(1)}, t) \cdots U_{i_{(n)}}(\mathbf{x}_{(n)}, t) \rangle. \tag{3}$$

Hence, Φ may be treated as a functional analogue of the characteristic function Φ_n for $n \to \infty$, defined, in the probability theory, as the inverse Fourier transform of f_n. E. Hopf derived evolution equation for the characteristic functional, cf. [2]. It is only one equation (not a hierarchy) which embodies the statistical properties of the fluid flow in a very concise form.

The objectives of the present work is to discuss the classical and new statistical Lie symmetries that were first found for the FK hierarchy [6] and are also present in the LMN hierarchy [9] and find corresponding symmetries for the Hopf equation. Lie one-point symmetry transformation is such transformation of the independent and dependent variables, $\mathbf{x}$ and $\mathbf{y}$, respectively, which does not change the functional form of a considered equation [1]

$$F(\mathbf{x}, \mathbf{y}, \underset{1}{\mathbf{y}}, \underset{2}{\mathbf{y}}, \ldots, \underset{p}{\mathbf{y}}) = 0 \quad \Leftrightarrow \quad F(\mathbf{x}^*, \mathbf{y}^*, \underset{1}{\mathbf{y}}^*, \underset{2}{\mathbf{y}}^*, \ldots, \underset{p}{\mathbf{y}}^*) = 0 \tag{4}$$

where the transformed variables $\mathbf{x}^* = g(\mathbf{x}, \mathbf{y}; \varepsilon)$ and $\mathbf{y}^* = h(\mathbf{x}, \mathbf{y}; \varepsilon)$ are functions of $\mathbf{x}$ and $\mathbf{y}$ and depend on a group parameter ε. The transformations can also be written in infinitesimal forms after a Taylor series expansion about $\varepsilon = 0$: $\mathbf{x}^* = \mathbf{x} + \xi(\mathbf{x}, \mathbf{y})\varepsilon + O(\varepsilon^2)$ and $\mathbf{y}^* = \mathbf{y} + \eta(\mathbf{x}, \mathbf{y})\varepsilon + O(\varepsilon^2)$. It follows from the Lie's first theorem that knowing the infinitesimal forms ξ and η uniquely determines the global form of the group transformation $g(\mathbf{x}, \mathbf{y}; \varepsilon)$ and $h(\mathbf{x}, \mathbf{y}; \varepsilon)$. With the use of infinitesimals invariant solutions of the considered Eq. (4) may be derived [1]. In fluid mechanics these solutions often represent attractors of the instantaneous fluctuating solutions of the Navier–Stokes equations, i.e. the characteristic turbulent scaling laws.

From the Lie symmetry analysis of the LMN hierarchy it followed that, surprisingly, the new symmetries are connected with intermittent laminar/turbulent flows [9]. The outcome of the symmetry analysis are invariant solutions for turbulence statistics and new possibilities to improve turbulence closures, such that invariance under the whole set of symmetries is accounted for.

2 Symmetries of the LMN Hierarchy

The LMN hierarchy, derived in Ref. [5] is an infinite set of equations for the multipoint PDFs where in the nth equation the $n+1$-point PDF is present

$$\begin{aligned} &\frac{\partial f_n}{\partial t} + \sum_{k=1}^{n} v_{i(k)} \frac{\partial f_n}{\partial x_{i(k)}} \\ &= -\frac{1}{4\pi} \sum_{k=1}^{n} \frac{\partial}{\partial v_{i(k)}} \int\int \left(\frac{\partial}{\partial x_{i(k)}} \frac{1}{\left|\mathbf{x}_{(k)} - \mathbf{x}_{(n+1)}\right|} \right) \left(v_{j(n+1)} \frac{\partial}{\partial x_{j(n+1)}} \right)^2 f_{n+1} \mathrm{d}\mathbf{v}_{(n+1)} \mathrm{d}\mathbf{x}_{(n+1)} \\ &- \sum_{k=1}^{n} \frac{\partial}{\partial v_{i(k)}} \left[\lim_{|\mathbf{x}_{(n+1)} - \mathbf{x}_{(k)}| \to 0} \nu \frac{\partial^2}{\partial x_{j(n+1)} \partial x_{j(n+1)}} \int v_{i(n+1)} f_{n+1} \mathrm{d}\mathbf{v}_{(n+1)} \right]. \end{aligned} \tag{5}$$

Symmetries of the LMN hierarchy were investigated in Ref. [9]. Therein, it was shown that the hierarchy is invariant under the classical symmetries of the Navier–Stokes, equations, in particular, time and space translations, Galilean invariance and, for $\nu = 0$ two scaling groups

$$\bar{T}_2: \quad t^* = t, \mathbf{x}^*_{(l)} = \mathrm{e}^{a_2}\mathbf{x}_{(l)}, \mathbf{v}^*_{(l)} = \mathrm{e}^{a_2}\mathbf{v}_{(l)}, \ f^*_n = \mathrm{e}^{-3na_2} f_n, \tag{6}$$

$$\bar{T}_3: \quad t^* = \mathrm{e}^{a_3}t, \mathbf{x}_{(l)} = \mathbf{x}_{(l)}, \mathbf{v}^*_{(l)} = \mathrm{e}^{-a_3}\mathbf{v}_{(l)}, \ f^*_n = \mathrm{e}^{3na_3} f_n. \tag{7}$$

with $l = 1, \ldots, n$, which for $\nu \neq 0$ reduce to one scaling group. Moreover, it was shown that the LMN hierarchy is invariant under new statistical symmetry groups observed in the Friedrich–Keller hierarchy as the scaling and translations of multipoint velocity correlations cf. Ref. [6]. In the LMN approach the symmetries have the following forms

$$f^*_n = \delta(\mathbf{v}_{(1)}) \cdots \delta(\mathbf{v}_{(n)}) + \mathrm{e}^{a_s}(f_n - \delta(\mathbf{v}_{(1)}) \cdots \delta(\mathbf{v}_{(n)})). \tag{8}$$

for the scaling and

$$f^*_n = f_n + \psi(\mathbf{v}_{(1)})\delta(\mathbf{v}_{(1)} - \mathbf{v}_{(2)}) \cdots \delta(\mathbf{v}_{(1)} - \mathbf{v}_{(n)}). \tag{9}$$

for the translation symmetry, where ψ is a function such that $\int \psi(\mathbf{v})\mathrm{d}\mathbf{v} = 0$ and δ is the Dirac delta function.

In Ref. [9] the particular case of a plane Poiseuille channel flow was considered. Both the scaling (8) and the translation symmetry (9) were taken into account. For the channel flow the symmetries have slightly different form. First, the scaling symmetry (8) transforms a PDF f_n of a turbulent signal into the PDF with delta function at $\mathbf{v} = 0$. Next, the translation symmetry with the function ψ defined as $\psi = (1 - \mathrm{e}^{a_s})[\delta(\mathbf{v}_{(1)} - \mathbf{U}_L(\mathbf{x}_{(1)})) - \delta(\mathbf{v}_{(1)})]$ where $\mathbf{U}_L(\mathbf{x}_{(k)}) = [U_0(1 - x^2_{2(k)}/H^2), 0, 0]$, U_0 is the velocity at the centerline and H is the channel half-width, transforms PDF into

$$f_n^* = \mathrm{e}^{a_s} f_n + (1 - \mathrm{e}^{a_s})\delta(\mathbf{v}_{(1)} - \mathbf{U}_L(\mathbf{x}_{(1)})) \dots \delta(\mathbf{v}_{(n)} - \mathbf{U}_L(\mathbf{x}_{(n)})). \tag{10}$$

To sum up, both symmetries, scaling and translation transform a PDF of a turbulent signal into the PDF of an intermittent laminar-turbulent flow. This would correspond to a situation where the flow in a channel is induced by a certain pressure difference ΔP, such that the resulting Reynolds number $Re = U_b H/\nu$ where U_b is the bulk velocity, is close to the critical value Re_{cr}. For certain range of Re both, laminar or turbulent solutions are possible with certain probability, leading to the PDF of the form given in Eq. (10). As discussed in Ref. [9] such interpretation of the symmetries has important consequences. First, it leads to certain conditions on the group parameter e^{a_s}. As this parameter is present in invariant solutions for turbulent statistics derived in [6] it may provide restrictions on the scaling parameters in these laws, such as e.g. the von Karman constant. Second, to properly describe physics, turbulence models should be invariant under the same set of symmetries as the exact equations for statistics (e.g. the FK hierarchy), hence, new symmetries should be included in these models. It may be expected that this is especially important in the case of models describing laminar-turbulent transition.

With the use of new symmetries series of invariant solutions for turbulence statistics can be derived [6]. In Ref. [9] three particular symmetries were taken into account: classical scaling of the Navier–Stokes equations, cf. (5), $y^* = \mathrm{e}^{k_2} y$, $\langle U \rangle^* = \mathrm{e}^{-k_2}\langle U \rangle$ where k_2 is an arbitrary constant, new scaling and translation symmetries of the mean velocity $\langle U \rangle^* = \mathrm{e}^{a_s}\langle U \rangle + C_1(1 - y^2/H^2)$, which follows from Eq. (10). Invariant solution can be found from the solution of the characteristic equation [6]

$$\frac{\mathrm{d}\langle U \rangle}{(a_s - k_2)\langle U \rangle + C_1(1 - y^2/H^2)} = \frac{\mathrm{d}y}{k_2 y} \tag{11}$$

which for $a_s = k_2$ gives

$$\langle U \rangle = \frac{C_1}{k_2} \ln(y) + \frac{C_1}{2k_2}\left(1 - \frac{y^2}{H^2}\right) + \mathscr{C} \tag{12}$$

where $\mathscr{C}$ is a constant. The formula above is, apparently, a sum of the turbulent and laminar velocity in the plane channel flow.

3 Symmetries of the Hopf Equation

Based on the Navier–Stokes equations, E. Hopf derived evolution equation for the characteristic functional [2]. It is only one equation (not a hierarchy) and all turbulence statistics can formally be calculated from the solution of the Hopf equation. The Hopf equation for velocity in the physical space reads [2]

$$\frac{\partial \Phi}{\partial t} = \int_R \mathrm{d}\mathbf{x}\, \widetilde{y}_k(\mathbf{x}) \left[i \frac{\partial}{\partial x_l} \frac{\delta^2 \Phi}{\delta y_l(\mathbf{x}) \delta y_k(\mathbf{x})} + \nu \nabla_x^2 \frac{\delta \Phi}{\delta y_k(\mathbf{x})} \right]. \tag{13}$$

where, in order to eliminate pressure functional Π from the equation, vector field $\widetilde{\mathbf{y}}$ such that $\mathbf{y}(\mathbf{x}) = \widetilde{\mathbf{y}}(\mathbf{x}) + \nabla\phi$ was introduced. The scalar ϕ is chosen such that $\widetilde{\mathbf{y}} = \mathbf{0}$ at the boundary B and the continuity equation is satisfied $\nabla \cdot \widetilde{\mathbf{y}} = 0$. In order to check the invariance of the Hopf functional equation under the scaling groups we first consider transformations of $n+1$-point characteristic functions. From the relation

$$\Phi_{n+1}^* = \int \mathrm{e}^{i\mathbf{v}_{(0)}^* \cdot \mathbf{y}_{(0)}^*} \ldots \mathrm{e}^{i\mathbf{v}_{(n)}^* \cdot \mathbf{y}_{(n)}^*} f_{n+1}^* \mathrm{d}\mathbf{v}_{(0)}^* \ldots \mathrm{d}\mathbf{v}_{(n)}^* \tag{14}$$

we find that the scaling symmetries (6) and (7) will hold if $\mathbf{y}_{(i)}^* = \mathrm{e}^{-k_2}\mathbf{y}_{(i)}$ and $\mathbf{y}_{(i)}^* = \mathrm{e}^{k_3}\mathbf{y}_{(i)}$ for each i, as in such a case the exponent $\mathbf{v}_{(i)}^* \cdot \mathbf{y}_{(i)}^* = \mathbf{v}_{(i)} \cdot \mathbf{y}_{(i)}$ remain unchanged and using (6) we obtain $\Phi_n^* = \Phi_n$. The same holds for the second scaling group (7), i.e. the n-point characteristic function is not transformed $\Phi_n^* = \Phi_n$. We expect that the same should hold for the limit $n \to \infty$ and the characteristic functional. In this case instead of the discrete kth variable $y_{i_{(k)}}$ we deal with $y_i(\mathbf{x})\mathrm{d}\mathbf{x}$. The sums are replaced by integrals in the continuum limit and hence $y_i\mathrm{d}\mathbf{x}$ should scale as $y_{i_{(k)}}$ in the discrete case, i.e. $y_i^*\mathrm{d}\mathbf{x}^* = \mathrm{e}^{-k_2} y_i \mathrm{d}\mathbf{x}$. Because $\mathrm{d}\mathbf{x}$ scales as $\mathrm{d}\mathbf{x}^* = \mathbf{x}\mathrm{e}^{3k_2}$ it follows that $y_i^* = \mathrm{e}^{-4k_2} y_i$.

To sum up, it can be shown that the Hopf functional equation (13) for $\nu = 0$ is invariant under the following transformation of variables

$$\bar{T}_2: \ \Phi^* = \Phi, \quad \mathbf{x}^* = \mathrm{e}^{k_2}\mathbf{x}, \quad t^* = t, \quad y_i^*\mathrm{d}\mathbf{x}^* = \mathrm{e}^{-k_2} y_i \mathrm{d}\mathbf{x}, \quad \mathbf{y}^* = \mathrm{e}^{-4k_2}\mathbf{y}, \tag{15}$$

$$\bar{T}_3: \ \Phi^* = \Phi, \quad \mathbf{x}^* = \mathbf{x}, \quad t^* = \mathrm{e}^{k_3} t, \quad y_i^*\mathrm{d}\mathbf{x}^* = \mathrm{e}^{k_3} y_i \mathrm{d}\mathbf{x}, \quad \mathbf{y}^* = \mathrm{e}^{k_3}\mathbf{y}. \tag{16}$$

For $\nu \neq 0$ instead of two scaling groups we obtain one scaling.

The difficulty is connected with the Galilei invariance which is broken. The functional Φ transforms under the Galilean invariance as follows [4]

$$\Phi^* = \langle \mathrm{e}^{i\int \mathbf{U}^*(\mathbf{x},t)\cdot\mathbf{y}^*(\mathbf{x})\mathrm{d}\mathbf{x}^*} \rangle = \langle \mathrm{e}^{i\int \mathbf{U}(\mathbf{x},t)\cdot\mathbf{y}(\mathbf{x})\mathrm{d}\mathbf{x}} \rangle \mathrm{e}^{i\int \mathbf{y}(\mathbf{x})\cdot\mathbf{U}_0\mathrm{d}\mathbf{x}}. \tag{17}$$

Hence, the transformation of Φ may be written as $\Phi^* = C([\mathbf{y}(\mathbf{x})])\Phi$, where $C([\mathbf{y}(\mathbf{x})])$ is a functional

$$C([\mathbf{y}(\mathbf{x})], t) = \mathrm{e}^{i\int \mathbf{y}(\mathbf{x})\cdot\mathbf{U}_0\mathrm{d}\mathbf{x}} \tag{18}$$

The nth derivative of the transformed functional Φ^* at $\mathbf{y}=0$ gives

$$\left.\frac{\delta^n \Phi^*}{\delta y_{i_{(0)}}(\mathbf{x})\cdots\delta y_{i_{(n-1)}}(\mathbf{x})}\right|_{\mathbf{y}=0} = i^n(U_{i_{(0)}}(\mathbf{x},t)+U_{0i_{(0)}})\cdots(U_{i_{(n-1)}}(\mathbf{x},t)+U_{0i_{(n-1)}}) \tag{19}$$

as expected for Galilean invariance. In the Galilean transformation, the space derivatives in Eq. (13) transform as $\partial/\partial x_i^* = \partial/\partial x_i$ and the integral over infinite space $\int \mathrm{d}\mathbf{x}^* = \int \mathrm{d}\mathbf{x}$. As the variables $\mathbf{y}^*=\mathbf{y}$, also the functional derivative remains unchanged $\delta/\delta(y_i(\mathbf{x}))^* = \delta/\delta y_i(\mathbf{x})$. The derivative $\partial/\partial t$ can be presented as

$$\frac{\partial}{\partial t} = \frac{\partial t^*}{\partial t}\frac{\partial}{\partial t^*} + \frac{\partial x_i^*}{\partial t}\frac{\partial}{\partial x_i^*} + \int \frac{\partial y^* \mathrm{d}\mathbf{x}^*}{\partial t}\frac{\delta}{\delta y(x)^*} = \frac{\partial}{\partial t^*} + U_{0i}\frac{\partial}{\partial x_i}, \tag{20}$$

The transformed functional equation (13) reads

$$C([\mathbf{y}(\mathbf{x})])\frac{\partial \Phi}{\partial t} = \int y_k(\mathbf{x})\left[i\frac{\partial}{\partial x_l}\frac{\delta^2 C([\mathbf{y}(\mathbf{x})])\Phi}{\delta y_l(\mathbf{x})\delta y_k(\mathbf{x})} + \nu\nabla_x^2\frac{\delta C([\mathbf{y}(\mathbf{x})])\Phi}{\delta y_k(\mathbf{x})} - \frac{\partial \Pi}{\partial x_k}\right]\mathrm{d}\mathbf{x}\,. \tag{21}$$

The functional derivative of Φ^* in (21) reads $C\delta\Phi/\delta y_k(\mathbf{x}) + \Phi\delta C/\delta y_k(\mathbf{x})$ and we note that Laplacian ∇_x^2 of the second term is zero as this term is not a function of $\mathbf{x}$. Hence, the last RHS term of equation (21) inside the integral reads $C([\mathbf{y}(\mathbf{x})])\nu\nabla_x^2\delta\Phi/\delta y_k(\mathbf{x})$. Further, the second functional derivative of Φ^* reads

$$C\frac{\delta^2\Phi}{\delta y_k(\mathbf{x})\delta y_l(\mathbf{x})} + \frac{\delta\Phi}{\delta y_k(\mathbf{x})}\frac{\delta C}{\delta y_l(\mathbf{x})} + \frac{\delta\Phi}{\delta y_l(\mathbf{x})}\frac{\delta C}{\delta y_k(\mathbf{x})} + \Phi\frac{\delta^2 C}{\delta y_k(\mathbf{x})\delta y_l(\mathbf{x})}. \tag{22}$$

Again, the derivative $\partial/\partial x_l$ of the last term is zero, as it does not depend on $\mathbf{x}$. In addition we also have

$$\frac{\partial}{\partial x_k}\left[\frac{\delta\Phi}{\delta y_l(\mathbf{x})}\frac{\delta C}{\delta y_k(\mathbf{x})}\right] = \frac{\delta C}{\delta y_k(\mathbf{x})}\frac{\partial}{\partial x_l}\frac{\delta\Phi}{\delta y_l(\mathbf{x})} = 0, \tag{23}$$

where the first equality follows from the fact that the derivative of C does not depend explicitly on $\mathbf{x}$ and the second, from the continuity condition. It can be seen from the definition of the Hopf functional (2) that its derivative with respect to $y_l(\mathbf{x})$ reads $\langle iU_l(\mathbf{x},t)\exp[i\int \mathbf{U}(\mathbf{x},t)\cdot\mathbf{y}(\mathbf{x})\mathrm{d}\mathbf{x}]\rangle$, hence differentiating once again with respect to x_l is zero as $\partial U_l/\partial x_l = 0$. Finally, the transformed Hopf equation reads

$$\begin{aligned} C([\mathbf{y}(\mathbf{x})])\frac{\partial\Phi}{\partial t} &= C([\mathbf{y}(\mathbf{x})])\int \widetilde{y}_k(\mathbf{x})\left[i\frac{\partial}{\partial x_l}\frac{\delta^2\Phi}{\delta y_l(\mathbf{x})\delta y_k(\mathbf{x})} + \nu\nabla_x^2\frac{\delta\Phi}{\delta y_k(\mathbf{x})}\right]\mathrm{d}\mathbf{x} \\ &\quad + \frac{\delta C}{\delta y_l(\mathbf{x})}\int y_k(\mathbf{x})\left[i\frac{\partial}{\partial x_l}\frac{\delta\Phi}{\delta y_k(\mathbf{x})}\right]\mathrm{d}\mathbf{x} \end{aligned} \tag{24}$$

As it is seen the last RHS term does not cancel. Hence, in the integral formulation of the Hopf equation the Galilei invariance is broken.

We will next consider the transformation of the Hopf functional under the statistical symmetries, formulated for the PDF's in Eqs. (8) and (9). E. Hopf proposed to present the solution of the Hopf functional as the infinite series expansion [2]

$$\Phi = 1 + C_1 + C_2 + \cdots, \tag{25}$$

where

$$C_n = \int K_{i_{(1)}\dots i_{(n)}}(\mathbf{x}_{(1)}, \dots, \mathbf{x}_{(n)}, t) y_{i_{(1)}}(\mathbf{x}_{(1)}) \cdots y_{i_{(n)}}(\mathbf{x}_{(n)}) \mathrm{d}\mathbf{x}_{(1)} \cdots \mathrm{d}\mathbf{x}_{(n)} \tag{26}$$

with functions

$$K_{i_{(1)}\dots i_{(n)}}(\mathbf{x}_{(1)}, \dots, \mathbf{x}_{(n)}, t) = \frac{i^n}{n!} \langle U_{i_{(1)}}(\mathbf{x}_{(1)}, t) \cdots U_{i_{(n)}}(\mathbf{x}_{(n)}, t) \rangle. \tag{27}$$

If we substitute the statistical symmetries of moments into the above equations we find that the scaling symmetry of multipoint velocity correlations, cf. Ref. [6] and Eq. (8), transforms the kernel functions as

$$K^*_{i_{(1)}\dots i_{(n)}} = \mathrm{e}^{k_s} K_{i_{(1)}\dots i_{(n)}}, \tag{28}$$

hence the series expansion reads

$$\Phi^* = 1 + \mathrm{e}^{k_s} \left(\Phi^1 + \Phi^2 + \cdots \right), \tag{29}$$

or

$$\Phi^* = 1 + \mathrm{e}^{k_s} (\Phi - 1). \tag{30}$$

The translation symmetry of multipoint correlations, cf. Ref. [6] and Eq. (9), translates the kernel functions by a constant which leads to the following translation of the nth term in the Taylor series expansion

$$\Phi^*_n = \Phi_n + \int C_{i_{(1)}\dots i_{(n)}} y_{i_{(1)}}(\mathbf{x}_{(1)}) \cdots y_{i_{(n)}}(\mathbf{x}_{(n)}) \mathrm{d}\mathbf{x}_{(1)} \cdots \mathrm{d}\mathbf{x}_{(n)}. \tag{31}$$

Hence, the translation symmetry of the characteristic functional Φ can be written in the following form

$$\Phi^* = \Phi + \Psi([\mathbf{y}(\mathbf{x})]), \tag{32}$$

where Ψ is a functional such that its nth functional derivative at the origin equals $n!C_{i_{(1)}\ldots i_{(n)}}$ and its functional derivatives does not depend explicitly on x or t which makes the Eq. (13) invariant under the transformation (32).

4 Conclusions

To sum up, it was argued that all methods for the full statistical description of turbulence, namely FK hierarchy, LMN hierarchy for PDFs and the Hopf characteristic functional equations are invariant under classical scaling symmetries of Navier–Stokes equations and additionally under the set of statistical symmetries: translation and scaling. Deriving transformations of the Hopf equation equivalent to the FK symmetries is a new contribution of the present work. Through the analysis of PDF equations the statistical translation and scaling were identified in Ref. [9] as connected with the (external) intermittency. Hence, the statistical symmetries indicate the fact that solutions of Navier–Stokes equations may have different character. Such transformation could only be observed in the statistical approach, hence the statistical symmetries were not found in the Lie group analysis of the Navier–Stokes equations. With the use of statistical symmetries series of invariant solutions for turbulence statistics were obtained in Ref. [6]. It can be expected that similar, new results could be obtained for PDFs based on the symmetries of LMN hierarchy. Moreover, the invariance under new statistical symmetries could be introduced into turbulence models to improve their predictions.

The Lie group analysis of infinite hierarchies of equations, such as LMN or FK cannot be performed with the use of common computer algebra systems. Hence, the symmetries of FK equations were rather guessed than calculated and their set may not be complete. A possibility would be to apply the Lie group method to one, Hopf functional equation in order to find possibly new statistical symmetries and next derive corresponding transformations in the LMN and FK hierarchy. For this purpose, the extended Lie group method, introduced in Refs. [7, 8] could be used. This issue is the subject of the current study.

Acknowledgments Support from the DFG under project WA 3097/3-1 is gratefully acknowledged.

References

1. Bluman, G., Cheviakov, A., Anco, S.: Applications of Symmetry Methods to Partial Differential Equations. Springer, New York (2009)
2. Hopf, E.: Statistical hydromechanics and functional calculus. J. Ration. Mech. Anal. **1**, 87–122 (1952)
3. Keller, L., Friedmann, A.: Differentialgleichungen für die turbulente Bewegung einer kompressiblen Flüssigkeit. In: Proceedings of the First International Congress for Applied Mechanics, pp. 395–405 (1924)

4. Frewer, M., Khujadze, G.: Comment on: Lie symmetry analysis of the Hopf functional-differential equation [Symmetry 2015, **7**(3), 1536], submitted to Symmetry (2015)
5. Lundgren, T.S.: Distribution functions in the statistical theory of turbulence. Phys. Fluids **10**, 969–975 (1967)
6. Oberlack, M., Rosteck, A.: New statistical symmetries of the multi-point equations and its importance for turbulent scaling laws. Discrete Contin. Dyn. Syst. Ser. S. **3**, 451–471 (2010)
7. Oberlack, M., Wacławczyk, M.: On the extension of Lie group analysis to functional differential equations. Arch. Mech. **58**, 597 (2006)
8. Wacławczyk, M., Oberlack, M.: Application of the extended Lie group analysis to the Hopf functional formulation of the Burgers equation. J. Math. Phys. **54**, 072901 (2013)
9. Wacławczyk, M., Staffolani, N., Oberlack, M., Rosteck, A., Wilczek, M., Friedrich, R.: Statistical symmetries of the Lundgren–Monin–Novikov hierarchy. Phys. Rev. E **90**, 013022 (2014)

Application of an Integral Fluctuation Theorem to Turbulent Flows

N. Reinke, D. Nickelsen, A. Engel and J. Peinke

1 Introduction

There is a long lasting discussion on universal properties of turbulence. The following questions arise: do turbulent properties change with the Reynolds number, or are they even dependent on the large scale properties of turbulence? An important feature would be that turbulence could be taken as universal below some scales. In this case, even for turbulent flows which are generated on a large scale by different processes, the same subgrid models can be used, an important aspect for numerical simulations. For large eddy simulations, it is essential to know the connections between larger scales and the unresolved subgrid turbulence. From this aspect it is important to get a profound understanding of the turbulent cascade, relating turbulent structures on different scales. Rigorous results on the turbulent cascade are still missing.

In the last decades there have been many works to achieve a better understanding of the turbulent cascade. Interesting new methods have been developed. Often these works are based on two-point correlations, although it was shown that three-point correlations are elementary fundamental for the description of the turbulent cascade [1, 2]. Such a three-point closure allows also to achieve a stochastic description of the cascade by Fokker–Planck equations, which finally enables a general

N. Reinke (✉) · J. Peinke
Institute of Physics and ForWind, University of Oldenburg, Carl-v-Ossietzky-Str. 9-11, Oldenburg, Germany
e-mail: nico.reinke@uni-oldenburg.de

J. Peinke
e-mail: joachim.peinke@uni-oldenburg.de

D. Nickelsen · A. Engel
Institute of Physics, University of Oldenburg, Carl-v-Ossietzky-Str. 9-11, Oldenburg, Germany
e-mail: d.nickelsen@uni-oldenburg.de

A. Engel
e-mail: andreas.engel@uni-oldenburg.de

J. Peinke et al. (eds.), *Progress in Turbulence VI*,
Springer Proceedings in Physics 165, DOI 10.1007/978-3-319-29130-7_3

N-point characterization of turbulence [3]. This stochastic approach to a more general description of turbulence has recently been related to new developments in non-equilibrium thermodynamics. Prominent among these developments are fluctuation theorems [4, 5], of which an integral fluctuation theorem (IFT) has been shown to hold for the stochastic approach to the turbulent cascade [6]. It provides a characterization of entropy fluctuations along the turbulent cascade in form of a generalized second law of thermodynamics.

In this paper we use the IFT to analyze different turbulent flows. Measurements of a free jet flow and wake flows behind a fractal grid for two different Reynolds numbers are presented in the next chapter. Afterwards, these data is analyzed and compared on the background of the above mentioned questions of universal features of turbulence.

2 Experimental Data

We analyzed velocity time series (a)–(c) based on different turbulent flows, namely, a free jet flow [2] and two fractal grid flows [7]. Data-sets have been selected with respect to two aspects. On the one hand, two data-sets of one flow with different Reynolds numbers and, on the other hand, data from two different flows with similar Reynolds numbers have been chosen. The velocity time series were measured in an air stream with single-wire probes. The data-sets consists of 3×10^6–1.25×10^7 samples. Flow characteristics, namely the Taylor Reynolds number Re_λ, the longitudinal integral length L, the Taylor length λ, the Einstein–Markov length l_{EM}, and the length of inertial range L/λ, are summarized in Table 1. The procedure proposed by Aronson and Löfdahl [8] is used to estimate λ and Re_λ; L is estimated by integrating the autocorrelation function [9] and l_{EM} is determined by using the relation $l_{EM} = 0.9\lambda$ proposed by Lück [10].

Figure 1 presents the energy density spectrum. Instead of using $E(k)$ we plot the spectral energy density as a function of the scale $r = 1/k$ normalized by λ. Note that we present and discuss all scales in units of λ. As usual, Taylor's hypothesis of frozen turbulence is used to transfer the measurements in time to those in space. The three data-sets show the common scaling behavior of homogeneous isotropic turbulence, $E(r) \propto r^{5/3}$.

Table 1 Flow characteristics of the measurements from three turbulent flows

Data-set	Re_λ (−)	L (mm)	λ (mm)	l_{EM} (mm)	L/λ (−)
(a) Free jet	166	63.1	6.56	5.90	9.6
(b) Fractal grid	153	53.5	7.32	6.59	7.3
(c) Fractal grid	740	51.2	8.15	7.34	6.3

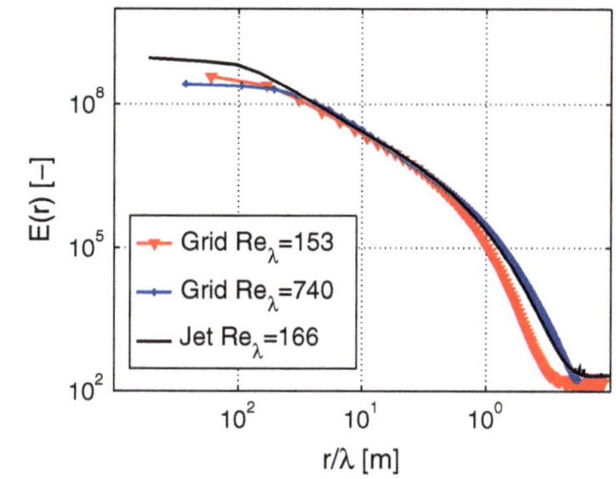

Fig. 1 Energy spectrum of fractal grid wake flows with $Re_\lambda = 153$; 740 and free jet flow with $Re_\lambda = 166$

3 Theoretical Framework

3.1 N-Scale Statistics

The turbulent cascade can be described by the statistics of velocity increments $u_r = u(x+r) - u(x)$. A general approach is the N-scale characterization of the turbulent cascade. It is given by the joint probability $p(u_r; u_{r+\Delta}; ...; u_{r+N\Delta})$ over N scales r [1, 3].

For $\Delta \geq l_{EM}$ the stochastic process can be expressed as a Markov-chain [1, 10]. The joint probability factorizes and derives from the conditional PDFs $p(u_r|u_{r+\Delta})$ [2, 11], reducing the N-scale statistics to three-point or two-scale statistics. The transition probability $p(u|u')$ can be described by a Kramers–Moyal expansion. (We abbreviate (u_r) with u and $(u_{r+\Delta})$ with u'). Due to the fact that the first two terms of the Kramers–Moyal expansion strongly dominate the expansion [1, 2], the evolution of the velocity increments for decreasing r can be expressed by the following Fokker-Planck equation

$$-\partial_r p(u|\tilde{u}) = -\partial_u \left[D^{(1)}(u,r)p(u|\tilde{u})\right] + \partial_{uu}\left[D^{(2)}(u,r)p(u|\tilde{u})\right] . \quad (1)$$

where $\tilde{u}$ denotes an increment at some fixed large scale $\tilde{r} > r$.

The remaining two Kramers–Moyal coefficients $D^{(1)}(u,r)$ and $D^{(2)}(u,r)$ are also called drift and diffusion coefficient, respectively. (We abbreviate $D^{(1)}(u,r)$ and $D^{(2)}(u,r)$ with $D^{(1,2)}$). These coefficients, together with the initial distribution $p(u_L, L)$, contain the complete stochastic information about the cascade.

$D^{(1,2)}$ are estimated by the moments

$$M^{(\kappa)}(u,\Delta,r) = \int_{-\infty}^{+\infty} (u-u')^\kappa \; p(u|u')\, du' ,$$
$$D^{(\kappa)}(u,r) = \lim_{\Delta\to 0} \frac{r}{\kappa!\Delta} M^{(\kappa)}(u,\Delta,r) . \quad (2)$$

Due to a subsequent optimization of $D^{(1,2)}$, small uncertainties in Eq. (2) (e.g. estimating limit) can be overcome. The optimization cost function is based on conditional PDFs, which are deduced on the one hand from experimental data and on the other hand from Kramers–Moyal coefficients [12, 13].

3.2 Integral Fluctuation Theorem as Criterion for $D^{(1,2)}$

The two Kramers–Moyal coefficients $D^{(1,2)}$ define the Markov cascade process. Markov processes are central in the emerging field of stochastic thermodynamics [4]. A typical example is the motion of colloidal particles subject to an external force [14]. In this non-equilibrium setting, the particles produce entropy as they move through the fluid. The entropy production ΔS can be defined for individual fluctuating trajectories by which it becomes a fluctuating quantity itself. Due to the nanoscopic setting, also negative values of ΔS are possible. The balance between fluctuations that produce or consume entropy is expressed by the IFT [5]

$$\langle e^{-\Delta S} \rangle = 1 \, , \tag{3}$$

where $\langle \dots \rangle$ is the expectation value over many fluctuating trajectories. Note that Eq. (3) implies that on average $\langle \Delta S \rangle > 0$, in agreement with the second law of thermodynamics.

For any Markov process an IFT for ΔS in the form Eq. (3) is known to hold. Thus, the IFT can be used to test the validity of the approximation defined by the Kramers–Moyal coefficients $D^{(1,2)}$ and the initial distribution $p(u_L, L)$ for the turbulent cascade. The entropy production for a single realization $u(\cdot)$ can be calculated from the functional

$$\Delta S[u(\cdot)] = - \int_L^{\lambda} \partial_r u(r) \, \partial_u \varphi\big(u(r), r\big) \, \mathrm{d}r - \ln \frac{p(u_\lambda, \lambda)}{p(u_L, L)} \tag{4}$$

with the non-equilibrium potential. Here $(\cdot)$ denotes the evolution of r from L to λ .

$$\varphi(u) = \ln D^{(2)}(u, r) - \int_{-\infty}^{u} \frac{D^{(1)}(u', r)}{D^{(2)}(u', r)} du' \, . \tag{5}$$

To see if the IFT, Eq. (3), holds for our experimental data, we use Eq. (4) to determine $\Delta S^{(i)} = \Delta S[u^{(i)}(\cdot)]$ from the estimated coefficients $D^{(1,2)}$ for different measured realizations $u^{(i)}(\cdot)$. The probability densities $p(u_L, L), \;\; p(u_\lambda, \lambda)$ are taken directly from the measured data. Averaging over different realizations, we test

$$\langle e^{-\Delta S} \rangle_n = \frac{1}{n} \sum_{i=1}^{n} e^{-\Delta S^{(i)}} \simeq 1 \, . \tag{6}$$

This relation holds only if, firstly, $D^{(1,2)}$ correctly capture the three-point statistics of the flow field and, secondly, if the number n of measured $u(\cdot)$ is large enough to include rare fluctuations with negative values for ΔS sufficiently often.

In [6], Nickelsen and Engel showed that Eq. (6) indeed holds for a free jet flow with $D^{(1,2)}$ estimated from the experimental data. Moreover, they found that fluctuations of u on integral scales result in positive values for ΔS, whereas negative ΔS arise from intermittent small-scale fluctuations. The IFT hence probes the estimation of $D^{(1,2)}$ particularly with regard to the appropriate balance between typical large-scale fluctuations and intermittent small-scale fluctuations, that is, it checks the correct modeling of the cascade features that cause small-scale intermittency.

4 Result

We estimate the Kramers–Moyal coefficients $D^{(1,2)}$ for all three above mentioned turbulent flows using the functional forms

$$r\,D^{(1)}(u,r) = a_1(r)\,u\ ,\quad r\,D^{(2)}(u,r) = b_0(r) + b_1(r)\,u + b_2(r)\,u^2\ . \qquad (7)$$

The results of the optimized estimation of $D^{(1,2)}$ are shown in Table 2.

With these estimates of $D^{(1,2)}$, we compute from Eq. (4) the entropy production $\Delta S[u(\cdot)]$ for different realizations of the three flows. Firstly, the validity of the IFT, Eq. (6), is confirmed for the turbulent flows considered here, as shown exemplarily in Fig. 2a, b. With increasing number n of samples a clear convergence to the value 1 can bee seen. Secondly, we can test the universality of turbulent features of our chosen flows. An universal turbulent cascade implies universal coefficients $D^{(1,2)}$.

To check this assumption, we combine realizations $u(\cdot)$ and probability densities $p(u_L, L)$, $p(u_\lambda, \lambda)$ from one flow with coefficients $D^{(1,2)}$ of another one. In particular, we find for the fractal grid that the IFT even holds if $D^{(1,2)}$ for flow (b) is combined with $u(\cdot)$ and $p(u_L, L)$, $p(u_\lambda, \lambda)$ from the measurement for another Re number (flow (c)), see Fig. 2c.

In a similar way, we analyze data for the flows shown in Fig. 2a, b which have been generated in a different way but are of similar Re_λ. In this case $\langle e^{-\Delta S}\rangle$ converges to a value clearly higher than one, as shown in Fig. 2d. This suggests that the three-point statistics of the free jet flow is not transferable to the statistics of the fractal grid flow, or expressing this in other terms, both flows have clear different statistical features.

Table 2 Coefficients $a_1(r)$ and $b_i(r)$ of the drift and diffusion terms for the experimental data-sets (a)–(c)

Data-set	a_1	b_0	b_1	b_2
(a) Free jet	−0.043r − 0.97	0.058r − 0.034	−0.0028r − 0.035	−0.0015r + 0.069
(b) Fractal grid	−0.047r − 0.86	0.075r − 0.035	−0.0039r − 0.037	−0.0018r + 0.057
(c) Fractal grid	−0.030r − 0.94	0.096r − 0.048	−0.0048r − 0.038	−0.0036r + 0.044

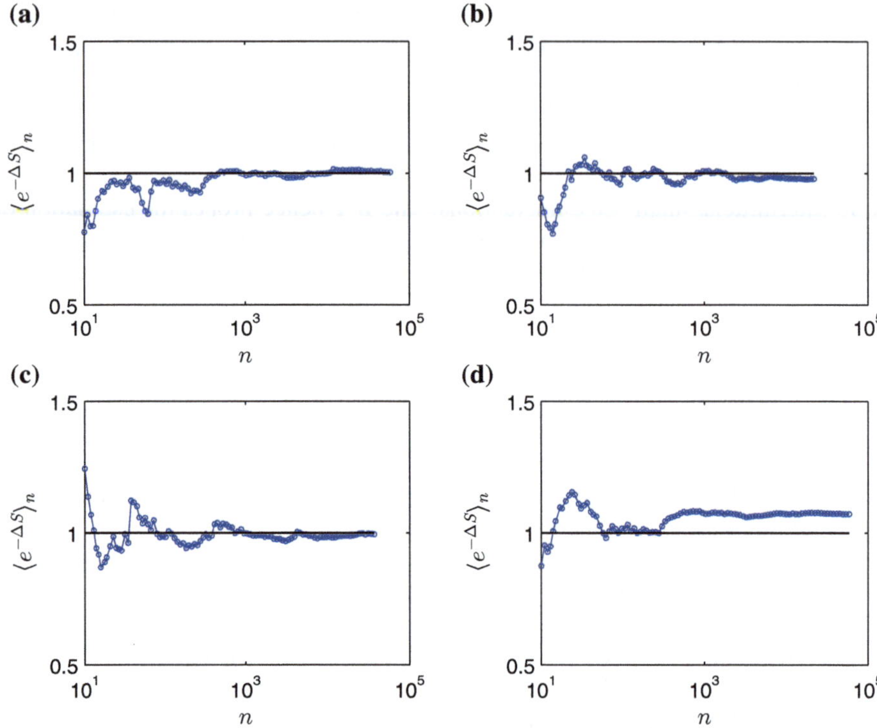

Fig. 2 Application of the IFT, Eq. (6), for increasing number of samples n to **a** the free jet flow with $Re_\lambda = 166$, **b** the fractal grid flow with $Re_\lambda = 153$, **c** $u(\cdot)$, $p(u_L)$ and $p(u_\lambda)$ from the fractal grid flow with $Re_\lambda = 740$ and $D^{(1,2)}$ from the fractal grid flow with $Re_\lambda = 153$, and **d** $u(\cdot)$, $p(u_L)$ and $p(u_\lambda)$ from the free jet flow with $Re_\lambda = 166$ and $D^{(1,2)}$ from the fractal grid flow with $Re_\lambda = 153$

To conclude, we established that the IFT is fulfilled for our three different turbulent flows. The crosswise checking of the IFT suggests that the stochastic cascade of fractal grid turbulence is universal in terms of being independent of Re_λ. Comparing the free jet with the fractal cascade at approximately the similar Re_λ, no universality is found. Establishing the link to small-scale intermittency, the statement is that within the class of fractal generated turbulence the signature of small-scale intermittency is universal, but is fundamentally different for the free jet flow.

References

1. Friedrich, R., Peinke, J.: Description of a turbulent cascade by a Fokker–Planck equation. Phys. Rev. Lett. **78**, 863 (1997)
2. Renner, C., Peinke, J., Friedrich, R., Chanal, O., Chabaud, B.: Universality of small scale turbulence. Phys. Rev. Lett. **89**, 124502 (2002)

3. Stresing, R., Peinke, J.: Towards a stochastic multi-point description of turbulence. New J. Phys. **12**, 103046 (2010)
4. Seifert, U.: Stochastic thermodynamics, fluctuation theorems and molecular machines. Rep. Prog. Phys. **75**, 126001 (2012)
5. Seifert, U.: Entropy production along a stochastic trajectory and an integral fluctuation theorem. Phys. Rev. Lett. **95**, 040602 (2005)
6. Nickelsen, D., Engel, A.: Probing small-scale intermittency with a fluctuation theorem. Phys. Rev. Lett. **110**, 214501 (2012)
7. Stresing, R., Peinke, J., Seoud, R.E., Vassilicos, J.C.: Defining a new class of turbulent flows. Phys. Rev. Lett. **104**, 194501 (2010)
8. Aronson, D., Löfdahl, L.: The plane wake of a cylinder. Measurements and inferences on the turbulence modeling. Phys. Fluids A **5**, 1433 (1993)
9. Batchelor, G.K.: The Theory of Homogeneous Turbulence. Cambridge Science Classic. Cambridge University Press, Cambridge (1953)
10. Lück, St., Renner, Ch., Peinke, J., Friedrich, R.: The Markov–Einstein coherence length—a new meaning for the Taylor length in turbulence. Phys. Lett. A **359**, 335–338 (2006)
11. Tutkun, M., Mydlarski, L.: Markovian properties of passive scalar increments in grid-generated turbulence. New J. Phys. **6**, 49 (2004)
12. Nawroth, A.P., Peinke, J., Kleinhans, D., Friedrich, R.: Improved estimation of Fokker-Planck equations through optimisation. Phys. Rev. E. **76**, 056102 (2007)
13. Kleinhans, D., Friedrich, R., Nawroth, A., Peinke, J.: An iterative procedure for the estimation of drift and diffusion coefficients of Langevin processes. Phys. Lett. A **346**, 42–46 (2005)
14. Blickle, V., Speck, T., Helden, L., Seifert, U., Bechinger, C.: Thermodynamics of a colloidal particle in a time-dependent nonharmonic potential. Phys. Rev. Lett. **96**, 070603 (2006)

A Marker for Studying the Turbulent Energy Cascade in Real Space

J.I. Cardesa and J. Jiménez

Abstract The equation for the kinetic energy based on the residual (sub-filter) velocity is written in several ways in which each term in the equation is Galilean invariant. A different expression for the inter-scale energy transfer term arises from each equation. The statistics of these different terms are studied, together with those of the subgrid-scale (SGS) dissipation. We report on the expression yielding the variance which best matches that of the SGS dissipation, and which is most negatively correlated with it. We argue why the term exhibiting these features is preferred over the others based on its physical meaning. Our study used direct numerical simulation data of homogeneous isotropic turbulence, and our conclusions were observed to be valid with both Gaussian and sharp spectral (low-pass) filters.

1 Introduction and Governing Equations

The subgrid-scale (SGS) dissipation is an example of a real-space quantity which means transfer of kinetic energy between resolved (large) and unresolved (small) scales. We define the large-scale velocity as a spacially low-pass filtered quantity as follows

$$\overline{u}_i\left(x_j\right) = \int u_i\left(x_j - r_j\right) G\left(r_j\right) dr_j, \tag{1}$$

so that the total velocity is decomposed into large- and small-scale components as

$$u_i = \overline{u}_i + u_i'. \tag{2}$$

J.I. Cardesa (✉) · J. Jiménez
School of Aeronautics, Universidad Politécnica de Madrid, 28040 Madrid, Spain
e-mail: ji.cardesa@upm.es

J. Jiménez
e-mail: jimenez@torroja.dmt.upm.es

J. Peinke et al. (eds.), *Progress in Turbulence VI*,
Springer Proceedings in Physics 165, DOI 10.1007/978-3-319-29130-7_4

The evolution of the kinetic energy in the filtered velocity field is then given by

$$\left(\frac{\partial}{\partial t} + \overline{u}_j \frac{\partial}{\partial x_j}\right) \frac{1}{2}\overline{u}_i\overline{u}_i = -\frac{\partial}{\partial x_j}\left(\overline{u}_j\overline{p} + \overline{u}_i\tau_{ij} - 2\nu\overline{u}_i\overline{S}_{ij}\right) - 2\nu\overline{S}_{ij}\overline{S}_{ij} + \overline{S}_{ij}\tau_{ij}, \quad (3)$$

where τ_{ij} is the SGS stress tensor $\overline{u_iu_j} - \overline{u}_i\overline{u}_j$ and $\overline{S}_{ij}\tau_{ij}$ is the SGS dissipation. Its statistics have been studied extensively [1, 3]. Here, we focus on an alternative equation, which is that of the kinetic energy contained in the residual scales. The reason behind this is that some terms within Eq. (3) are not, by themselves, Galilean invariant. Another reason is that $\overline{S}_{ij}\tau_{ij}$ contains the triple product $\overline{u}_i\overline{u}_j\overline{S}_{ij}$ which features interactions between large scales only. We are looking for an inter-scale transfer term arising uniquely from interactions between filtered and residual terms. Among the ways of writing the evolution equation for the kinetic energy of the residual (small-scale) velocities, we report on the following four:

$$(\partial_t + \overline{u}_j\partial_j)\frac{1}{2}u'_iu'_i = -\partial_j\left(u'_jp' - 2\nu u'_iS'_{ij}\right) - 2\nu S'_{ij}S'_{ij} \underbrace{+ u'_i\partial_j(\tau_{ij}) - u'_iu'_j\overline{S}_{ij} - u'_iu'_jS'_{ij}}_{T_A} \quad (4)$$

$$(\partial_t + \overline{u}_j\partial_j)\frac{1}{2}u'_iu'_i = -\partial_j\left(u'_jp' - 2\nu u'_iS'_{ij} - u'_i\tau_{ij}\right) - 2\nu S'_{ij}S'_{ij} \underbrace{- S'_{ij}\tau_{ij} - u'_iu'_j\overline{S}_{ij} - u'_iu'_jS'_{ij}}_{T_B} \quad (5)$$

$$(\partial_t + \overline{u}_j\partial_j)\frac{1}{2}u'_iu'_i = -\partial_j\left(u'_jp' - 2\nu u'_iS'_{ij} + \tfrac{1}{2}u'_iu'_iu'_j\right) - 2\nu S'_{ij}S'_{ij} \underbrace{+ u'_i\partial_j(\tau_{ij}) - u'_iu'_j\overline{S}_{ij}}_{T_C} \quad (6)$$

$$(\partial_t + \overline{u}_j\partial_j)\frac{1}{2}u'_iu'_i = -\partial_j\left(u'_jp' - 2\nu u'_iS'_{ij} - u'_i\tau_{ij} + \tfrac{1}{2}u'_iu'_iu'_j\right) - 2\nu S'_{ij}S'_{ij} \underbrace{- S'_{ij}\tau_{ij} - u'_iu'_j\overline{S}_{ij}}_{T_D}. \quad (7)$$

There are additional ways of writing the right-hand side, but they lead to terms which are not, by themselves, Galilean invariant. These four equations give rise to four different inter-scale energy transfer terms outside of the divergence, which we label $T_{A\rightarrow D}$. They should be the small-scale counterparts of $\tau_{ij}\overline{S}_{ij}$. Our task will be to compare some simple statistics of these four resulting terms, with the aim of choosing the most suitable one for our needs.

2 Results and Conclusions

We computed the inter-scale transfer terms $T_{A\rightarrow D}$ from the direct numerical simulation of forced homogeneous isotropic turbulence found in the databases of the Johns Hopkins University [2]. Three 1024^3 velocity fields were used, separated by half a large-eddy turnover time between them. The Re_λ reported is 433. Entire fields of the three velocity components were downloaded and Fourier transformed, in order to filter and differentiate in wave-number space—thus taking full advantage of the simulation's spectral accuracy. We used a Gaussian and a sharp spectral filter. The filter widths were [20, 28, 40, 57, 80, 113, 160, 226, 320, 453, 640, 905]η/π, with η being the Kolmogorov length scale.

The dependence of the mean and standard deviation of terms $T_{A \to D}$ as well as that of $-\tau_{ij}\overline{S}_{ij}$ on the filter width is shown in Fig. 1a, c for the Gaussian filter, and in Fig. 1b, d for the sharp spectral filter. The mean values of $T_{A \to D}$ for a given filter type are equal to each other. This is consistent with the fact that they all differ by

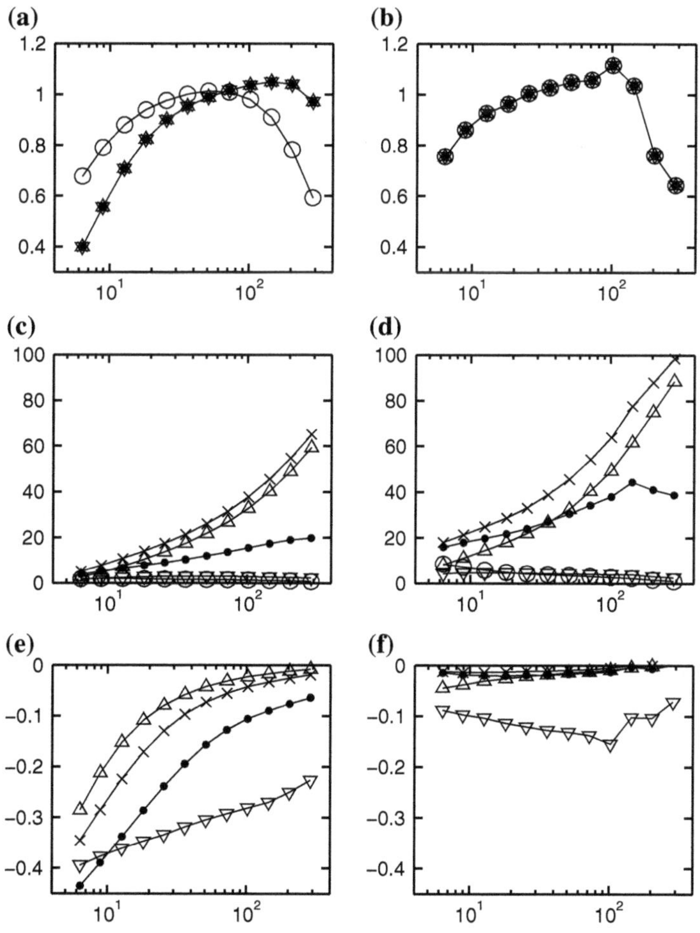

Fig. 1 Statistics of the transfer terms $T_{A \to D}$ obtained with a Gaussian filter (*left*) and a sharp spectral filter (*right*), as a function of filter width Δ. The horizontal axis is Δ/η on all figures. Symbol legend: $\triangle$ T_A, $\times$ T_B, $\triangledown$ T_C, $\bullet$ T_D, $\bigcirc$ $-\overline{S}_{ij}\tau_{ij}$. **a, b** Mean of the transfer terms normalised by the mean total viscous dissipation. **c, d** Standard deviation of the transfer terms normalised by the mean total viscous dissipation. **e, f** Spatial correlation between $\tau_{ij}\overline{S}_{ij}$ and $T_{A \to D}$

a divergence term which vanishes on average in homogeneous flows. However, the mean that the terms converge to changes depending on the filter type. For both filter types, it remains of the same order as the mean total viscous dissipation. A significant effect of the filter type is the fact that $\langle -\tau_{ij}\overline{S}_{ij}\rangle$ is different from the mean of terms $T_{A\rightarrow D}$ for the Gaussian filter. Yet these are all equal for the sharp spectral filter, which is a *projection* contrary to the Gaussian filter. From Fig. 1c, d we conclude that the standard deviation of T_C is the most similar to that of $\tau_{ij}\overline{S}_{ij}$, whereas those of T_A, T_B and T_D are larger. This lead us to investigate the spatial correlation between quantities $T_{A\rightarrow D}$ with $\tau_{ij}\overline{S}_{ij}$ found with the same filter width and type. It seemed reasonable to expect, at least on average, that there exists a negative correlation between $\tau_{ij}\overline{S}_{ij}$ and the quantities $T_{A\rightarrow D}$. Such a negative correlation would imply that when $\tau_{ij}\overline{S}_{ij}$ acts as a sink in the equation for the kinetic energy of the large-scale velocity field, terms $T_{A\rightarrow D}$ behave as sources in the kinetic energy of the small-scale velocity field—and vice-versa. These correlations are shown on Fig. 1e, f. It is clear from them that the most negative correlation is given by T_C, indicating that $\tau_{ij}\overline{S}_{ij}$ and T_C behave more like "communicating vessels" than the combination of $\tau_{ij}\overline{S}_{ij}$ with T_A, T_B or T_D.

From the results outlined so far, we are left with the conclusion that T_C features the best-behaved statistics in the sense that they are the most consistent with the observed statistics of $\tau_{ij}\overline{S}_{ij}$. Our conclusion so far is based solely on the data. We now try to get to the same conclusion by looking back at the equations where terms $T_{A\rightarrow D}$ appear. T_A and T_B are found to be the same as T_C and T_D with the additional $u'_i u'_j S'_{ij}$ product. This product can be written entirely as part of the divergence—using continuity, and so it contributes nothing to the mean but affects only the standard deviation. As seen on Fig. 1c, d, $u'_i u'_j S'_{ij}$ increases the standard deviation. Furthermore, this triple product is a coupling of small-scale quantities interacting with each other, and hence cannot convey the meaning of an energy transfer resulting from large- and small-scale interaction. So T_A and T_B could be discarded based on that criterion alone. In order to chose the best expression—for our purpose—between T_C and T_D, we see that the equation with T_D includes the term $u'_i\tau_{ij}$ inside the divergence. The latter term contains a coupling between filtered (large) and residual (small) quantities which is outside of T_D. Such coupling terms within the equation with T_C have been entirely confined to the inter-scale transfer term, so that all large- and small-scale interactions lie inside T_C. To summarise, all the products coupling large and small scales in Eq. (6) are inside T_C, while no additional term within T_C is the result of small-scale quantities interacting with each other. It is the only expression out of Eqs. (4)→(7) which achieves this, and for this reason we chose it as our preferred way of writing the evolution equation of the small-scale kinetic energy.

Acknowledgments This work was supported by the European Research Council Multiflow grant ERC-2010.AdG-20100224.

References

1. Aoyama, T., Ishihara, T., Kaneda, Y., Yokokawa, M., Itakura, K., Uno, A.: Statistics of energy transfer in high-resolution direct numerical simulation of turbulence in a periodic box. J. Phys. Soc. Jpn. **74**(12), 3202–3212 (2005)
2. Li, Y., Perlman, E., Wan, M., Yang, Y., Meneveau, C., Burns, R., Chen, S., Szalay, A., Eyink, G.: A public turbulence database cluster and applications to study Lagrangian evolution of velocity increments in turbulence. J. Turbul. **9**(31), 1–29 (2008)
3. Piomelli, U., Cabot, W.H., Moin, P., Lee, S.: Subgrid-scale backscatter in turbulent and transitional flows. Phys. Fluids A **3**(7), 1766–1771 (1991)

Scale Energy of Turbulence Based on Two-Point Velocity Correlation

Fujihiro Hamba

Abstract The energy density in the scale space was introduced on the basis of the two-point velocity correlation toward better understanding inhomogeneous turbulence and improving turbulence models. The transport equation for the energy density was derived for inhomogeneous turbulence. Using the DNS of homogeneous isotropic turbulence, the transport equation in the scale space was evaluated and compared with that in the wavenumber space. It was shown that the energy density can be used to examine the energy transfer in the scale space.

1 Introduction

In order to better understand inhomogeneous turbulence, it must be useful to examine the energy transport not only in the physical space but also in the wavenumber space. Instead of the energy spectrum in the wavenumber space, the second-order velocity structure function $\langle (u_i'(\mathbf{x}+\mathbf{r}) - u_i'(\mathbf{x}))^2 \rangle$ was treated as the scale energy and its transport in the $\mathbf{r}$ space was discussed [1–4]. However, its meaning in the limit of $r \to \infty$ for inhomogeneous turbulence is not clear because it corresponds to the sum of the energies at very distant two points as $\langle u_i'^2(\mathbf{x}+\mathbf{r}) \rangle + \langle u_i'^2(\mathbf{x}) \rangle$.

In this work, we propose another definition of the scale energy based on the two-point velocity correlation $\langle u_i'(\mathbf{x}) u_i'(\mathbf{x}+\mathbf{r}) \rangle$. In the limit of $r \to 0$ it corresponds to the turbulent energy at a single point $\mathbf{x}$. We expect that the energy density based on the two-point velocity correlation is suitable in discussing the energy transfer in the scale space. We derive the transport equation for the energy density in inhomogeneous turbulence. As a first step we examine the transport equation using the direct numerical simulation (DNS) data of homogeneous isotropic turbulence. Analysis of inhomogeneous turbulence such as channel flow will be done in future work.

F. Hamba (✉)
Institute of Industrial Science, The University of Tokyo, Komaba, Meguro-ku, Tokyo 153-8505, Japan
e-mail: hamba@iis.u-tokyo.ac.jp

J. Peinke et al. (eds.), *Progress in Turbulence VI*,
Springer Proceedings in Physics 165, DOI 10.1007/978-3-319-29130-7_5

2 Energy Density in Scale Space

The transport equation for the energy spectrum $E(k)$ for homogeneous isotropic turbulence is given by

$$\frac{\partial}{\partial t}E(k) = T(k) - \varepsilon(k), \tag{1}$$

$$T(k) = 2\pi k^2 \iint d\mathbf{p}d\mathbf{q}S(\mathbf{k}, \mathbf{p}, \mathbf{q}), \quad \varepsilon(k) = 2\nu k^2 E(k). \tag{2}$$

The energy transfer can be discussed partly because the energy spectrum is the energy density in the wavenumber space and satisfies $K = \int_0^\infty dk E(k)$.

For inhomogeneous turbulence, it is not always possible to apply Fourier transform. Instead, we can treat the two-point velocity correlation $Q_{ii}(\mathbf{x}, \mathbf{r}) = \left\langle u_i'(\mathbf{x})u_i'(\mathbf{x}+\mathbf{r})\right\rangle$ to examine the energy transfer in the scale space like the Kármán-Howarth equation. The correlation $Q_{ii}(\mathbf{x}, \mathbf{r})/2$ can be considered the energy whose scale is larger than r and it does not exactly represent the energy density in the scale space. Instead of $Q_{ii}(\mathbf{x}, \mathbf{r})/2$, we introduce the following quantity as the scale energy:

$$E(\mathbf{x}, \mathbf{r}) = -\frac{1}{2}\frac{\partial}{\partial r}Q_{ii}(\mathbf{x}, \mathbf{r}) = -\frac{1}{2}\frac{\partial}{\partial r}\langle u_i'(\mathbf{x})u_i'(\mathbf{x}+\mathbf{r})\rangle, \tag{3}$$

where $r = |\mathbf{r}|$. Since this energy satisfies $K = \int_0^\infty dr E(\mathbf{x}, \mathbf{r})$, it represents the energy density in r space.

The transport equation for $E(\mathbf{x}, \mathbf{r})$ can be written as

$$\begin{aligned}
\frac{D}{Dt}E(\mathbf{x}, \mathbf{r}) = {} & \frac{1}{2}\frac{\partial}{\partial r}\langle u_k'(\mathbf{x})u_i'(\mathbf{x}+\mathbf{r})\rangle \left(\frac{\partial}{\partial x_k}U_i(\mathbf{x}) + \frac{\partial}{\partial x_i}U_k(\mathbf{x})\right) \\
& + \frac{1}{2}\frac{\partial}{\partial r}\left[\langle u_k'(\mathbf{x}+\mathbf{r})u_i'(\mathbf{x})\rangle \frac{\partial}{\partial x_k}(U_i(\mathbf{x}+\mathbf{r}) - U_i(\mathbf{x}))\right] \\
& + \nu\frac{\partial}{\partial r}\left\langle \frac{\partial}{\partial x_k}u_i'(\mathbf{x})\frac{\partial}{\partial x_k}u_i'(\mathbf{x}+\mathbf{r})\right\rangle + \frac{\partial}{\partial x_k}\left(\frac{1}{2}\frac{\partial}{\partial r}\langle u_k'(\mathbf{x})u_i'(\mathbf{x})u_i'(\mathbf{x}+\mathbf{r})\rangle\right) \\
& + \frac{\partial}{\partial x_i}\left(\frac{1}{2}\frac{\partial}{\partial r}\langle p'(\mathbf{x})u_i'(\mathbf{x}+\mathbf{r}) + p'(\mathbf{x}+\mathbf{r})u_i'(\mathbf{x})\rangle\right) + \nu\frac{\partial^2}{\partial x_k \partial x_k}E(\mathbf{x}, \mathbf{r}) \\
& + \frac{1}{2}\frac{\partial}{\partial r}\frac{\partial}{\partial r_k}[(U_k(\mathbf{x}+\mathbf{r}) - U_k(\mathbf{x}))\langle u_i'(\mathbf{x})u_i'(\mathbf{x}+\mathbf{r})\rangle] \\
& + \frac{1}{2}\frac{\partial}{\partial r}\frac{\partial}{\partial r_k}\langle(u_k'(\mathbf{x}+\mathbf{r}) - u_k'(\mathbf{x}))u_i'(\mathbf{x})u_i'(\mathbf{x}+\mathbf{r})\rangle.
\end{aligned} \tag{4}$$

By integrating each term from $r = 0$ to ∞, Eq. (4) is reduced to the transport equation for the turbulent energy K. On the right-hand side of Eq. (4) the first and second terms correspond to the energy production, the third term to the dissipation, and the fourth to sixth terms to the diffusion in the K equation. The remaining seventh and eighth terms represent the energy transfer in the $\mathbf{r}$ space. By evaluating these terms we could examine in which scale the processes of energy production, dissipation, etc. occur in wall bounded flows. This transport equation is expected to give an insight into improving turbulence models. For homogeneous isotropic turbulence the above transport equation is rewritten as

$$\frac{\partial}{\partial t}E(r) = T_E(r) - \varepsilon_E(r), \tag{5}$$

$$T_E(r) = \frac{1}{2}\frac{\partial}{\partial r}\frac{\partial}{\partial r_k}\langle(u'_k(\mathbf{x}+\mathbf{r}) - u'_k(\mathbf{x}))u'_i(\mathbf{x})u'_i(\mathbf{x}+\mathbf{r})\rangle, \tag{6}$$

$$\varepsilon_E(r) = -\nu\frac{\partial}{\partial r}\left\langle\frac{\partial}{\partial x_k}u'_i(\mathbf{x})\frac{\partial}{\partial x_k}u'_i(\mathbf{x}+\mathbf{r})\right\rangle. \tag{7}$$

3 Analysis Using Homogeneous Isotropic Turbulence DNS

In this work, we examine the energy transfer in the scale space for homogeneous isotropic turbulence as a first step to investigating the transport equation given by Eq. (4). We carry out DNS of decaying homogeneous isotropic turbulence using 512^3 grid points. The initial spectrum is set to $E(k) \propto k^4 \exp(-2(k/k_p)^2)$ where $k_p = 3.5$. We will show data at $t = 3$ where the Reynolds number R_λ is 75. At $t = 3$ the spectrum is as broad as that in the case of forced isotropic turbulence and the dependence on the initial condition has been almost forgotten.

Figure 1a shows the energy spectrum $E(k)$ as a function of k. Since the Reynolds number is not very high, the inertial range where $E(k) \propto k^{-5/3}$ is narrow and is located at $k \sim 20$. Figure 1b shows the energy density $E(r)$ as a function of r. The inertial range given by $E(r) \propto r^{-1/3}$ is located at $r \sim 0.15$.

Next we compare the energy transfer between the wavenumber and scale spaces. Figure 2a shows terms in the transport equation in the wavenumber space given by Eq. (1). The transfer term $T(k)$ is negative in the low wavenumber region and

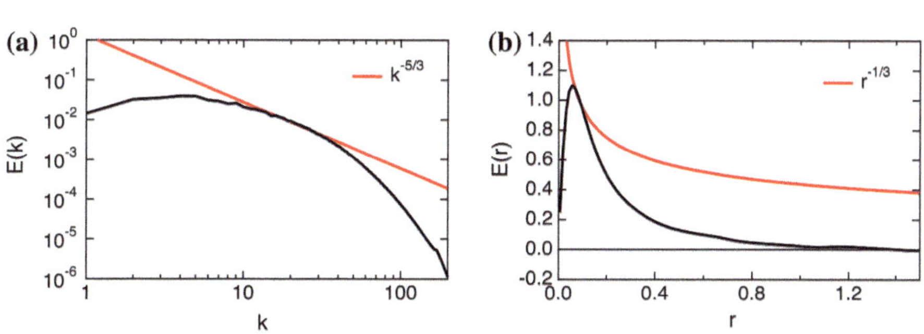

Fig. 1 Profiles of **a** energy spectrum $E(k)$ and **b** energy density in the scale space $E(r)$

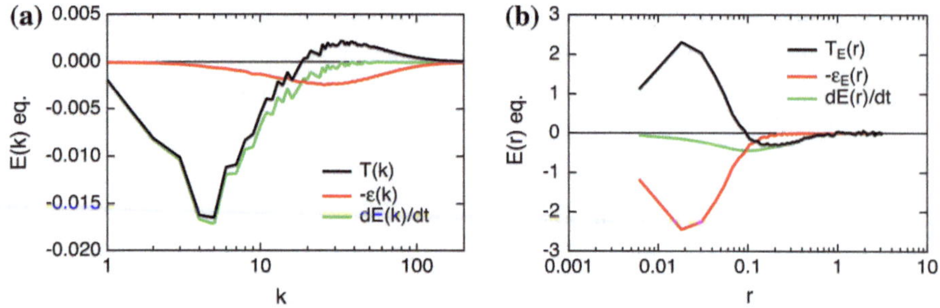

Fig. 2 Profiles of **a** terms in the transport equation for $E(k)$ given by Eq. (1) and **b** terms in the transport equation for $E(r)$ given by Eq. (5)

positive in the high wavenumber region, representing the forward energy cascade. At the dissipation range the transfer and dissipation terms are balanced to each other. Figure 2b shows terms in the transport equation in the scale space given by Eq. (5) as functions of r. The small (large) scale region in Fig. 2b corresponds to the high (low) wavenumber region in Fig. 2a. The transfer term $T_E(r)$ is negative in the large scale region and positive in the small scale region, representing the energy transfer from the large scale to the small scale. Although the amplitude of terms is different between Fig. 2a, b, the tendency of the energy transfer is the same. The profiles shown in Fig. 2 suggest that the energy density defined as Eq. (3) can be used to examine the energy transfer in the scale space.

4 Conclusions

The energy density in the scale space was introduced on the basis of the two-point velocity correlation. The transport equation for the energy density was derived for inhomogeneous turbulence. The energy transfer in the scale space was evaluated using the DNS of homogeneous isotropic turbulence. It was shown that the energy density is useful for examining the energy transfer in the scale space.

References

1. Marati, N., Casciola, C.M., Piva, R.: Energy cascade and spatial fluxes in wall turbulence. J. Fluid Mech. **521**, 191–215 (2004)
2. Davidson, P.A.: Turbulence. Oxford University Press, Oxford (2004)
3. Hill, R.J.: Exact second-order structure-function relationships. J. Fluid Mech. **468**, 317–326 (2002)
4. Cimarelli, A., De Angelis, E., Casciola, C.M.: Paths of energy in turbulent channel flows. J. Fluid Mech. **715**, 436–451 (2013)

A Numerical Study of the Shear-Less Turbulent/Non-turbulent Interface

G. Cocconi, A. Cimarelli, B. Frohnapfel and E. De Angelis

Abstract A DNS simulation of an interface between a decaying turbulent flow without mean shear and a quiescent non-turbulent region is presented here. The analysis of the instantaneous fields highlight a complex multi-scale behaviour at the turbulent/non-turbulent interface. According to previous results in the literature, by analysing the enstrophy budgets it is possible to observe that enstrophy propagate from the turbulent region mainly by inviscid process while viscous diffusion is relevant only at the interface.

1 Introduction

Turbulent/non-turbulent interfaces (TNTI) can be found in a number of flows of engineering interest where they are responsible for some of the key features of the flow [1]. One of the still debated points on the propagation of turbulent fronts is the relevant mechanism of entrainment of irrotational flow at the interface. Several evidences based on local scaling of turbulent properties at the TNTI seems to suggest that vorticity is initiated via small-scale viscous interactions along the convoluted surface of the interface [1, 2, 4]. Here we try to asses the problem by analysing propagation mechanics of enstrophy.

G. Cocconi (✉) · B. Frohnapfel
ISTM - Institute of Fluid Mechanics, Karlsruhe Institute of Technology, Karlsruhe, Germany
e-mail: giacomo.cocconi@kit.edu

A. Cimarelli
CIRI - Centro Interdipartimentale di Ricerca Industriale Aeronautica, Università di Bologna, Forlì, Italy

E. De Angelis
DIN - Dipartimento Ingegneria Industriale Università di Bologna, Bologna, Italy

J. Peinke et al. (eds.), *Progress in Turbulence VI*,
Springer Proceedings in Physics 165, DOI 10.1007/978-3-319-29130-7_6

2 DNS Simulation and Results

First a DNS simulation of homogeneous isotropic turbulence is generated by using a pseudo-spectral code with a discretization of 256^3 Fourier modes. Then two identical fields are put one next to each other in order to obtain a continuous field with twice the original size in the x direction. In such field, velocity fluctuations are multiplied by a continuous function $p(x) \in [1, 0]$, which damps the velocity to 0 in half the domain (in a similar way as seen in [3] for a shear-less mixing), generating the initial TNTI. The function p(x) had been defined as

$$p(x) = \frac{1}{2}\left[1 + tanh\left(a\frac{x}{L}\right) tanh\left(a\frac{x - L/2}{L}\right) tanh\left(a\frac{x - L}{L}\right)\right], \quad (1)$$

where L is the domain length while a is a parameter affecting the steepness of the mean velocity gradient generated by the function. After the introduction of the TNTI the flow is let freely decay while the turbulent front propagates in the irrotational region. In order to asses the effect of the choice of the parameter a three different values for it were used corresponding to $a = 12\pi$, $a = 20\pi$ and $a = 40\pi$, corresponding to three different steepnesses of the initial enstrophy profile. For each of these three parameters 20 independent decay runs where performed which constitute the statistical data-set for the present study. The average initial Reynolds number based on the Taylor microscale is $Re_\lambda = 50$ and the initial ratio between the discretization step and the Kolmogorov scale η is $\Delta x/\eta = 1.44$. The interface detection is performed using an enstrophy threshold [1] with the threshold set to be 2 % of the mean enstrophy in the turbulent core at a given time. A sample of the resulting TNTI is shown in Fig. 1 for the case $a = 40\pi$ at $t/t_0 \approx 4$ where t_0 is defined as the integral time scale of the flow at the beginning of the decay, here the cut-out of the color map represent the enstrophy threshold used to determine the mean interface position.

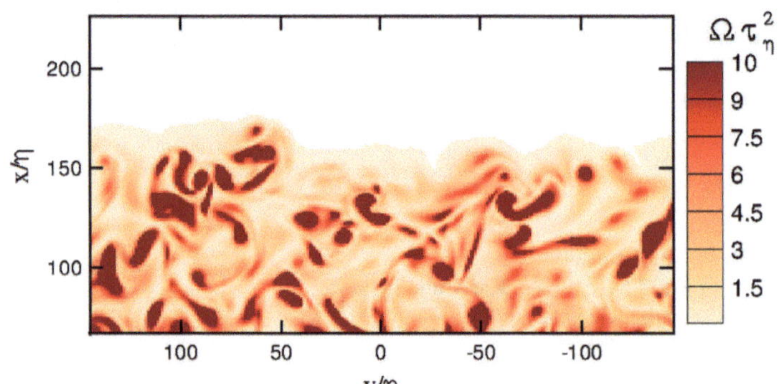

Fig. 1 Instantaneous enstrophy field for $a = 40\pi$ and $t/t_0 \approx 4$

The decay rate of the kinetic energy in the turbulent core for all the three cases simulated tends to the same power-law decay $E \propto t^{-7/6}$, implying that the turbulent region is only marginally affected by the differences in the imposed initial conditions. The location of the turbulent front is proportional to $\sqrt{t}$, highlighting a possible relevance of the viscous diffusion mechanisms on the propagation of the interface. Also from the point of view of the propagation rates of the interface the qualitative behaviour of the TNTI was not affected by the choice of the initial condition. The same is true for other statistics. The balance equation of enstrophy $\Omega = \omega_i \omega_i / 2$ is given by

$$\frac{\partial \langle \Omega \rangle}{\partial t} = \langle \omega_i \omega_j s_{ij} \rangle - \nu \left\langle \frac{\partial \omega_i}{\partial x_j} \frac{\partial \omega_i}{\partial x_j} \right\rangle - \frac{\partial \langle \Omega u \rangle}{\partial x} + \nu \frac{\partial^2 \langle \Omega \rangle}{\partial x^2}, \tag{2}$$

where the first right-hand side term represent the vortex stretching, the second the viscous dissipation, the third term is the advection of enstrophy and the last is the viscous diffusion. Such budget of enstrophy is depicted in Fig. 2 where it is shown how the bulk of the flow is dominated by the dissipation of enstrophy, partially counteracted only by the vortex stretching while the others contributes are almost negligible. As result enstrophy decays everywhere except at the interface, where the small increment in enstrophy is mainly sustained by the inviscid advection of enstrophy. In the right panel of Fig. 2 is depicted the ensemble average of the advection, together with some of its instantaneous samples at $t/t_0 \approx 4$. From this result, it can be observed that the growth of enstrophy at the interface is mainly determined by an inertial transfer of enstrophy, which is drawn by the advection in the turbulent region between -90η and -30η and then it is released close to the interface. Despite the scatter in the dataset, such behaviour is observable for all the single realisations. The

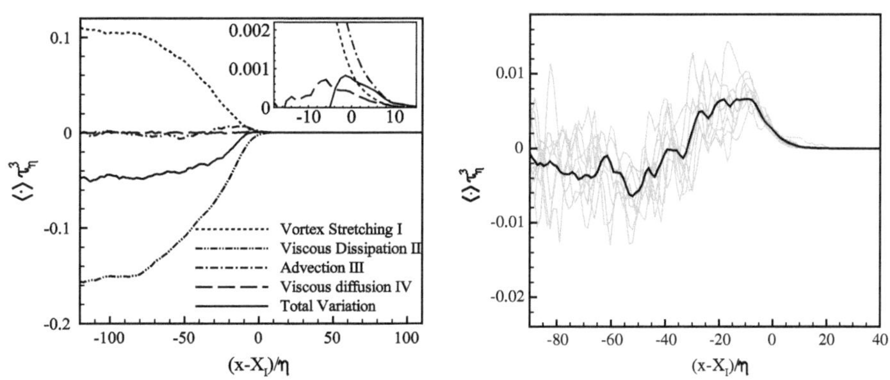

Fig. 2 *Left* enstrophy budget as a function of the distance from the average interface position X_I at $t/t_0 \approx 4$; *inset* magnification of the interface region. *Right* ensemble average of the advection from the entire data set (*black line*) and advection for some of the single realisations composing the data-set (*grey line*), at $t/t_0 \approx 4$

budget also shows that the viscous diffusion of enstrophy is negligible in most part of the flow. Only at the interface, in the region between -15η and -10η, it becomes positive and contributes in the spreading of enstrophy in the irrotational region.

3 Conclusions

A simulation of a shear-less turbulent non-turbulent interface was performed by introducing an interface in a decaying homogeneous isotropic turbulent flow. Thc analysis of the enstrophy budget in such flow showed how the bulk of the turbulent region is dominated mainly by the viscous dissipation of enstrophy followed by the vortex stretching. The viscous diffusion is non-negligible only in the region between -15η and -10η, while is practically zero everywhere else in the flow. The advection of enstrophy appears to be the feeding mechanism for the propagation of the turbulent front. Inviscid mechanisms draw enstrophy from the turbulent region between 30η and 90η behind the interface, and release it at the interface leading to a positive temporal variation and hence to the propagation of the turbulent front.

References

1. da Silva, C.B., Hunt, J.C.R., Eames, I., Westerweel, J.: Interfacial layers between regions of different turbulence intensity. Annu. Rev. Fluid Mech. **46**, 567–590 (2014)
2. Taveira, R.R., da Silva, C.B.: Characteristics of the viscous superlayer in shear free turbulence and in planar turbulent jets. Phys. Fluids **26**, 021702 (2014)
3. Tordella, D., Iovieno, M., Bailey, P.R.: Sufficient condition for gaussian departure in turbulence. Phys. Rev. E **77**, 016309 (2008)
4. Wolf, M., Holzner, M., Lüthi, B., Krug, D., Kinzelbach, W., Tsinober, A.: Effects of mean shear on the local turbulent entrainment process. J. Fluid Mech. **731**, 95–116 (2013)

The Imbalance Between Enstrophy Production and Destruction in Homogeneous Isotropic Unsteady Turbulence

P.C. Valente, R. Onishi and C.B. da Silva

Abstract We show in direct numerical simulations of homogeneous isotropic non-stationary turbulence that there is a systematic and significant imbalance between enstrophy production and its destruction which is concomitant with the previously observed imbalance between the non-linear energy cascade to fine scales and its dissipation (Valente, Onishi, da Silva, Phys Rev E 90(023003), 2014, [12]). However, contrary to the former, the imbalance between enstrophy production and destruction is affected by the 'cascade time-lag', i.e. the time it takes for the energy injected on the large-scales to reach the fine-scales.

1 Introduction

One of the great challenges in turbulence modelling has been the prediction of the rate of turbulent kinetic energy dissipation in statistically unsteady turbulent flows [7]. Perhaps the main underlying cause for this difficulty is the breakdown of Kolmogorov's postulate of local equilibrium [4, 5], at least for the range of Reynolds numbers typical of industrial and geophysical flows [12]. Local equilibrium, i.e. Kolmogorov's notion that "within short time intervals this regime [that of small-scale fluctuations that are approximately spatially isotropic] can be regarded as being stationary, even when the flow as a whole is non-stationary" [4], is essential to relate the small-scale dissipation with the inertial-range non-linear energy flux. In particular

P.C. Valente (✉) · C.B. da Silva
LAETA, IDMEC, Instituto Superior Técnico, Universidade de Lisboa, Av. Rovisco Pais, 1049-001 Lisboa, Portugal
e-mail: pedro.cardoso.valente@tecnico.ulisboa.pt

C.B. da Silva
e-mail: carlos.silva@tecnico.ulisboa.pt

R. Onishi
Center for Earth Information Science and Technology, Japan Agency for Marine-Earth Science and Technology, 3173-25 Showa-machi, Kanazawa-ku, Yokohama, Kanagawa 236-0001, Japan
e-mail: onishi.ryo@jamstec.go.jp

J. Peinke et al. (eds.), *Progress in Turbulence VI*,
Springer Proceedings in Physics 165, DOI 10.1007/978-3-319-29130-7_7

it can be shown that it leads to a balance between the energy cascade in the inertial-range and the kinetic energy dissipation. This balance, commonly known as Kolmogorov's 'four-fifth's law' (due to the pre-factor appearing in the isotropic form of the balance first derived by Kolmogorov) is quintessential in many turbulence models and underlies many theoretical developments in the field.

There are two main viewpoints to explain the origin of the imbalance between energy cascade and dissipation. One viewpoint is that of 'finite Reynolds number effects' whereby any observed departures from a balance between energy cascade and dissipation are due to an insufficient separation between large and small scales which is required for the onset of a true inertial range [1, 10]. An alternative viewpoint put forth by Kraichnan [6] and subsequently developed by Lumley [7] and Yoshizawa [3, 16] is that of a non-negligible time for the energy to 'cascade' from the large-scales to the small-scales where it is dissipated. Note that the two viewpoints of 'finite Reynolds number' and 'cascade time-lag' are fundamentally incompatible since for large Reynolds numbers the increasingly large separation between large- and small-scales would imply an increasing energy cascade delay leading to a growing imbalance between the energy cascade and the dissipation [2, 7] whereas in the case of the imbalance being a 'finite Reynolds number effect' it would, in contrast, become negligible at high Reynolds numbers [10].

2 Results

We present further results from our direct numerical simulations (DNS) of homogeneous isotropic turbulence driven by a periodic forcing following a square-wave protocol [12]. Previously we have shown that these statistically unsteady DNSs manifest a systematic and significant imbalance between the maximum non-linear energy cascade flux to fine scales, Π, and the kinetic energy dissipation rate, ε (see Figs. 1 and 2). Our data also indicated that the maximum energy cascade flux (Π) remains approximately proportional to the kinetic energy, $\sim u^2$, over the turnover time, ℓ/u (ℓ is an integral scale and u the root-mean-square velocity), i.e. $C_\Pi \equiv \Pi\ell/u^3 \approx$ constant, whereas the normalised dissipation followed a power-law of the local (Taylor-microscale) Reynolds number, $C_\varepsilon \equiv \varepsilon\ell/u^3 \sim Re_\lambda^{-\alpha}$ (see Fig. 1), which bears resemblance to the non-equilibrium behaviour found in the near-field of grid-generated decaying turbulence [13–15].

Here we use our data to investigate the balance between vortex stretching, $\overline{\omega_i\omega_j s_{ij}}$, and enstrophy destruction, $2\nu\overline{\nabla\omega_i\nabla\omega_i}$ (ω_i is the vorticity, s_{ij} the strain rate tensor and $\overline{\omega_i\omega_i}$ is the enstrophy; the overbars indicate averages over the whole computational domain) when there is a significant imbalance between Π and ε. These are the two quantities appearing in the evolution equation for the dissipation in a homogeneous flow.

$$\frac{d\varepsilon}{dt} = \nu\overline{\omega_i\omega_j s_{ij}} - 2\nu^2\overline{\nabla\omega_i\nabla\omega_i} \tag{1}$$

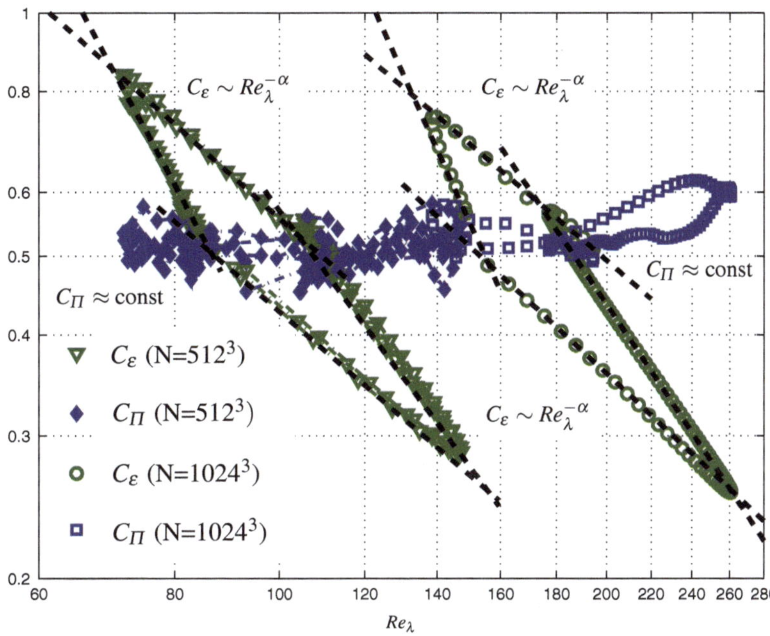

Fig. 1 Normalised maximum energy cascade flux $C_\Pi \equiv \Pi\ell/u^3$ and normalised energy dissipation $C_\varepsilon \equiv \varepsilon\ell/u^3$ as functions of the local Reynolds number, Re_λ, throughout the transients caused by the periodic forcing (N is the number of collocation points in the DNSs)

(recall that in a homogeneous incompressible flow $\varepsilon = \nu\overline{\omega_i\omega_i}$). In Fig. 2 we plot the evolution of $\overline{\omega_i\omega_j s_{ij}} - 2\nu\overline{\nabla\omega_i\nabla\omega_i}$ together with $\Pi - \varepsilon$ for one power input cycle out of the three cycles computed. It can be readily seen that the plotted quantities follow each other indicating that they may be physically related. This is perhaps not surprising since $\overline{\omega_i\omega_j s_{ij}} = \int_0^\infty k^2 T(k)\,dk$ and $2\nu\overline{\nabla\omega_i\nabla\omega_i} = 2\nu\int_0^\infty k^4 E(k)\,dk$, whereas $\Pi = \int_{k_c}^\infty T(k)\,dk$ and $\varepsilon = 2\nu\int_0^\infty k^2 E(k)\,dk$ (k_c is the wavenumber of the zero crossing of the transfer function, i.e. $T(k = k_c) = 0$). Therefore this would lead one to expect that $\overline{\omega_i\omega_j s_{ij}} \sim \Pi$ and $\nu\overline{\nabla\omega_i\nabla\omega_i} \sim \varepsilon$. One can go one step further and directly relate $\overline{\omega_i\omega_j s_{ij}}$ with Π and $\overline{\nabla\omega_i\nabla\omega_i}$ with ε by introducing two assumptions. First, one could assume that the energy and transfer spectra are self-similar for wavenumbers larger than k_c with the appropriate choice of normalising variables. In particular one could keep Kolmogorov's choice of ε and ν as the normalising variables for $E(k)$, but choose the Π and ν as the normalising variables for $T(k)$ since the data show that $\Pi \neq \varepsilon$ throughout most of the transients and Π is directly obtained from $T(k)$. Second, one could assume that the contribution of the wavenumbers smaller than k_c to the integrals leading to $\overline{\omega_i\omega_j s_{ij}}$ and $2\nu\overline{\nabla\omega_i\nabla\omega_i}$ is negligible. Introducing the ansätze $E(k) \sim \varepsilon^{2/3}k^{-5/3}f(k\eta)$ and $T(k) \sim \Pi\eta g(k\eta)$ and evaluating the integrals for $k > k_c$ (effectively neglecting the remaining infrared contribution) leads to

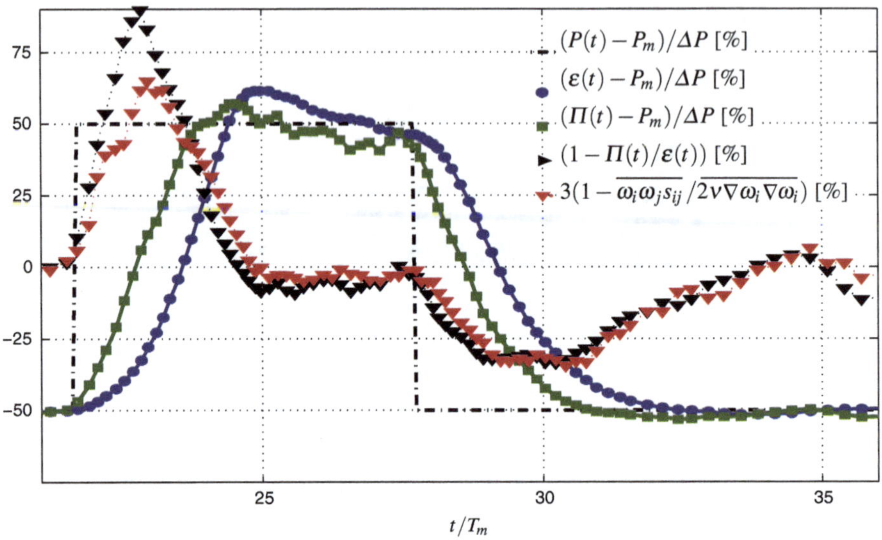

Fig. 2 The imbalance between Π and ε is followed closely by the imbalance between vortex stretching $\overline{\omega_i\omega_j s_{ij}}$ and destruction $\overline{2\nu\nabla\omega_i\nabla\omega_i}$ for the DNS of non-stationary turbulence with varying power input. The data are normalised by the mean and amplitude of the cycle, $P_m = (P|^{\max} + P|^{\min})/2$ and $\Delta P = P|^{\max} - P|^{\min}$ as described in the legend of the figure. The abscissae are normalised by the average turnover time T_m. A factor of 3 in the imbalance between enstrophy production and destruction is added for comparison purposes

$$\frac{d\varepsilon}{dt} \sim \Pi\sqrt{\varepsilon/\nu}\underbrace{\int_{k_c\eta}^{\infty}(k\eta)^2 g(k\eta)\,dk\eta}_{C_{\varepsilon_1}} - \varepsilon\sqrt{\varepsilon/\nu}\underbrace{\int_{k_c\eta}^{\infty}2(k\eta)^4 f(k\eta)\,dk\eta}_{C_{\varepsilon_2}}, \qquad (2)$$

where the integrals denoted as C_{ε_1} and C_{ε_2} would be invariant in time due to the hypothesised self-similarity. This model can be directly compared with the K-ε model where $d\varepsilon/dt = C^{\star}_{\varepsilon_1} P\,\varepsilon/K - C^{\star}_{\varepsilon_2}\varepsilon^2/K$ [8] which differs in two main things, (i) the rate of change of dissipation is proportional to $P - \varepsilon$ rather than $\Pi - \varepsilon$ and (ii) the choice of time scale is ε/K (which is proportional to the turnover time within Kolmogorov's phenomenology) rather than $\sqrt{\varepsilon/\nu}$. We do not wish to make a case for preferring the model equation (2) to the standard equation in the K-ε model. However, our DNS data do indicate that $\overline{\omega_i\omega_j s_{ij}} - 2\nu\overline{\nabla\omega_i\nabla\omega_i}$ seems to be more closely related to $\Pi - \varepsilon$ rather than to $P - \varepsilon$, regardless of the particular choice of time-scale. It would be interesting to test this finding in other unsteady turbulent flows.

Concerning the choice of time-scale, we note that the use of $\sqrt{\varepsilon/\nu}$ introduces a Reynolds number dependence in the dissipation (or enstrophy) equation, which disagrees with the phenomenological argument by Tennekes and Lumley [11], the so-called Tennekes and Lumley balance. We refer the readers to Refs. [8, 9] where this issue is thoroughly discussed. Nevertheless, what our data also show is that throughout the transients the transfer spectra deform (see Fig. 4 in Ref. [12]), hindering the possibility of a self-similar shape which is required to arrive to Eq. (2). The deformation of the transfer spectra, which we showed not to cause the imbalance between Π and ε [12], do seem to strongly influence the imbalance between $\overline{\omega_i\omega_j s_{ij}}$ and $2\nu\overline{\nabla\omega_i\nabla\omega_i}$. It is likely that these deformations of the spectra are a manifestation of the cascade time-lag and therefore it may be possible to model their effect using the ideas put forth by Lumley [2, 7] and Yoshizawa [3, 16].

3 Conclusions

In summary, we revisited the balance between vortex stretching and enstrophy destruction in statistically unsteady homogeneous turbulence and found that their imbalance follows closely the imbalance between the maximum energy cascade flux and the dissipation. A simple model for the evolution equation of the kinetic energy dissipation illustrated that the maximum cascade flux may be a better surrogate for the vortex stretching than the turbulence production (or external power input) and highlights the importance of revisiting the debate of the appropriate time-scale for the dissipation equation as well as considering the effect of the cascade time-lag. Further efforts in obtaining higher Reynolds number experimental and numerical data in different unsteady flows are needed together with reinforced efforts in modelling the dissipation equation.

References

1. Antonia, R.A., Burattini, P.: Approach to the 4/5 law in homogeneous isotropic turbulence. J. Fluid Mech. **550**, 175–184 (2006)
2. Bos, W.J.T., Shao, L., Bertoglio, J.-P.: Spectral imbalance and the normalized dissipation rate of turbulence. Phys. Fluids **19**, 045101 (2007)
3. Horiuti, K., Tamaki, T.: Nonequilibrium energy spectrum in the subgrid-scale one-equation model in large-eddy simulation. Phys. Fluids **25**, 12104 (2013)
4. Kolmogorov, A.N.: The local structure of turbulence in incompressible viscous fluid for very large reynolds numbers. Dokl. Akad. Nauk. SSSR **30**(4) (1941). (Reprinted in 1991, Proc. R. Soc. Lond. **434**, 9–13)
5. Kolmogorov, A.N.: Dissipation of energy in the locally isotropic turbulence. Dokl. Akad. Nauk. SSSR **32**(1) (1941). (Reprinted in 1991, paper 47 in Selected Works of A.N. Kolmogorov, vol. I: Mathematics and Mechanics)
6. Kraichnan, R.: On Kolmogorov's inertial-range theories. J. Fluid Mech. **62**(2), 305–330 (1974)
7. Lumley, J.L.: Some comments on turbulence. Phys. Fluids A **4**(2), 203–211 (1992)
8. Rubinstein, R., Clark, T.T., Livescu, D., Luo, L.-S.: Time-dependent isotropic turbulence. J. Turb. **5**, 11 (2004)
9. Speziale, C.G., Bernard, P.S.: The energy decay in self-preserving isotropic turbulence revisited. J. Fluid Mech. **241**, 645–667 (1992)
10. Tchoufag, J., Sagaut, P., Cambon, C.: Spectral approach to finite Reynolds number effects on Kolmogorov's 4/5 law in isotropic turbulence. Phys. Fluids **24**, 015107 (2012)
11. Tennekes, H., Lumley, J.L.: A First Course in Turbulence. MIT Press, Cambridge (1972)

12. Valente, P.C., Onishi, R., da Silva, C.B.: The origin of the imbalance between energy cascade and dissipation in turbulence. Phys. Rev. E **90**, 023003 (2014)
13. Valente, P.C., Vassilicos, J.C.: Universal dissipation scaling for nonequilibrium turbulence. Phys. Rev. Lett. **108**, 214503 (2012)
14. Valente, P.C., Vassilicos, J.C.: The energy cascade in grid-generated non-equilibrium decaying turbulence (submitted) (2014). arXiv:1307.5901
15. Vassilicos, J.C.: Dissipation in turbulent flows. Annu. Rev. Fluid Mech. **47**, 95–114 (2015)
16. Yoshizawa, A.: Nonequilibrium effect of the turbulent-energy-production process on the inertial-range spectrum. Phys. Rev. E **49**, 5 (1994)

From Time to Space and Back: Convection and Wave Velocities in Turbulent Shear Flows

B. Ganapathisubramani and R. de Kat

Abstract To obtain spatial spectra, experimentalists map temporal spectra using Taylor's hypothesis [18]. This transfer function relates wavenumber to frequency using a convection velocity, a velocity that proves difficult to define and use. Adding to this difficulty, convection and wave velocities have been treated the same, despite the obvious difference between them. Convection velocities are velocities at which a specific structure moves—group velocity—and wave velocities are velocities at which a single wave moves—phase velocity. The ideal mapping function to go between wavenumber and frequency is the wavenumber-frequency spectrum that shows a range of wave velocities can contribute to a single wavenumber or a single frequency, and this range can differ significantly between them. To try to capture the influence of this range, experimentalists have applied various peak or average wave velocities with varying success. However, none account for the spread in wave velocities directly. In this paper we propose a two-point cross-spectral approach that uses a distribution of wave velocities to reconstruct the wavenumber-frequency plane. This plane can then be integrated to obtain the spatial spectrum. We verify the technique on particle image velocimetry data set of a turbulent boundary layer, and we obtain a transfer function from this data set. The transfer function is applied to hot-wire data at a comparable Reynolds, and comparison of the newly proposed technique with the classic Taylor's hypothesis approach shows that Taylor's hypothesis hold for larger frequencies and wavenumbers the smaller frequencies and wavenumbers (long temporal and spatial scales) there are significant differences.

1 Introduction

To infer spatial characteristics from temporal signals, experimentalists have long used a time-space mapping that originates from Taylor's hypothesis of frozen turbulence [18]. Taylor assumes that—at a point—the change of turbulent velocity fluctuations in time can be directly related to their spatial change—via the mean [convection]

B. Ganapathisubramani (✉) · R. de Kat
Faculty of Engineering and the Environment, University of Southampton, Southampton, UK
e-mail: g.bharath@soton.ac.uk

J. Peinke et al. (eds.), *Progress in Turbulence VI*,
Springer Proceedings in Physics 165, DOI 10.1007/978-3-319-29130-7_8

velocity. Lin [15] shows that, while the hypothesis holds in isotropic homogenous convective turbulence, for (wall-bounded) shear flows the hypothesis breaks down and is restricted to a limited range of wavenumbers (or frequencies).

In order to still be able to use Taylor's hypothesis to map temporal spectra in to space, experimentalists focus on finding an appropriate convection velocity (or velocities) to replace the mean velocity. They use two-point (cross-)correlation of time signals and apply (time-)filtering to get frequency dependent convection velocities [3, 4, 8, 9, 12, 19]. An alternative method is to use a cross-spectral approach where the convection velocity is determined by the phase angle between two-point cross-spectra [2, 10, 17], which is similar to the spectral filtering techniques, where both are intended to overcome the dependancy of the convection velocity on the time delay and separation distance.

Using particle image velocimetry (PIV), large regions of flow can be sampled at the same time and these large fields-of-view (FOVs) combined with high-repetition-rate cameras and lasers allows to directly investigate Taylor's hypothesis and convection velocities [6, 7].

Because experimentalists are keen to compare data with computations (and vice versa) and find the appropriate mapping from time-spectra to space-spectra, they shifted their focus from correlation based techniques to spectral approaches [5, 13, 14, 16].

Most of these studies find that the convection of specific frequencies (waves) is frequency dependent and their results suggest that there exists a range of these 'convection' velocities, which they try to account for using averages or peaks of this distribution. Adding to the confusion is the difference between convection velocities (group velocities) of turbulent velocity fluctuations and the wave velocity (phase velocity) of the waves that form the turbulent velocity fluctuations, and this difference has become blurred. As a result, the efforts of finding a fix for Taylor's hypothesis by applying different 'convection' velocities are problematic.

Therefore, we determine a subset of the ideal mapping—the wavenumber-frequency spectrum—for a turbulent boundary layer using PIV data and show that wave velocities have a range of values that need to be taken into account. We will also propose a new way of estimating the wavenumber-frequency spectrum using a cross-spectral approach, which allows us to map spectra from wavenumber to frequency or vice versa without the issues related to defining and determining convection velocities. This new approach is verified on the same data that is used to obtain the subset of the ideal mapping. Finally, a complete transfer is determined from our PIV data and this transfer function is applied to a hot-wire data set at a comparable Reynolds number to show the new approach and the classic Taylor's hypothesis approach differ.

2 Experimental Setup and Methodology

A time resolved PIV experiment is performed in a streamwise wall-normal plane in a turbulent boundary layer in the water tunnel at Cambridge University Engineering Department. The test section of the water tunnel is 0.4 m deep and 0.8 m wide and

8 m long. The flow is tripped with a glass rod at the beginning of the test-section and particle image velocimetry measurements are performed 4.5 m downstream of this trip. At this location, nominal flow conditions are the following, the free-stream velocity is $U_\infty = 0.67\,\mathrm{m\,s^{-1}}$, the boundary layer thickness is $\delta = 0.1\,\mathrm{m}$, the friction velocity based on a Clauser plot is $U_\tau = 0.027\,\mathrm{m\,s^{-1}}$, and the corresponding Reynolds number is $Re_\tau = 2.7 \times 10^3$.

Particle image pairs are captured and processed using LaVision software DaVis 7.2. The field of view (FOV) of the measurements that covered an area of 17 × 4.5 cm ($1.7\delta \times 0.45\delta$) with a spatial resolution of $h^+ = 10$ ($l^+ = 20$) and a temporal resolution of $\Delta t^+ = 0.7$. A total of 25,000 velocity fields were acquired spanning a time interval of more than 300 δ/U_∞.

3 Wave Velocities

Particle image velocimetry data is—as is experimental data in general—limited in domain length and resolution. To minimise effects of truncation and account for differences in resolution, we determine power spectral density by taking the following windowed and truncated two-dimensional Fourier transform of the velocity fluctuations:

$$\hat{u}(\xi_x, f) = \frac{1}{\sqrt{XT\int_0^X\int_0^T w(x,t)^2 dtdx}} \int_0^X \int_0^T w(x,t)u'(x,t)e^{-i2\pi(ft+\xi_x x)}dtdx \tag{1}$$

where X is the domain length, T the sample time interval length, w the weighting function. After which, the power spectral density is estimated by:

$$\boldsymbol{\Phi}_{uu}(\xi_x, f) = \hat{u}^*\hat{u} \tag{2}$$

where $\hat{u}^*$ is the conjugate of $\hat{u}$.

Figure 1 shows power spectral densities with wavenumber and frequency of streamwise and wall-normal velocity fluctuations at $y/\delta = 0.1$. Wavenumber is shown in the horizontal direction and frequency in the vertical direction. This wavenumber-frequency mapping is what is required in order to obtain either the wavenumber or frequency spectrum by integrating this plane along one direction. The figure clearly shows the inclined nature of the spectrum suggesting a convecting flow. There are several ways to determine this wave velocity. The inclination of the contours provides a wave velocity that we can use to map time to space. At this wall-normal location, it can be shown that a best fit regression line would infact have a slope that is equal to the mean streamwise velocity at this plane. Other solutions to define a single wave velocity linking frequency to wavenumber have been proposed (for example, [13] traced the ridge). All these wave velocity definitions assume that

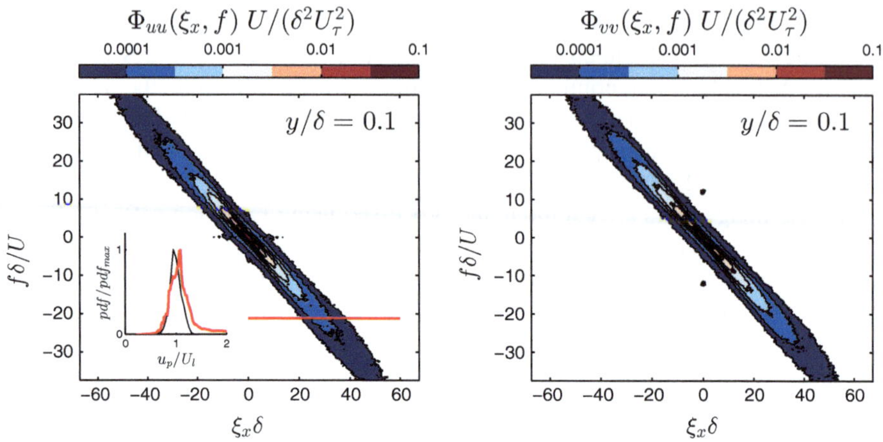

Fig. 1 Wavenumber frequency power spectral density of streamwise and wall-normal velocity fluctuations, $\boldsymbol{\Phi}_{uu}$ and $\boldsymbol{\Phi}_{vv}$ at $y/\delta = 0.1$. The inset shows the PDF of phase velocity (*red*) at a single frequency at this wall-normal location. Also shown in the inset is the PDF of instantaneous velocity fluctuations (*black*)

the peak of the power (or energy) spectral density is the best representative of the phase velocity. However, the actual spectra show a distribution around this peak and, therefore, [5] proposed a weighted average of the energy spectrum to define average wave velocities and as such take into account the whole distribution of energy over all frequencies f for a given wavenumber ξ_x or the whole distribution of energy over all wave numbers for a given frequency.

The above-mentioned methods still only takes a single wave velocity to capture a range of wave velocities. The inset in Fig. 1a shows a PDF of wave velocities at a single frequency at $y/\delta = 0.1$. The figure clearly shows that there is wide distribution of these wave velocities even at a single frequency. The figure also shows the PDF of streamwise velocity fluctuations and it can be seen that the range of wave velocities is larger than the range of instantaneous velocity fluctuations. Moreover, the distribution of wave velocities is in fact skewed to the positive side. Therefore, any averaging (weighted or otherwise) or peak tracing methods will not really capture the true extent of wave velocities. Therefore, it is necessary to capture this range entirely in order to map the frequencies to wavenumbers accurately.

4 Estimating the Wavenumber-Frequency Spectrum

A mapping from wavenumber to frequency (and vice versa) can be determined by using a cross-spectral approach and it requires two-point temporal measurements. Spatial spectra are determined for two different time instances and the difference in spatial phase between the two spectra is a measure for the distance a wave has traveled during the time in between the two instances [1]. Here we use a cross-spectral approach to determine this phase difference by using temporal data. We start with the one-dimensional truncated weighted Fourier transform and determine the auto-spectral power density in terms of frequency from a time series of data. We then calculate the cross-spectrum between two different spatial locations separated by a small distance (Δx), from which the phase spectrum is obtained. Fourier transforms describe pure sinusoidal waves for a specific wavenumber or frequency, therefore, the resulting phase shift can be used to estimate a corresponding frequency or wavenumber, which we will refer to as the pseudo-wavenumber, $\tilde{\xi}_x$. Using this pseudo-frequency or pseudo-wavenumber and the corresponding cross-spectral power density, a transfer function can be created by unfolding multiple cross-spectra in pseudo-wavenumber or pseudo-frequency direction, binning them, and normalising them into a probability density distributions in the unfolding direction (integral equal to one for each unfolding line). The wavenumber-frequency spectrum is then obtained by multiplying these probability density distributions with the respective auto-spectral power density of the wavenumber or frequency that is being unfolded. Finally, integration in frequency direction results in the wavenumber spectrum.

The estimated wavenumber-frequency spectra based on the above mentioned approach is shown in Fig. 2. Comparison of this figure with Fig. 1 (which shows the true spectra) reveals some differences. Most notably, the spectra appear to have a waist near zero frequency/wavenumber and this can be related to the truncation of the PIV data. A larger sampling window will eliminate this truncation effect. The spectra at higher frequency/wavenumber appears to have more noise compared to the true spectra. This again is a limitation of the technique where the velocity fluctuations at those small spatial and temporal scales become comparable to measurement noise. Despite these two sources of uncertainty, it is clear that the estimated frequency-wavenumber spectrum has a remarkbale similarity the true spectrum. This similarity

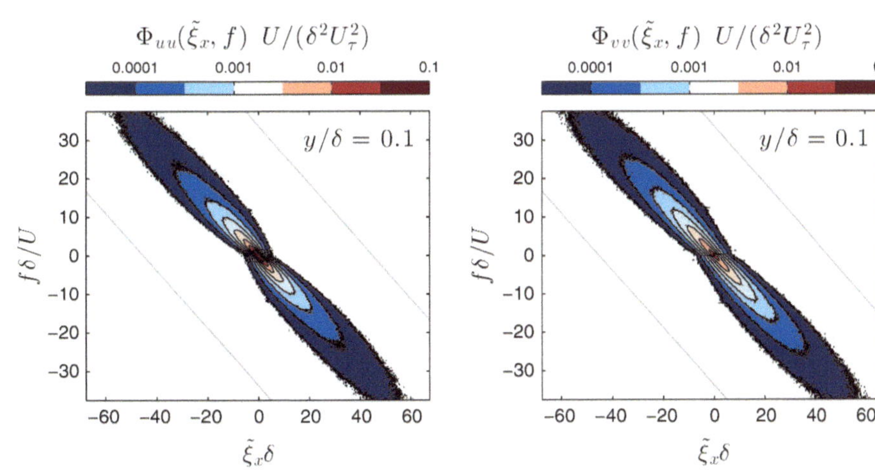

Fig. 2 Reconstructed wavenumber frequency power spectral density of streamwise and wall-normal velocity fluctuations

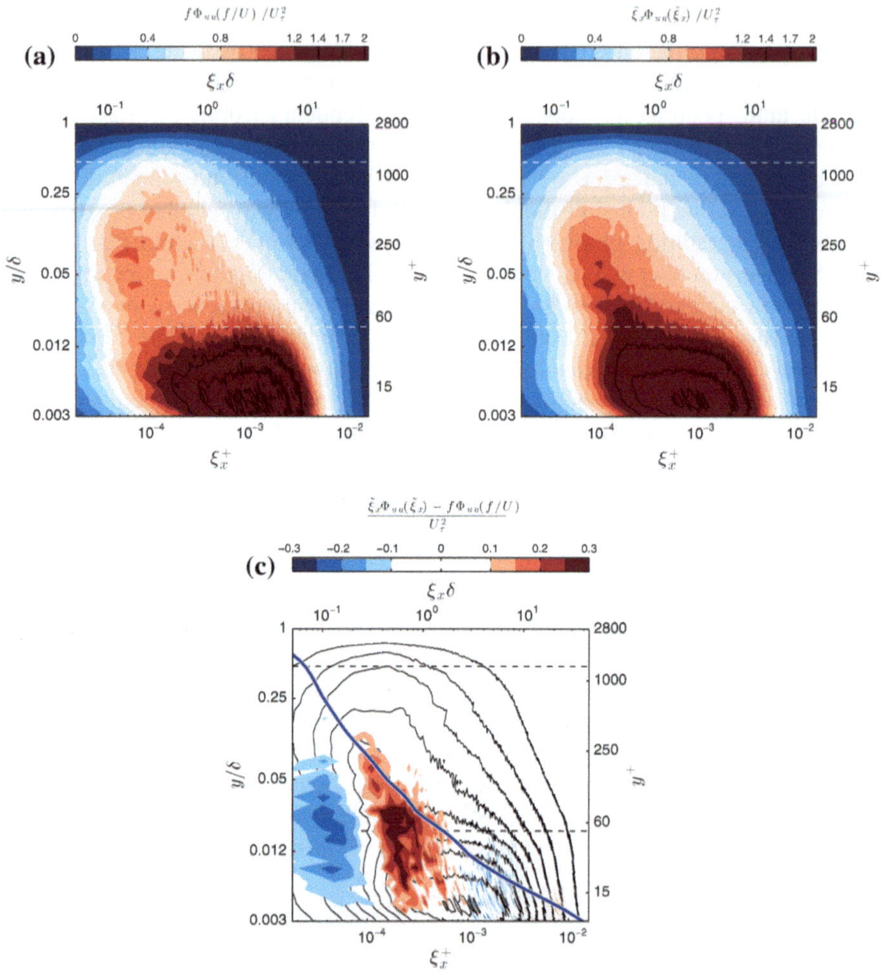

Fig. 3 Premultiplied mapped power density spectra of hot-wire data at a Reynolds number of $Re_\tau = 2820$ from [11] with wall-normal location, y/δ. *Dashed white* lines demarcate the region where mapping data is directly available. Outside this region we correct the closest available transfer function to conform to the local mean velocity. **a** Classic Taylor's hypothesis mapping using the local mean velocity. **b** Our proposed mapping based on cross-sectra. **c** Difference between Taylor's hypothesis and the mapped spectra with the line showing $\xi_x U = dU/dy$

is even higher in the spectra of the wall-normal velocity fluctuations where the truncation effects are minimal (i.e. the PIV FOV as well as the time-series captures all the energetic spatial and temporal scales). This estimated frequency-wavenumber spectra can now be integrated across all frequencies to obtain the wavenumber spectrum. Similar mapping functions can be obtained across different locations (i.e. wall-normal or cross stream directions) in a shear flow to obtain the variation of the wavenumber spectrum across the flow.

Since the PIV data is limited in resolution and length, we apply the mapping function we obtained from our current PIV data to hot-wire data at a comparable Reynolds number [11]. Figure 3a shows the spectrogram (i.e. wavenumber spectra across wall-normal locations) of streamwise velocity from hotwire measurements. The wavenumber (or the length scale) is obtained by applying Taylor's hypothesis. The same hotwire data is used in conjunction with the mapping generated for different wall-normal locations using our method to calculate the new wavenumber spectrum. Figure 3b shows the resulting spectrogram. The white lines show the wall-normal extent over which we have the data for the mapping function and hence any comparison between the two spectrograms should only be made over this range of wall-normal locations. It can clearly be seen from the two figures that the spectrogram that uses Taylor's hypothesis is significantly different from the one obtained using the mapping.

To examine the differences in detail, the difference between Taylor's hypothesis mapping and our proposed mapping is shown in Fig. 3c. It can be seen that there are large difference between the spectra (up to 40 %) at low wavenumbers. This is consistent with the prediction of [15], who indicated that Taylor's hypothesis is really only applicable for wavenumbers for which $\xi_x U \gg dU/dy$ (where U is the local mean streamwise velocity). Figure 3c shows a line that represents Lin's prediction, $\xi_x U = dU/dy$. We see that his prediction demarcates the region where the mappings differ significantly.

5 Conclusions

We proposed a two-point cross-spectral approach that uses a distribution of wave velocities (or the distribution of cross-spectral phase difference) to reconstruct the entire wavenumber-frequency plane. This plane can then be integrated to obtain the spatial or temporal spectrum depending on the input data. We verified this technique on particle image velocimetry data set of a turbulent boundary layer, and obtained a transfer function from this data set. The transfer function is applied to hot-wire data at a comparable Reynolds number. A comparison of the newly proposed technique with the classical Taylor's hypothesis approach shows that Taylor's hypothesis hold for larger wavenumbers (small spatial scales), for smaller wavenumbers (large spatial scales) there are significant differences. It is hoped that this technique can now be employed in other shear-flows as well as in wall-bounded flows at higher Reynolds numbers in order to examine the energy content in spatial scales without invoking Taylor's hypothesis.

Acknowledgments We gratefully acknowledge the support from UK Engineering and Physical Sciences Research Council (EPSRC) through Grant EP/I004785/1. The research leading to these results has also received funding from the European Research Council under the European Union's Seventh Framework Programme (FP7/2007–2013)/ERC Grant agreement no 277472-WBT

References

1. Buxton, O.H.R., de Kat, R., Ganapathisubramani, B.: The convection of large and intermediate scale fluctuations in a turbulent mixing layer. Phys. Fluids **25**(125), 105 (2013)
2. Cedenese, A., Romano, G.P., Di Felice, F.: Experimental testing of Taylor's hypothesis by L.D.A. in highly turbulent flow. Exp. Fluids **11**, 351–358 (1991)
3. Davies, P.O.A.L., Fisher, M.J.: Statistical properties of the turbulent velocity fluctuations in the mixing region of a round subsonic jet. AASU report 233, University of Southampton (1963)
4. Davies, P.O.A.L., Fisher, M.J., Barrat, M.J.: The characteristics of the turbulence in the mixing region of a round jet. J. Fluid Mech. **15**, 337–367 (1963)
5. del Álamo, J.C., Jiménez, J.: Estimation of turbulent convection velocities and corrections to Taylor's approximation. J. Fluid Mech. **640**, 5–26 (2009)
6. Dennis, D.J., Nickels, T.B.: On the limitations of Taylor's hypothesis in constructing long structures in a turbulent boundary layer. J. Fluid Mech. **614**, 197–206 (2008)
7. Elsinga, G.E., Poelma, C., Schröder, A., Geisler, R., Scarano, F., Westerweel, J.: Tracking of vortices in a turbulent boundary layer. J. Fluid Mech. **697**, 273–295 (2012)
8. Fisher, M.J., Davies, P.O.A.L.: Correlation measurements in a non-frozen pattern of turbulence. J. Fluid Mech. **18**, 97–116 (1964)
9. Goldschmidt, V.W., Young, M.F., Ott, E.S.: Turbulent convective velocities (broadband and wavenumber dependent) in a plane jet. J. Fluid Mech. **105**, 327–345 (1981)
10. Harrison, M.: Pressure fluctuations on the wall adjacent to a turbulent boundary layer. Technical Report 1260, David Taylor Model Basin (1958)
11. Hutchins, N., Nickels, T.B., Marusic, I., Chong, M.S.: Hot-wire spatial resolution issues in wall-bounded turbulence. J. Fluid Mech. **635**, 103–136 (2009)
12. Krogstad, P., Kaspersen, J.H., Rimestad, S.: Convection velocities in a turbulent boundary layer. Phys. Fluids **10**(4), 949–957 (1998)
13. LeHew, J., Guala, M., McKeon, B.J.: A study of the three-dimensional spectral energy distribution in a zero pressure gradient turbulent boundary layer. Exp. Fluids **51**, 997–1012 (2011)
14. LeHew, J., Guala, M., McKeon, B.J.: Time-resolved measurements of coherent structures in the turbulent boundary layer. Exp. Fluids **54**, 1508 (2013)
15. Lin, C.C.: On Taylor's hypothesis and the acceleration terms in the Navier-Stokes equations. Q. Appl. Maths. **10**(4), 295–306 (1953)
16. Monty, J.P., Chong, M.S.: Turbulent channel flow: comparison of streamwise velocity data from experiments and direct numerical simulation. J. Fluid Mech. **633**, 461–474 (2009)
17. Romano, G.P.: Analysis of two-point velocity measurements in near-wall flows. Exp. Fluids **20**, 68–83 (1995)
18. Taylor, G.I.: The spectrum of turbulence. Proc. R. Soc. Lond. **164**(919), 476–490 (1938)
19. Wills, J.A.B.: Convection velocities in turbulent shear flows. J. Fluid Mech. **20**, 417–432 (1964)

Concurrent Scale Interactions in the Far-Field of a Turbulent Mixing Layer

O.R.H. Buxton and B. Ganapathisubramani

Abstract The interaction between large- and small-scale fluctuations in turbulent flow is not only of great fundamental interest but an understanding of these interactions is fundamental to the modelling of the sub-grid scale (SGS) stresses in a large eddy simulation (LES). Particle image velocimetry (PIV) data is acquired with two different spatial resolutions, simultaneously, in the self-similar region of a turbulent planar mixing layer. The SGS activity is observed to be amplified by concurrent large-scale low momentum fluctuations and attenuated by high momentum fluctuations with both the sign and magnitude of the large-scale fluctuations being of significance. Further, regions in which the orientation of the large-scale Reynolds stress tensor is aligned perpendicularly to the direction of the mean shear of the flow are shown to lead to an increased level of small-scale activity.

1 Introduction

Turbulence is known to be a multi-scale problem, in which energy is transferred from the mean flow into turbulent kinetic energy at large-scales and dissipated into heat at the small-scales via a mean cascade of energy from the large to the small-scales [3, 6]. According to the phenomenology of Kolmogorov [3] the small-scales of turbulent flows are universal, but recently a distinct interaction between the large and small-scales has been observed. These results have primarily been driven by research in wall bounded flows, in particular the modulation of the near wall events by large-scale outer fluctuations [4, 5].

The study of [1] was the first (and perhaps the only) that examined the interaction between large and small scales in various different shear flows, including wall-bounded and free shear flows. Correlations were made between the low pass

O.R.H. Buxton (✉)
Department of Aeronautics, Imperial College London, London, UK
e-mail: o.buxton@imperial.ac.uk

B. Ganapathisubramani
Aerodynamics and Flight Mechanics Research Group,
University of Southampton, Southampton, UK

J. Peinke et al. (eds.), *Progress in Turbulence VI*,
Springer Proceedings in Physics 165, DOI 10.1007/978-3-319-29130-7_9

filtered time series data with the envelope of the high frequency component and a significant degree of coupling between the scales across all shear flows was observed. This coupling between the scales is maximised when the high frequency and low frequency signals are concurrent.

Here we examine the interactions between large- and small-scale velocity fluctuations in the self-similar region of a turbulent mixing layer. PIV experiments are performed at two different spatial resolutions, one that captures the range from integral scale (L) to Taylor microscale (λ) and the other that captures the range from Taylor microscale to the Kolmogorov scale (η), simultaneously.

2 Experimental Methods

A planar mixing layer was produced by placement of a perforated metal sheet, 50 % open area ratio, on one side of a splitter plate of length 1.25 m and thickness, $h = 20$ mm, in a recirculating water tunnel facility. A customised PIV setup was constructed in order to capture data at two different spatial resolutions simultaneously consisting of four 2048 × 2048 pixel resolution CCD cameras. Three of the cameras were mounted below the floor of the water tunnel facility and were fitted with 105 mm lenses and the final camera was mounted above the water channel and fitted with a 50 mm lens, all imaging a light sheet in the streamwise—cross-stream plane.

The spatial resolution for the large-scale field of view (FOV) is $0.4\lambda \times 0.4\lambda$, with adjacent vectors separated by 0.2λ due to a 50 % overlap. The resolution for the small-scale fields of view is $1.2\eta \times 1.2\eta$ (which is comparable to the thickness of the laser sheet), with adjacent vectors separated by 0.62η. Figure 1a illustrates simultaneous velocity fields from both the large- and small-scale FOVs. It should be

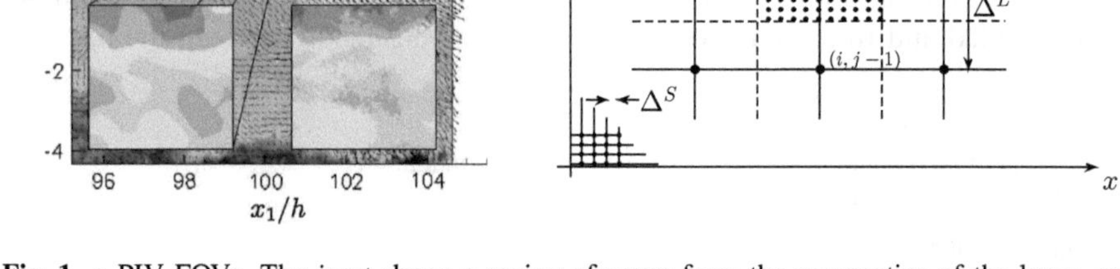

Fig. 1 **a** PIV FOVs. The *inset* shows a region of space from the perspective of the large- and small-scale FOVs. **b** Schematic of concurrent large-scale and small-scale FOV stencils

noted that the small-scale FOV is located close to the centreline but just to the high speed side ($x_2 \gtrsim 0$) of the cross-stream location of the peak Reynolds stresses and is situated within the self-similar region of the flow. The Reynolds number based on the Taylor length scale was $Re_\lambda = 260$ and mean velocity on the centre-line was $0.29\,\text{ms}^{-1}$ at this location. A full description of the experimental methods, including an uncertainty quantification and schematic figure is presented in [2].

3 Discussion and Statistical Methodology

At the measurement location it was observed that the Taylor microscale, defined as $\lambda^2 = 2u'^2 / \left\langle \left(\frac{\partial u_1}{\partial x_1}\right)^2 \right\rangle$, was the length scale that divided the dissipation spectrum into two sectors that each accounted for $\approx$50 % of the total dissipation rate. The large-scale FOV was thus filtered (running mean filter) to remove all components at length scales less than λ, leaving $u_1^{L\lambda}$, and the small-scale FOVs were similarly filtered to remove all components at length scales greater than λ, leaving $u_1^{S\lambda}$. In this manuscript the superscript L denotes that a quantity is taken from the large-scale FOV, superscript S denotes a quantity that is taken from one of the small-scale FOVs, velocities are decomposed into a mean component (upper case letter) and fluctuating component (lower case) and subscript 1 denotes the streamwise direction.

Figure 1b illustrates the overlap of the small-scale and large-scale FOV stencils. A "conditional window" is defined extending $\pm\Delta^L/2$ in both the x_1 and x_2 directions from the (i, j)th node of the large-scale stencil and encompasses $N \approx 100$ samples from the small-scale FOV. First and second order moments of the small-scale velocity fluctuations within this "conditional window" are then computed and statistics of these moments conditioned on the concurrent large-scale fluctuation, $u_1^{L\lambda}$, are produced.

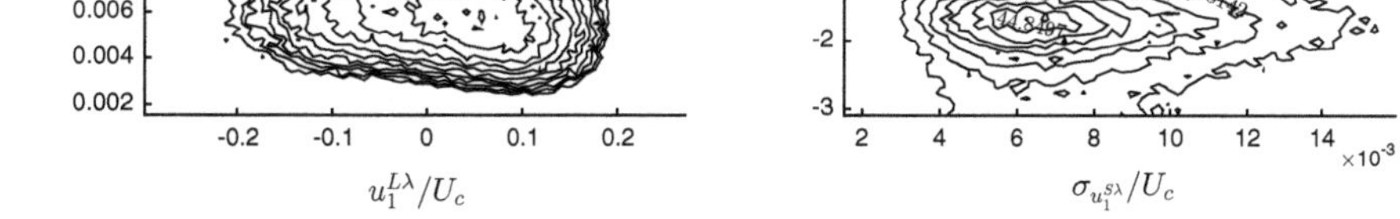

Fig. 2 **a** Joint pdf between the variance of the small-scale velocity fluctuations, $u_1^{S\lambda}$, and $u_1^{L\lambda}$. **b** Joint pdf between the variance of the small-scale velocity fluctuations and the "Reynolds stress angle" relative to the mean velocity gradient

Figure 2a shows the joint probability density function (pdf) between the variance of the small-scale fluctuations, $\sigma_{u_1^{S\lambda}}$, and $u_1^{L\lambda}$. It can be seen that there is a clear slope of the contours from high magnitude positive $u_1^{L\lambda}$ fluctuations concurrent with a low variance of the small-sale fluctuations towards high magnitude negative $u_1^{L\lambda}$ fluctuations and a concurrent high variance of the small-scale fluctuations. The dependence of the "roughness" of the small-scale fluctuations on $u_1^{L\lambda}$ can be explained with a convective argument. The location of the small-scale FOV to the high speed side of the peak Reynolds stresses means that, statistically, positive $u_2^{L\lambda}$ fluctuations will convect with them "rougher" small-scale turbulence. Since we are in a self preserving free shear flow $u_1^{L\lambda}$ is negatively correlated with $u_2^{L\lambda}$ ($\rho_{u_1 u_2} = -0.581$) due to the positivity of the turbulent kinetic energy production term, $\mathscr{P} = -\langle u_1 u_2 \rangle \partial \langle U_1 \rangle / \partial x_2$, thus we might expect to see negative $u_1^{L\lambda}$ fluctuations concurrent to "rougher" small-scale turbulence. However, the dependence of $\sigma_{u_1^{S\lambda}}$ on the magnitude of $u_1^{L\lambda}$ in addition to the sign confuses this picture and is suggestive of a true scale modulation effect.

Furthermore, Fig. 2b shows the effect of the orientation of the large-scale velocity fluctuation on the concurrent small-scale activity. A "Reynolds stress angle", θ_{uv}, is defined as the arctangent of $u_2^{L\lambda}/u_1^{L\lambda}$. Additionally a "mean strain angle" is defined as the arctangent of the non-dimensionalised mean velocity gradient $\partial \langle U_1 \rangle / \partial x_2$. The figure shows the joint pdf between the difference between the mean strain angle and the modified Reynolds stress angle (such that $\theta_{\frac{\partial \langle U_1 \rangle}{\partial x_2}} - \theta_{uv}{}' \in \{-\pi, \pi\}$) and the variance of the small-scale fluctuations. It can be seen that the small-scale activity is significantly amplified when the large-scale fluctuation is perpendicular to the mean strain, i.e. $\theta_{\frac{\partial \langle U_1 \rangle}{\partial x_2}} - \theta_{uv}{}' = \pm\pi/2$. The highest small-scale activity is observed for a positive value of θ_{uv} which would be described as an "ejection" in wall bounded terminology. This is further evidence that the amplification of the small-scale activity cannot be explained by a convective argument alone and that a genuine scale interaction is present in self-similar free shear flows.

References

1. Bandyopadhyay, P., Hussain, A.: The coupling between scales in shear flows. Phys. Fluids **27**(9), 2221–2228 (1984)
2. Buxton, O.: Fine scale features of turbulent shear flows. PhD thesis (2011)
3. Kolmogorov, A.: The local structure of turbulence in incompressible viscous fluid for very large Reynolds numbers. C.R. Acad. Sci. URSS **30**(4), 301–305 (1941)
4. Mathis, R., Hutchins, N., Marusic, I.: Large-scale amplitude modulation of the small-scale structures in turbulent boundary layers. J. Fluid Mech. **628**, 311–337 (2009)
5. Priyadarshana, P., Klewicki, J., Treat, S., Foss, J.: Statistical structure of turbulent-boundary-layer velocity—vorticity products at high and low Reynolds numbers. J. Fluid Mech. **570**, 307–346 (2007)
6. Richardson, L.: Atmospheric diffusion shown on a distance-neighbour graph. Proc. R. Soc. Lond. A **110**(756), 709–737 (1926)

Euler Angle and Axis—"Fingerprints" of a Subgrid-Scale Stress Model

Zixuan Yang and Bing-Chen Wang

Abstract The concepts of Euler angle and axis are utilized to investigate the relative rotation between the eigenframes of the deviatoric subgrid-scale (SGS) stress tensor $-\tau_{ij}^d$ and the resolved strain rate tensor $\bar{S}_{ij}$. Both Euler angle and axis are "natural invariants" of fluid tensors, which uniquely describe the relative rotation between eigenframes of two tensors. The Euler angle and axis can be regarded as "fingerprints" of a SGS stress model and have a profound implication for structural modeling of the SGS stress tensor. As an application, three SGS models are tested in the context of turbulent channel flows. The proposed Euler angle and axis are proven to be effective for demonstrating geometrical properties of a SGS stress model.

1 Introduction

The rotational dynamics between two orthonormal triads are essential for studying the relative attitude between two or more rigid bodies. The methodology of attitude dynamics can be introduced into SGS stress modeling to provide an interesting alternative approach for investigating the local geometrical property of the constitutive relationship. In order to study the attitude of the eigenframe of $-\tau_{ij}^d$ with respect to that of $\bar{S}_{ij}$, Tao et al. [1] adopted the so-called axis-azimuth representation, which however, is not uniquely defined, but rather depends on the specific method for decomposing the relative rotation between two eigenframes.

Based on Euler's theorem (1775), Wang et al. [2] suggested that the relative rotation of the eigenframe of $-\tau_{ij}^d$ with respect to that of $\bar{S}_{ij}$ can be uniquely quantified using the Euler angle χ about the Euler axis $\boldsymbol{q}$ (see Fig. 1). They further tested several SGS models using their proposed representation. However, their results were not compared against any direct numerical simulation (DNS) or experimental results, and as such, their findings on model performances were not conclusive. In view of

Z. Yang · B.-C. Wang (✉)
Department of Mechanical Engineering, University of Manitoba,
Winnipeg, MB R3T 5V6, Canada
e-mail: BingChen.Wang@ad.umanitoba.ca

J. Peinke et al. (eds.), *Progress in Turbulence VI*,
Springer Proceedings in Physics 165, DOI 10.1007/978-3-319-29130-7_10

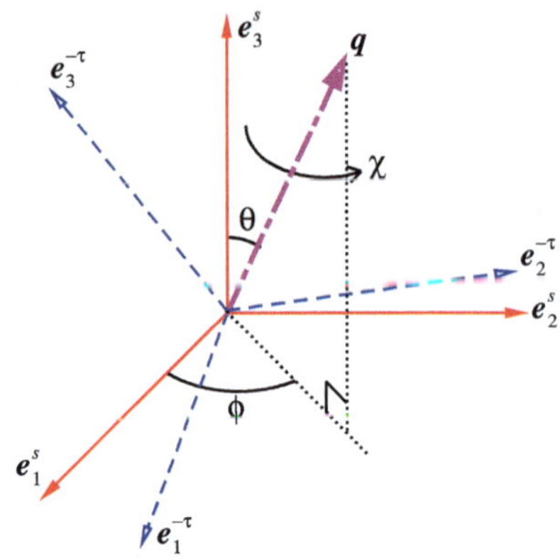

Fig. 1 Sketch of Euler angle and axis between eigenframes of $-\tau_{ij}^d$ and $\bar{S}_{ij}$. The colatitude and longitude with respect to the eigenframe of $\bar{S}_{ij}$ are denoted by θ and ϕ, respectively

this, we test the Euler angle and axis of different SGS stress models in the context of turbulent channel flows using the a posteriori approach and validate the results predicted by large eddy simulations (LES) against the DNS results.

2 Definitions of Euler Angle and Axis

To study the geometrical property of SGS stress models, we use $\boldsymbol{E}^S = [\boldsymbol{e}_1^S, \boldsymbol{e}_2^S, \boldsymbol{e}_3^S]$ (eigenframe of $\bar{S}_{ij}$) as the observer-referenced (or absolute) frame, and treat $\boldsymbol{E}^{-\tau} = [\boldsymbol{e}_1^{-\tau}, \boldsymbol{e}_2^{-\tau}, \boldsymbol{e}_3^{-\tau}]$ (eigenframe of $-\tau_{ij}^d$) as the object-referenced (or relative) frame. Here, $\boldsymbol{e}_i^{-\tau}$ and $\boldsymbol{e}_i^S$ are eigenvectors corresponding to eigenvalues $\lambda_i^{-\tau}$ and λ_i^S, respectively, and the eigenvalues are sorted in the descending order of their absolute values, i.e. $|\lambda_1^{-\tau}| \geq |\lambda_2^{-\tau}| \geq |\lambda_3^{-\tau}|$ and $|\lambda_1^S| \geq |\lambda_2^S| \geq |\lambda_3^S|$.

The Euler angle and axis are natural invariants of the rotation matrix $\boldsymbol{R}$, which can be calculated as $\boldsymbol{R} = \boldsymbol{E}^{-\tau}(\boldsymbol{E}^S)^T$. The Euler angle χ is related to the trace of $\boldsymbol{R}$ as $\chi = \arccos[(\mathtt{tr}(\boldsymbol{R}) - 1)/2]$, and the Euler axis $\boldsymbol{q} = [q_1, q_2, q_3]^T$ can be determined as $q_i = \varepsilon_{ijk} R_{jk}/(2 \sin \chi)$, where ε_{ijk} represents the Levi-Civita symbol.

3 Numerical Algorithm and Test Cases

In order to test the proposed method, DNS and LES of turbulent channel flows are conducted using a pseudo-spectral method code. Variables are expanded into Fourier series in x–z planes and into Chebyshev polynomial series in the y direction, where x, y and z represent streamwise, wall-normal and spanwise directions, respectively. Time advancement is achieved by using a 3rd order time-splitting method.

DNS and LES of turbulent channel flows have been conducted for $Re_\tau = u_\tau h/\nu = 595$, where u_τ, h and ν represent wall frictional velocity, half channel height and kinematic viscosity of the fluid, respectively. The computational domain size is $L_x \times L_y \times L_z = 2\pi h \times 2h \times \pi h$. The number of grid points is $N_x \times N_y \times N_z = 384 \times 256 \times 384$ for DNS, and is $64 \times 96 \times 64$ for LES. We have tested three SGS stress models, which include the dynamic Smagorinsky model (DSM) of Lilly [3], dynamic mixed model (DMM) of Morinishi and Vasilyev [4] and dynamic nonlinear model (DNM) of Wang et al. [5]

4 Results Analysis

Figure 2 compares the profiles of mean Euler angle $\langle\chi\rangle$ of different SGS stress models. As shown by the DNS results, the mean Euler angle reaches 45° asymptotically as the wall is approached. In the far-wall region for $y^+ > 100$, the value of $\langle\chi\rangle$ predicted by DNS keeps approximately constant around 72°.

In sharp contrast to the DNS results, the value of $\langle\chi\rangle$ predicted by the DSM is 0° identically. The reason $\langle\chi\rangle \equiv 0°$ holds strictly for the DSM is that this model assumes a linear constitutive relationship between $-\tau_{ij}^d$ and $\bar{S}_{ij}$, characterized by the unique condition $\boldsymbol{R} \equiv \boldsymbol{I}$ and $\chi \equiv 0°$. As a result, the equation for determining the Euler axis $\boldsymbol{q}$ becomes singular as the denominator becomes zero and the Euler axis of the DSM is undefined. Both DMM and DNM correctly predict the wall value of the Euler angle, i.e. $\langle\chi\rangle_w = 45°$. In the far-wall region, the DMM under-predicts $\langle\chi\rangle$ as 57°, while the DNM makes a better prediction in comparison with DNS results.

Figure 3 compares the PDF of Euler angle $P(\chi)$ in the viscous sublayer ($y^+ = 1$), buffer layer ($y^+ = 11$) and logarithmic layer ($y^+ = 500$). The DNS results on $P(\chi)$ is approximately a delta function at $\chi = 45°$ in the viscous sublayer, such that the mean value of χ is also 45°. As the distance from the wall increases, the probability of $\chi = 90°$ increases. In the logarithmic layer, mode $\chi = 90°$ becomes predominant. Both DMM and DNM predict $P(\chi)$ as a delta function at $\chi = 45°$ in the viscous sublayer, which is consistent with the DNS result. In the logarithmic layer, the DMM fails to capture the peak at $\chi = 45°$ and significantly over-predicts the probability

Fig. 2 Profiles of plane- and time-averaged Euler angle

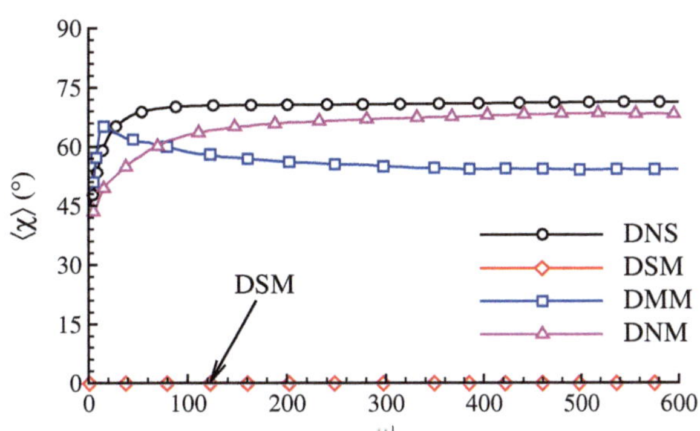

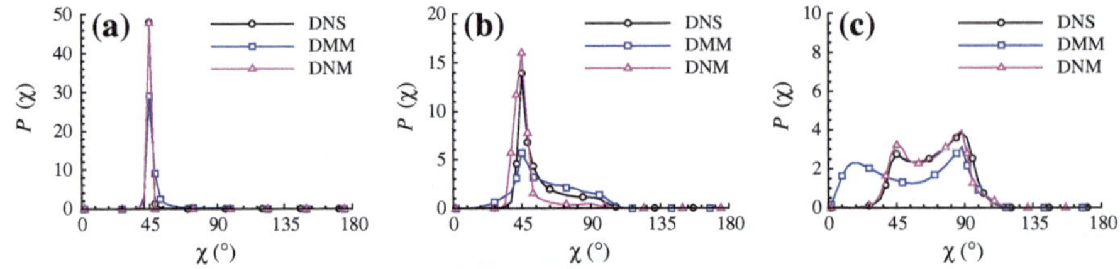

Fig. 3 PDF of Euler angle $P(\chi)$ in different layers of the turbulent channel flow: **a** viscous sublayer, $y^+ = 1$, **b** buffer layer, $y^+ = 11$ and **c** logarithmic layer, $y^+ = 500$

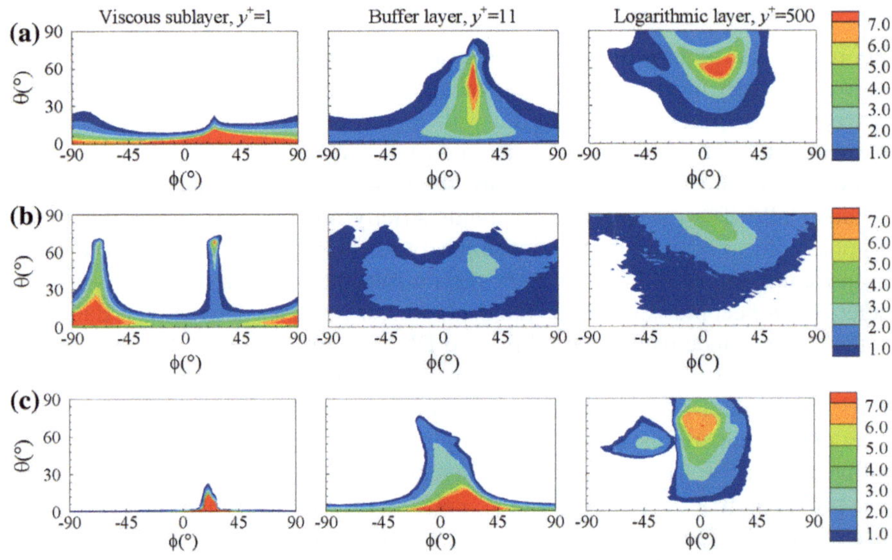

Fig. 4 Contours of pre-multiplied JPDF $(\theta, \phi) \sin\theta$ related to the orientation of Euler axis predicted by **a** DNS, **b** DMM and **c** DNM. Contours corresponding to low probability ($P(\theta, \phi) \sin\theta < 0.5$) are clipped

of modes for $\chi \leq 30°$. The PDF of Euler angle predicted by the DNM agrees well with the DNS results in all three layers. As a result, the mean Euler angle predicted by the DNM shown in Fig. 2 is also close to DNS results.

Figure 4 compares the DNS and LES results on the pre-multiplied JPDF of (θ, ϕ) that quantifies the orientation of the Euler axis. As shown in Fig. 4a, the most probable mode predicted by DNS in the viscous sublayer is located at $\theta = 0°$ with ϕ being an arbitrary angle. The most probable modes obtained from DNS are $(\theta, \phi) = (41°, 23°)$ and $(62°, 15°)$ in the buffer layer and logarithmic layer, respectively.

As shown in Fig. 4b, the DMM predicts a mode for $(\theta, \phi) = (70°, 23°)$ in the viscous sublayer, which however, is not shown in the contours of $P(\theta, \phi) \sin\theta$ obtained from DNS. In the buffer layer and logarithmic layer, the most probable modes predicted by the DMM are $(\theta, \phi) = (52°, 29°)$ and $(88°, 2°)$, respectively, which deviate from the DNS results. From Fig. 4c, it is observed that the colatitude θ of the most probable mode predicted by the DNM is 0° in the viscous sublayer, which is consistent with the DNS results. However, in the buffer layer, the predominant colatitude value is still 0°, indicating that the DNM over-estimates the wall effect on the Euler axis. In the logarithmic layer, the most probable Euler axis predicted by the DNM corresponds to $(\theta, \phi) = (70°, 2°)$, which does not agree well with the DNS results. The above results on Euler axis indicate that as SGS models, the ability of both DMM and DNM is still limited in terms of their prediction of advanced geometrical properties of the SGS stress tensor.

5 Conclusions

As natural invariants of the SGS stress and resolved strain rate tensors, the proposed Euler angle and axis are proven to be effective for demonstrating the geometrical properties of a SGS stress model and for evaluating its capability of predicting the near-wall anisotropic effects. The mean Euler angle based on the DNS results is 45° in the viscous sublayer, and reaches 72° in the center of the channel. The mean Euler angle predicted by the DSM is 0° identically, indicating that the DSM fails to mimic the near-wall effect on the geometrical property of the SGS stress tensor. The Euler angle predicted by the DNM agrees well with the DNS results. In contrast, the Euler angle predicted by the DMM deviates from the DNS results significantly in the logarithmic layer. The colatitude of the most probable Euler axis predicted by DNS is 0° in the viscous sublayer, which is correctly captured by the DNM, but in the buffer layer and logarithmic layer, the capability of both DMM and DNM is limited in terms of their prediction of the orientation of the Euler axis.

References

1. Tao, B., Katz, J., Meneveau, C.: Statistical geometry of subgrid-scale stresses determined from holographic particle image velocimetry measurements. J. Fluid Mech. **457**, 35–78 (2002)
2. Wang, B.-C., Yee, E., Bergstrom, D.J.: Geometrical description of subgrid-scale stress tensor based on Euler axis/angle. AIAA J. **44**, 1106–1110 (2006)
3. Lilly, D.K.: A proposed modification of the Germano subgrid scale closure method. Phys. Fluids A. **4**(3), 633–635 (1992)
4. Morinishi, Y., Vasilyev, O.V.: A recommended modification to the dynamic two-parameter mixed subgrid scale model for large eddy simulation of wall bounded turbulent flow. Phys. Fluids. **13**(11), 3400–3410 (2001)
5. Wang, B.-C., Bergstrom, D.J.: A dynamic nonlinear subgrid-scale stress model. Phys. Fluids. **17**(3), 035109 (2005)

References

Low-Cost Energy-Preserving RK Schemes for Turbulent Simulations

Francesco Capuano, Gennaro Coppola and Luigi de Luca

Abstract A low-cost numerical strategy is presented for energy-preserving simulations of incompressible flows. This method consists in an explicit Runge–Kutta scheme in which the divergence and advective forms of the convective term are suitably alternated within the sub-steps. As a result, the conservation properties of the skew-symmetric form are recovered at a reduced computational cost. The performances and the accuracy of the method are proved by numerical simulations.

1 Introduction

Global kinetic energy $e = \int_{\Omega} u_i^2/2\,dV$ is a well-known invariant of the incompressible Navier-Stokes equations, in the limit of vanishing viscosity and on a periodic domain Ω. Numerical methods that aim to reproduce this property on a discrete level are highly desirable in Direct and Large-Eddy Simulation of turbulent flows, and can be obtained by adopting suitable spatial and temporal discretization operators [1]. For the former, semi-discrete conservation is often achieved by using the so-called skew-symmetric splitting of the convective term

$$(Skew.)_i \equiv \frac{1}{2}\frac{\partial u_j u_i}{\partial x_j} + \frac{1}{2}u_j\frac{\partial u_i}{\partial x_j}, \tag{1}$$

which is defined as a suitable average of the divergence and advective formulations, whose errors on energy conservation are equal and of opposite sign [2]. This choice, however, has the drawback of being roughly twice as expensive as standard divergence or advective forms alone. Moreover, exact conservation of spatial discretizations is lost when a non-conservative time-integration method is used. In this regard, implicit schemes are required [3], with an additional charge on the cost of a single

F. Capuano (✉) · G. Coppola · L. de Luca
Dipartimento di Ingegneria Industriale, University of Naples Federico II, Naples, Italy
e-mail: francesco.capuano@unina.it

G. Coppola
e-mail: gcoppola@unina.it

J. Peinke et al. (eds.), *Progress in Turbulence VI*,
Springer Proceedings in Physics 165, DOI 10.1007/978-3-319-29130-7_11

time step. A compromise between computational cost and energy-conservation is thus attractive. In this work, a novel explicit time-advancement strategy that retains the conservation properties of skew-symmetric-based schemes at a reduced computational cost is proposed and tested numerically.

2 The Novel Runge–Kutta Strategy

The proposed strategy consists in integrating the semi-discretized Navier-Stokes equations in time by means of a modified, s-stage explicit Runge–Kutta method:

$$\mathbf{u}^{n+1} = \mathbf{u}^n - \Delta t \sum_{i=1}^{s} b_i \mathbf{C}_i(\mathbf{u}_i)\mathbf{u}_i + \Delta t \mathbf{G}\mathbf{p}^{n+1}, \tag{2}$$

$$\mathbf{u}_i = \mathbf{u}^n - \Delta t \sum_{j=1}^{i-1} a_{ij} \left(\mathbf{C}_j(\mathbf{u}_j)\mathbf{u}_j - \mathbf{G}\mathbf{p}_j\right), \tag{3}$$

where $\mathbf{u}$ and $\mathbf{p}$ are the discrete velocity and pressure vectors, $\mathbf{G} \in R^{N_\mathbf{u} \times N_p}$ and $\mathbf{C} \in R^{N_\mathbf{u} \times N_\mathbf{u}}$ are the discrete gradient and convective operators, respectively, and a_{ij} and b_i are the Runge–Kutta coefficients. Pressure is to be solved from a Poisson equation. The operator $\mathbf{C}$ in Eqs. (2–3) is indexed by a suffix, meaning that can be expressed in either divergence (D) or advective (A) form within the stages, in any of the possible sequences available (e.g., for a four-stage scheme, ADDA, DADA, ...). Centered finite-difference discretizations on a periodic, equally spaced grid will be considered here. The velocity components $\mathbf{u}$ and pressure $\mathbf{p}$ are stored at the same points.

New schemes with different order of accuracy on solution (S) and energy (E) can be obtained by coupling the classical order conditions to a set of special equations [4]. The latter can be derived by nullifying the first- and second-order terms in the expression of the energy-conservation error arising from Eqs. (2–3)

$$\frac{\Delta E}{\Delta t} = \overbrace{-\mathbf{u}^T \left[\sum_i b_i \tilde{\mathbf{C}}_i\right] \mathbf{u}}^{1^{\text{st}} \text{ order term}} + \underbrace{\frac{\Delta t}{2}\mathbf{u}^T \left[\sum_{ij} 2 b_i a_{ij} \left(\tilde{\mathbf{C}}_i \tilde{\mathbf{C}}_j + \tilde{\mathbf{C}}_{ij}\right) + g_{ij} \tilde{\mathbf{C}}_i^T \tilde{\mathbf{C}}_j\right] \mathbf{u}}_{2^{\text{nd}} \text{ order term}} + O\left(\Delta t^2\right), \tag{4}$$

where $\mathbf{u} = \mathbf{u}^n$, $g_{ij} = b_i a_{ij} + b_j a_{ji} - b_i b_j$, $\tilde{\mathbf{C}}_{ij} = \tilde{\mathbf{C}}_i \left(\tilde{\mathbf{C}}_j \left(\mathbf{u}^n\right) \mathbf{u}^n\right)$. In Eq. (4), the operator $\tilde{\mathbf{C}}$ is assumed to be evaluated at $\mathbf{u}^n$ and projected with the pressure gradient to yield a divergence-free velocity field. The total discrete energy $E = \|\mathbf{u}\|^2/2$ and its error $\Delta E = E^{n+1} - E^n$ have also been defined. It can be shown that first- and second-order conditions on energy conservation lead to one and three additional constraints, respectively. In four-stage methods, 10 degrees of freedom are available, hence either 4S1E(4) or 3S2E(4) schemes can be derived. The number in brackets is a cost-coefficient which is proportional to the total number of derivatives to be evaluated. In this notation, a four-stage RK scheme in skew-symmetric form would be indicated as 4S4E(8). The following are Butcher arrays of two RK schemes with good conservation properties:

ADAD/DADA – 4S1E(4)

$$\begin{array}{|cccc} 0 & & & \\ \frac{1}{2} & 0 & & \\ 0 & \frac{1}{2} & 0 & \\ 0 & 0 & 1 & 0 \\ \hline \frac{1}{6} & \frac{1}{3} & \frac{1}{3} & \frac{1}{6} \end{array}$$

ADDA/DAAD – 3S2E(4)

$$\begin{array}{|cccc} 0 & & & \\ \frac{1}{3} & 0 & & \\ 0 & 1 & 0 & \\ \frac{1}{3} & 0 & \frac{1}{3} & 0 \\ \hline \frac{1}{8} & \frac{3}{8} & \frac{1}{8} & \frac{3}{8} \end{array} .$$

The 4S1E(4) scheme is the classical RK4. It can be shown that, when the sequences ADAD/DADA are used, this scheme satisfies the first-order condition on energy, while remaining fourth-order accurate on solution. The 3S2E(4) scheme is novel and has been obtained by solving the nonlinear system constituted by the four order conditions for third-order accuracy on solution, together with four conditions for second-order accuracy on energy conservation. The resulting two-parameter family has been optimized to yield as many null coefficients as possible.

3 Numerical Results

As a first test, an order of accuracy analysis is carried out to confirm the theoretical results obtained in the previous sections. The inviscid Navier-Stokes equations are integrated in a square region of size $2\pi \times 2\pi$, discretized with 16×16 mesh points. A centered second-order finite-difference scheme is used for convection; for time-advancement, the various Runge–Kutta schemes developed in Sect. 2 are used. Figure 1 (left) shows the time-step convergence of the energy error for three Runge–Kutta methods measured at $t_f = 10$, where the error measure $\varepsilon(t) = (E(t) - E_0)/E_0$ has been defined. The predicted orders of accuracy for the new schemes are confirmed numerically, proving the consistency of the theoretical framework.

As a turbulent test-case, direct numerical simulation of a temporal plane jet is performed. A standard pseudo-spectral code is used, with the diffusive term being computed exactly. Second-order finite-difference schemes are mimicked by using modified wavenumbers. The Navier-Stokes equations are discretized using 256^3 grid points on a cubic domain of length $4H$, where H is the plane jet thickness. The initial condition is based on a hyperbolic tangent profile with a velocity difference ΔU. The time step is computed from the CFL condition, with a maximum CFL number of 0.9. The Reynolds number $\mathrm{Re} = H\Delta U/\nu$ is equal to 3200. The time-evolution of energy is shown in Fig. 1 (right). Both divergence and advective forms lead to early blow-up, due to violation of energy conservation, while the new schemes perform

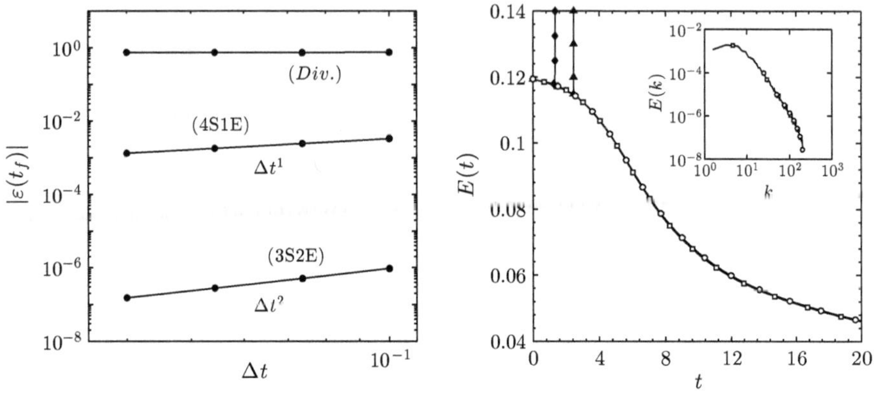

Fig. 1 *(Left)* Time-step convergence of the relative error on energy conservation for the inviscid random flow case. *(Right)* Time evolution of energy and spectra at $t = 5$ (*inset* plot) for the turbulent plane jet for five methods. —— $(Skew.)$, ▲ $(Adv.)$, ♦ $(Div.)$, □ (4S1E), ○ (3S2E)

remarakably well. The beneficial properties are retained also in wavenumber space, as shown by the energy spectrum within the inset plot.

In both tests, the novel procedure allowed to roughly halve the CPU time spent for the computation of the nonlinear term.

4 Conclusions

A novel time-advancing strategy for the incompressible Navier-Stokes equations has been proposed. The RK-based technique has proved to reproduce the energy-conserving properties of the skew-symmetric form while halving the CPU time requested for the computation of the nonlinear term.

The authors gratefully acknowledge Dr. Guillaume Balarac for performing the plane jet simulations with the alternating procedure.

References

1. Koren, B., Abgrall, R., Bochev, P., Frank, J., Perot, B.: Physics-compatible numerical methods. J. Comput. Phys. **257**, 1039 (2014)
2. Capuano, F., Coppola, G., Balarac, G., de Luca, L.: Energy preserving turbulent simulations at a reduced computational cost. J. Comput. Phys. **298**, 480–494 (2015)
3. Sanderse, B.: Energy-conserving Runge–Kutta methods for the incompressible Navier-Stokes equations. J. Comput. Phys. **233**, 100–131 (2013)
4. Capuano, F., Coppola, G., de Luca, L.: An efficient time advancing strategy for energy-preserving simulations. J. Comput. Phys. **295**, 209–229 (2015)

Assessment of Subfilter Scalar Dissipation Rate and Mixture Fraction Variance Models

J. Ventosa-Molina, O. Lehmkuhl, C.D. Pérez-Segarra and A. Oliva

Abstract The Flamelet/Progress-Variable (FPV) model is able to describe diffusion flames in the flamelet regime by parametrising the chemical problem as a function of a limited number of variables, namely the mixture fraction Z and the progress-variable c. The aim of the present paper is to study the effect of different mixture fraction variance Z_v and subgrid scalar dissipation rate χ_{sgs} models in the prediction of turbulent combustion phenomena, particularly for diffusion flames. Both parameters model scalar mixing at the subgrid level, which in turn controls the combustion process. Four different models are compared: an Equilibrium model, a mixture fraction Variance Transport Equation (VTE) model, a Transport Equation for the Second moment of the mixture fraction (STE), and a Scalar Dissipation Rate transport equation (SDR-TE).

1 Introduction and Mathematical Formulation

Evaluation of subgrid mixture fraction variance and subfilter scalar dissipation rate is a critical issue in flamelet and Conditional Moment Closure (CMC) models. Both parameters model scalar mixing at the subgrid level, which in turn controls the combustion process. A scale similarity model [8] was proposed, which assumes local equilibrium between production and dissipation of the variance at the small scales. However, this model presented accuracy limitations when applied to technically relevant flow configurations [3]. Non-equilibrium modelling can be achieved by constructing a variance transport equation (VTE). Alternatively, through the

J. Ventosa-Molina · O. Lehmkuhl · C.D. Pérez-Segarra · A. Oliva (✉)
Centre Tecnològic de Transferència de Calor (CTTC),
Universitat Politècnica de Catalunya (UPC), C/Colom 11, 08222 Terrassa, Spain
e-mail: cttc@cttc.upc.edu

O. Lehmkuhl
Termo Fluids, S.L., Av. Jacquard 97, 1-E, 08222 Terrassa, Spain
e-mail: termofluids@termofluids.com

J. Peinke et al. (eds.), *Progress in Turbulence VI*,
Springer Proceedings in Physics 165, DOI 10.1007/978-3-319-29130-7_12

definition of a second moment of any statistical value, the variance can be computed as $Z_v = \widetilde{Z^2} - \widetilde{Z}^2$, which requires a transport equation for $\widetilde{Z^2}$ (STE). However, both equations require a closure for the subfilter dissipation rate. Either through an algebraic expression using a turbulent time scale [2, 3] or by using a transport equation for the scalar dissipation rate [4].

In the flamelet regime thermochemical changes occur normal to the flame front. The FPV model in its steady form (SFPV) allows tabulating chemical kinetics and thermodynamical properties through an equation of state (EoS), $\psi = \{\rho, \dot{w}_c, \alpha, \ldots\} = \Gamma_\psi(Z, c)$, where Z represents the mixing process and c accounts for the evolution of the combustion process. However, in LES simulations solution variables are Favre filtered quantities. Consequently, turbulence chemistry interactions are characterised through a presumed joint PDF. Hence, a β-pdf for the statistical distribution of the mixture fraction and a δ-pdf for the progress-variable have been assumed, $\widetilde{P}(Z, c|_{st}) = \beta(Z; \widetilde{Z}, Z_v)\delta(c|_{st} - c|^*_{st})$. Thus, the EoS for LES simulations becomes $\widetilde{\psi} = \widetilde{\Gamma}_\psi(\widetilde{Z}, Z_v, \widetilde{c})$.

Besides from computing the transport equations for the Favre filtered mixture fraction $\widetilde{Z}$ and progress-variable $\widetilde{c}$, the description of the turbulent combustion state in these models requires modelling the subfilter variance Z_v and the filtered scalar dissipation rate $\widetilde{\chi}_Z$. To model Z_v, it may be transported (VTE) or computed through its definition using $\widetilde{Z^2}$ (STE). On the one hand, the VTE can be constructed as

$$\overline{\rho}\frac{DZ_v}{Dt} = \frac{\partial}{\partial x_i}\left(\overline{\rho}(\widetilde{\alpha}_Z + \alpha_{t,Z})\frac{\partial Z_v}{\partial x_i}\right) + 2\overline{\rho}(\widetilde{\alpha}_Z + \alpha_{t,Z})\frac{\partial \widetilde{Z}}{\partial x_i}\frac{\partial \widetilde{Z}}{\partial x_i} - \overline{\rho}\widetilde{\chi}_Z \tag{1}$$

$$\widetilde{\chi}_Z = 2\widetilde{\alpha}_Z\widetilde{\frac{\partial Z}{\partial x_i}\frac{\partial Z}{\partial x_i}} = 2\widetilde{\alpha}_Z\frac{\partial \widetilde{Z}}{\partial x_i}\frac{\partial \widetilde{Z}}{\partial x_i} + \varepsilon_Z \tag{2}$$

where ε_Z is the subfilter dissipation rate. On the other hand, the STE reads

$$\overline{\rho}\frac{D\widetilde{Z^2}}{Dt} = \frac{\partial}{\partial x_i}\left(\overline{\rho}(\widetilde{\alpha}_Z + \alpha_{t,Z})\frac{\partial \widetilde{Z^2}}{\partial x_i}\right) - \overline{\rho}\widetilde{\chi}_Z \tag{3}$$

The subfilter scalar dissipation rate can be closed through a turbulent mixing timescale τ [2, 3] and a model constant $C_Z = 4$

$$\varepsilon_Z = \frac{C_Z}{\tau}Z_v = C_Z\frac{\nu_t}{\Delta^2}Z_v \tag{4}$$

Assuming equilibrium of production and destruction of Z_v at the small scales, the last two terms in Eq. (1) cancel out. Thus, ε_Z becomes

$$\varepsilon_Z = 2\alpha_{t,Z}\frac{\partial \widetilde{Z}}{\partial x_i}\frac{\partial \widetilde{Z}}{\partial x_i} \tag{5}$$

With the local equilibrium assumption (LEA) Z_v cannot be computed from Eq. (1). Therefore, a scale similarity model was proposed [8]

$$\overline{\rho} Z_v = C_{var} \Delta^2 \overline{\rho} |\nabla \widetilde{Z}| \tag{6}$$

where C_{var} is calculated using the *Leonard term Expansion Dynamic* model [1].

A last approach is to consider a transport equation, see Eq. (7), for the filtered scalar dissipation rate $\widetilde{\chi}_Z$ (SDR-TE) [4], together with Eq. (3). This approach comes at the cost of having to model several unclosed terms and evaluating several computationally expensive terms.

$$\begin{aligned}
\overline{\rho}\frac{D}{Dt}\left(\widetilde{|\nabla Z|^2}\right) = & \frac{\partial}{\partial x_i}\left(\overline{\rho}(\widetilde{\alpha}_Z + \alpha_{t,Z})\frac{\partial \widetilde{|\nabla Z|^2}}{\partial x_i}\right) - 2\overline{\rho}\left(\frac{\partial \widetilde{u}_i}{\partial x_j}\frac{\partial \widetilde{Z}}{\partial x_i}\frac{\partial \widetilde{Z}}{\partial x_j}\right) - 2\overline{\rho}\,\widetilde{\alpha}_Z\left(\frac{\partial^2 \widetilde{Z}}{\partial x_i \partial x_j}\right)^2 \\
& - \frac{2}{\overline{\rho}}\frac{\partial \overline{\rho}}{\partial x_i}\frac{\partial \widetilde{Z}}{\partial x_i}\left(\frac{\partial}{\partial x_j}\left(\overline{\rho}\,\widetilde{\alpha}_Z\frac{\partial \widetilde{Z}}{\partial x_j}\right)\right) + 2\frac{\partial \overline{\rho}\,\widetilde{\alpha}_Z}{\partial x_i}\frac{\partial \widetilde{Z}}{\partial x_i}\left(\frac{\partial^2 \widetilde{Z}}{\partial x_j^2}\right) + 2\frac{\partial \widetilde{Z}}{\partial x_i}\frac{\partial \widetilde{Z}}{\partial x_j}\left(\frac{\partial^2 \overline{\rho}\,\widetilde{\alpha}_Z}{\partial x_i \partial x_j}\right) \\
& + \overline{\rho}\frac{32\nu_t}{\Delta^2}\left(\widetilde{|\nabla Z|^2} - |\nabla \widetilde{Z}|^2\right) - 12 C_{var}\overline{\rho}\frac{\widetilde{\alpha}_Z}{Z_v}\left(\widetilde{|\nabla Z|^2} - |\nabla \widetilde{Z}|^2\right)^2
\end{aligned} \tag{7}$$

2 Numerical Simulation and Discussion

Numerical computations are performed using the general purpose unstructured and parallel object-oriented CFD code TermoFluids [5]. Turbulent fluxes are modelled using a standard eddy diffusivity model, specifically the WALE model [7]. A jet

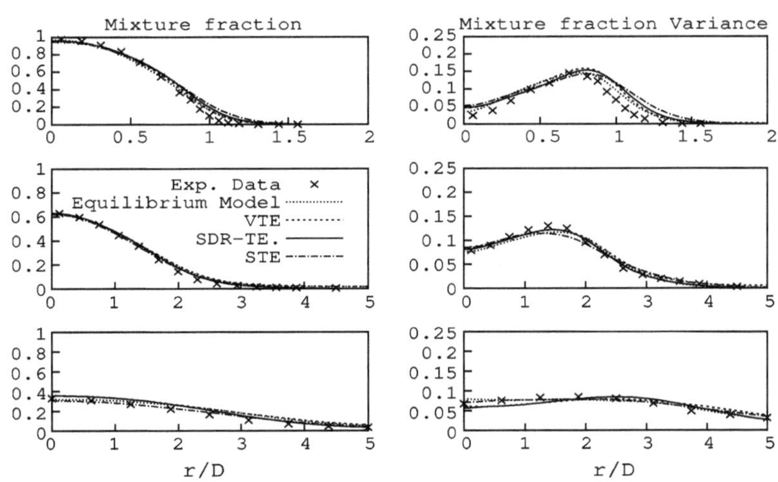

Fig. 1 Radial distribution of the mixture fraction and its rms. *Top* $y/D = 5$, *middle* $y/D = 20$, *bottom* $y/D = 40$. The fuel jet diameter is $D = 8$ mm

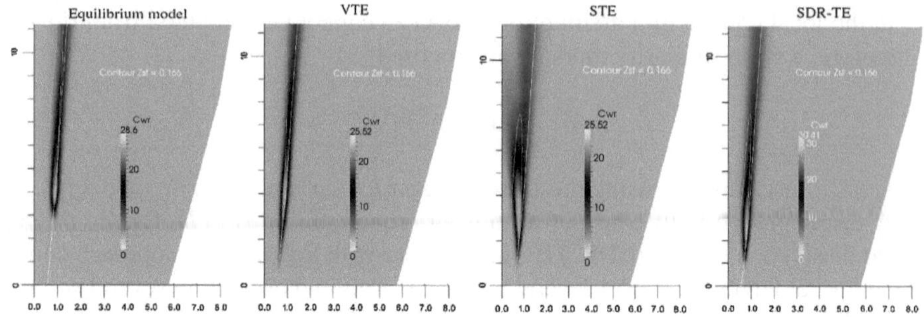

Fig. 2 Time averaged progress-variable reaction rate $[{}^{\mathrm{kg}}/_{\mathrm{m}^3 s}]$ for the different models

diffusion flame with a Reynolds number of 15,200, namely the DLR Flame A [6], is used as a test case. A mesh of 1.8M CV has been used. The SFPV model [9] is used to model chemistry-turbulence interactions. The GRI 3.0 mechanism is used to generate the database, and differential diffusion effects are considered. The progress variable is here defined as $c = Y_{CO} + Y_{CO_2} + Y_{H_2O} + Y_{H_2}$.

As it can be seen in Fig. 1, the radial distribution of the mixture fraction, both mean and rms values, are in good agreement with the experimental data [6], without significant differences between the four different models. For other quantities such as the progress-variable and velocities, there are also good agreement. Nonetheless, the four models predict flame stabilisation, and onset of reactions, at different axial position as depicted in Fig. 2. In the experiment, the flame is anchored to the fuel jet nozzle rim. Although good agreement downstream is observed, the stabilisation of the flame is not correctly captured. VTE and SDR-TE models show a better agreement, which indicates a better description of subgrid mixing. The Equilibrium model, having the longest stabilisation distance, clearly indicates that there is an imbalance between production and destruction of variance at the subgrid level.

Acknowledgments The author thankfully acknowledges the computer resources, technical expertise and assistance provided by the Barcelona Supercomputing Center—Centro Nacional de Supercomputación.

References

1. Balarac, G., Pitsch, H., Raman, V.: Development of a dynamic model for the subfilter scalar variance using the concept of optimal estimators. Phys. Fluids **20**, 035,114 (2008)
2. Ihme, M., Pitsch, H.: Prediction of extinction and reignition in nonpremixed turbulent flames using a flamelet/progress variable model 2. Application in LES of Sandia flames D and E. Combust. Flame **155**, 90–107 (2008)
3. Kaul, C., Raman, V., Knudsen, E., Richardson, E., Chen, J.: Large eddy simulation of a lifted ethylene flame using a dynamic nonequilibrium model for subfilter scalar variance and dissipation rate. Proc. Combust. Inst. **34**, 1289–1297 (2013)

4. Knudsen, E., Richardson, E., Doran, E., Pitsch, N., Chen, J.: Modeling scalar dissipation and scalar variance in large eddy simulation: Algebraic and transport equation closures. Phys. Fluids **24**, 055,103 (2012)
5. Lehmkuhl, O., Pérez Segarra, C., Borrell, R., Soria, M., Oliva, A.: Termofluids: a new parallel unstructured CFD code for the simulation of turbulent industrial problems on low cost PC cluster. Proc. Par. CFD Conf. pp. 1–8 (2007)
6. Meier, W., Barlow, R., Chen, Y.L., Chen, J.Y.: Raman/Rayleigh/LIF measurements in a turbulent CH4/H2/N2 jet diffusion flame: experimental techniques and turbulence-chemistry interaction. Combust. Flame **123**, 326–343 (2000)
7. Nicoud, F., Ducros, F.: Subgrid-scale stress modeling based on the square of the velocity gradient tensor. Flow Turbul. Combust. **62**, 183–200 (1999)
8. Pierce, C., Moin, P.: A dynamic model for subgrid-scale variance and dissipation rate of a conserved scalar. Phys. Fluids **10**, 3041 (1998)
9. Pierce, C., Moin, P.: Progress-variable approach for large-eddy simulation of non-premixed turbulent combustion. J. Fluid Mech. **504**, 73–97 (2004)

Backward Energy Transfer and Subgrid Modeling Approaches in Wall-Turbulence

A. Cimarelli and E. De Angelis

Abstract We report here results from a Large Eddy Simulation (LES) of a turbulent channel flow at a friction Reynolds number $Re_\tau = 550$ performed with a new subgrid modeling approach proposed by the same authors in Cimarelli et al., Phys. Fluids, 26, 055103 (2014), [1]. This subgrid scale model aims at reproducing the double feature of energy sink and source of the small scales of wall flows which become relevant when large filter lengths are adopted. Here we report a further analysis of the model by considering the instantaneous behavior of events of backward and forward energy transfer.

1 Introduction

One of the most important features of subgrid models in Large Eddy Simulation (LES) should be their ability to accurately reproduce the energy transfer between resolved and unresolved scales. In wall-turbulent flows, this physical process has been found to behave very differently depending on the amount of scales removed by the filter and on the wall-distances considered, see e.g. Härtel et al. [2]. In particular, for large filter lengths, in the production region a net energy flux takes origin in the subgrid scales to feed the resolved motion. As recently shown in Cimarelli et al. [3], this reverse energy transfer is a crucial mechanism characterizing wall-turbulence since is responsible for the formation of the commonly observed very long streaks of velocity fluctuations. The presence of an up-scale energy transfer leads to overwhelming difficulties for LES, since energy should emerge from the unknown subgrid scales to drive the quasi-coherent dynamics of the resolved motion. Several attempts to account for backscatter have been proposed by using random forcing or deterministic approach, see e.g. Carati et al. [4] and references therein. Here we show the results obtained by an alternative approach based on the coupling of the classical linear formulation of eddy viscosity with the nonlinear anisotropic features

A. Cimarelli (✉) · E. De Angelis
DIN, Università di Bologna, Via Fontanelle 40, 47121 Forlì, Italy
e-mail: andrea.cimarelli2@unibo.it

J. Peinke et al. (eds.), *Progress in Turbulence VI*,
Springer Proceedings in Physics 165, DOI 10.1007/978-3-319-29130-7_13

of the velocity increment tensor. This modeling approach, proposed in Cimarelli and De Angelis [1], has been found to properly account for backscatter and to reproduce the complex features of energy transfer in wall-flows.

2 The Subgrid Scale Model

From the classical picture of turbulence, the continuous spectrum of fluctuations is divided into three main ranges: large production scales, inertial energy transfer scales and small dissipative scales. This spectral view of turbulence is based on the energy cascade concept where the rate of energy transfer in the so-called inertial sub-range, is constant and proportional to the viscous dissipation $\langle \varepsilon \rangle$. This result, when homogeneous isotropic turbulence is considered, is one of the very few exact theories on turbulence and for this reason many LES models take inspiration from it. Unfortunately, when dealing with wall-turbulent flows the overall picture is drastically modified. As shown in Cimarelli et al. [3], a reverse energy flux from the small scales of the near-wall region systematically takes place driving the formation of large coherent structures further away from the wall. Again in Cimarelli et al. [3], a new simple equation for the energy transfer has been also developed and shown able to capture the switch between source-like and sink-like scales which induces both forward and reverse energy cascade. This equation differs from the Kolmogorov one on the energy cascade since takes into account the scale-dependent nonlinear dynamics of turbulent production beside the linear isotropic behavior of turbulent dissipation. By taking into account this double linear/nonlinear feature, the subgrid stress model proposed in Cimarelli and De Angelis [1] reads,

$$\tau_{ij}(x_s, t) = C_\Delta \delta \bar{u}_i \delta \bar{u}_j - 2\nu_T \bar{S}_{ij}. \tag{1}$$

Hence, the subgrid stress tensor τ_{ij} is written as a function of the nonlinear velocity increment tensor, $\delta \bar{u}_i \delta \bar{u}_j$, where $\delta u_i = u_i(x_s + \Delta_s) - u_i(x_s - \Delta_s)$ is the velocity increment at position x_s and separation Δ_s given by the grid resolution and C_Δ is a coefficient dynamically determined. In addition to the increment tensor an eddy viscosity term is used, $2\nu_T \bar{S}_{ij}$, where $\bar{S}_{ij} = 1/2(\partial \bar{u}_i/\partial x_j + \partial \bar{u}_j/\partial x_i)$ is the resolved strain rate tensor and ν_T is the eddy viscosity computed with the dynamic procedure proposed by Germano et al. [5] and refined by Lilly [6].

3 Results

The LES of a turbulent channel flow has been performed by solving the Navier-Stokes equations written in terms of wall-normal velocity and vorticity discretized using a spectral method, Fourier in the homogeneous x and z directions and Chebyshev in the wall-normal y direction. Time advancement is carried out with a partially implicit, fourth-order Crank–Nicholson/Runge–Kutta scheme. The nominal friction Reynolds number of the simulation is $Re_\tau = 550$ and the resolution adopted is very low corresponding to $\Delta x^+ = 108$ and $\Delta z^+ = 54$ in the homogeneous directions. In Cimarelli and De Angelis [1], the subgrid stress model (1) has been tested for different Reynolds numbers, grid resolutions and filter type. From a statistical point of view the proposed subgrid scale model has been found able to reproduce with a good accuracy the main features of the flow. Here we want to extend this analysis by considering the instantaneous realizations of the LES flow. In particular, since the goal of the present model is the correct description of the energetics of the flow we consider the topology of backward energy transfer events.

In LES, the net energy exchange between large resolved and small subgrid scales is quantified by the so-called *subgrid dissipation*, $\varepsilon_{sgs}(x_s, t) = \tau'_{ij}\bar{S}'_{ij}$, where $'$ denotes fluctuating quantities. Positive values of ε_{sgs} represent a forward energy cascade from large resolved to small subgrid scales. On the contrary, negative values of ε_{sgs} describe events of reverse energy cascade from small subgrid to large resolved scales. In Fig. 1, the probability density function of ε_{sgs} for two distances from the wall, $y^+ = 13$ and $y^+ = 130$ respectively, is shown. Strong backscatter events appear to be significant near the wall leading to a net contribution that overcomes the one due to forward events. Interestingly, also in the putative overlap layer where the net contribution of the small subgrid scales is of energy draining, backscatter events are still present but weaker. Indeed, the distribution is narrower and skewed toward forward energy transfer events. Let us now consider the topological distribution of the forward and backward events. Indeed, it is generally thought that the occurrence of large energy transfer between the resolved and unresolved scales is strongly correlated with some of the common turbulent structures that characterize wall-bounded flows [7]. In Fig. 2, the isosurfaces of forward and backward energy transfer events are shown. The main feature of these phenomena is that occur at relatively small scales. This is one of the main source of problem for modelling. In particular, we measure that the events of forward and backscatter form structures whose lengths are of the order

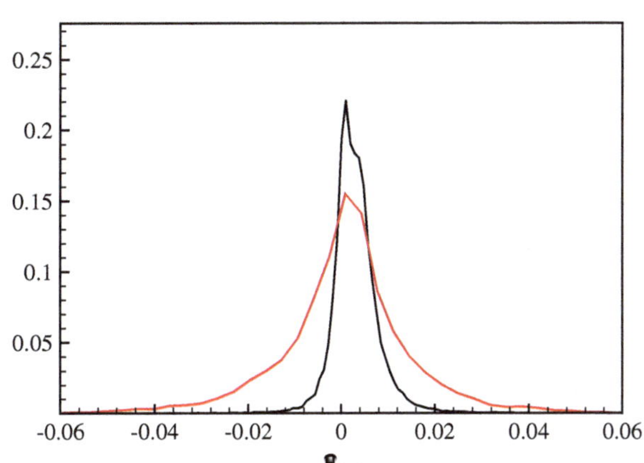

Fig. 1 Probability density function of ε_{sgs} at $y^+ = 13$ (*red line*) and $y^+ = 130$ (*black line*)

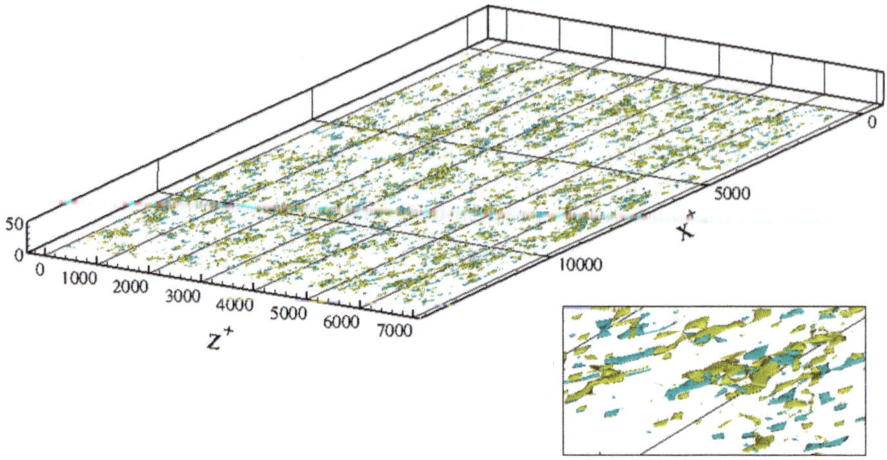

Fig. 2 Isosurfaces of forward (*yellow*) and backward (*light blue*) energy transfer events, $\varepsilon_{sgs} = 0.02$ and $\varepsilon_{sgs} = -0.02$ respectively

$\Delta x^+ \sim 200$, $\Delta y^+ \sim 20$ and $\Delta z^+ \sim 50$. Furthermore, it appears that backward and forward scatter usually occur in close proximity of each other, the backscatter event being generally surrounded by a region of significant forward scatter. Accordingly to Piomelli et al. [7], this close proximity indicates that the separation into resolved scales and subgrid scales is not isolating physically distinct structures. In other terms, it appears that forward and backscatter events arise from different parts of the same coherent motion.

References

1. Cimarelli, A., De Angelis, E.: The physics of energy transfer toward improved subgrid-scale models. Phys. Fluids **26**, 055103 (2014)
2. Härtel, C., Kleiser, L., Unger, F., Friedrich, R.: Subgrid-scale energy transfer in the near-wall region of turbulent flows. Phys. Fluids **6**, 3130–3143 (1994)
3. Cimarelli, A., De Angelis, E., Casciola, C.M.: Paths of energy in turbulent channel flows. J. Fluid Mech. **715**, 436–451 (2013)
4. Carati, D., Ghosal, S., Moin, P.: On the representation of backscatter in dynamic localization models. Phys. Fluids **7**, 606–616 (1995)
5. Germano, M., Piomelli, U., Moin, P., Cabot, W.H.: A dynamic subgrid-scale eddy viscosity model. Phys. Fluids A **3**, 1760–1765 (1991)
6. Lilly, D.K.: A proposed modification of the Germano subgrid-scale closure method. Phys. Fluids A **4**, 633–635 (1992)
7. Piomelli, U., Yu, Y., Adrian, R.J.: Subgrid-scale energy transfer and near-wall turbulence structure. Phys. Fluids **8**, 215–224 (1996)

Building Proper Invariants for Eddy-Viscosity Models

F.X. Trias, A. Gorobets and A. Oliva

Abstract Direct numerical simulations of the incompressible Navier-Stokes equations are limited to relatively low-Reynolds numbers. Therefore, dynamically less complex mathematical formulations are necessary for coarse-grain simulations. Regularization and eddy-viscosity models for LES are examples thereof. They rely on differential operators that should capture well different flow configurations (laminar and 2D flows, near-wall behavior, transitional regime ...). Most of them are based on the combination of invariants of a symmetric second-order tensor that is derived from the gradient of the resolved velocity field. In the present work, they are presented in a framework where the models are represented as a combination of elements of a 5D phase space of invariants. In this way, new models can be constructed by imposing appropriate restrictions in this space.

1 Theory: A 5D Phase Space for Eddy-Viscosity Models

The essence of turbulence are the smallest scales of motion. They result from a subtle balance between convective transport and diffusive dissipation. Numerically, if the grid is not fine enough, this balance needs to be restored by a turbulence model. The success of a turbulence model depends on the ability to capture well this (im)balance. In this regard, many eddy-viscosity models for LES have been proposed in the last decades. In order to be frame invariant, most of them rely on differential operators that are based on the combination of invariants of a symmetric second-order tensor (with the proper scaling factors). To make them locally dependent such tensors are derived from the gradient of the resolved velocity field, $G \equiv \nabla \overline{u}$. This is a second-order

F.X. Trias (✉) · A. Gorobets · A. Oliva
Heat and Mass Transfer Technological Center, Technical University of Catalonia
ETSEIAT, C/Colom 11, 08222 Terrassa, Spain
e-mail: cttc@cttc.upc.edu

A. Gorobets
Keldysh Institute of Applied Mathematics, 4A, Miusskaya Sq., Moscow 125047, Russia

J. Peinke et al. (eds.), *Progress in Turbulence VI*,
Springer Proceedings in Physics 165, DOI 10.1007/978-3-319-29130-7_14

traceless tensor, $tr(G) = \nabla \cdot \overline{u} = 0$. Therefore, it contains 8 independent elements and it can be characterized by 5 invariants (3 scalars are required to specify the orientation in 3D). This set of five invariants can be defined as follows

$$\{Q_G, R_G, Q_S, R_S, V^2\}, \tag{1}$$

where $Q_G = -1/2tr(G^2)$, $R_G = det(G)$ and $Q_S = -1/2tr(S^2)$, $R_S = det(S)$ are the second and third invariants of G and S, respectively. In this case, first invariants are zero, i.e. $P_G = tr(G) = \nabla \cdot u = 0$ and $P_S = tr(S) = \nabla \cdot u = 0$. Finally, $V^2 = tr(S^2\Omega^2)$, where $S = 1/2(G + G^T)$ and $\Omega = 1/2(G - G^T)$ are the symmetric and the skew-symmetric parts of the gradient tensor, G. Starting from the classical Smagorinsky model [3] that reads

$$\nu_e^{Smag} = (C_S\Delta)^2|S(\overline{u})| = 2(C_S\Delta)^2(-Q_S)^{1/2}, \tag{2}$$

most of the eddy-viscosity models for LES are based on invariants of second-order tensors that are derived from the gradient tensor, G. Therefore, it seems natural to re-write them in terms of the 5D phase space defined in (1). For instance, the WALE [4] and the Vreman's model [5] read

$$\nu_e^W = (C_W\Delta)^2 \frac{(2/3Q_G^2 + Z^2)^{3/2}}{(-2Q_S)^{5/2} + (2/3Q_G^2 + Z^2)^{5/4}}, \; \nu_e^{Vr} = (C_{Vr}\Delta)^2 \sqrt{\frac{Q_G^2 + 4Z^2}{2(Q_\Omega - Q_S)}} \tag{3}$$

respectively, where $Q_\Omega = Q_G - Q_S$ and $Z^2 = V^2 - 2Q_SQ_\Omega$. Other eddy-viscosity models that can be re-written in terms of the above-defined invariants are the R_S-based model proposed by Verstappen [1] and the σ-model proposed in [2]. The major drawback of the Smagorinsky model is that the differential operator it is based on does not vanish in near-wall regions (see Table 1). It is possible to build models based on invariants that do not have this limitation. Examples thereof are the WALE, the Vreman's, the R_S-based and the σ-model (see also Table 1).

Table 1 Top: near-wall behavior and units of the five basic invariants in the 5D phase space given in (1) together with the invariants $Q_\Omega = Q_G - Q_S$ and $Z^2 = V^2 - 2Q_SQ_\Omega$. Bottom: near-wall behavior of the Smagorinsky, the WALE, the Vreman's, the R_S-based and the σ-models

	Invariants						
	Q_G	R_G	Q_S	R_S	V^2	Q_Ω	Z^2
Wall-behavior	$\mathcal{O}(y^2)$	$\mathcal{O}(y^3)$	$\mathcal{O}(y^0)$	$\mathcal{O}(y^1)$	$\mathcal{O}(y^0)$	$\mathcal{O}(y^0)$	$\mathcal{O}(y^2)$
Units	$[T^{-2}]$	$[T^{-3}]$	$[T^{-2}]$	$[T^{-3}]$	$[T^{-4}]$	$[T^{-2}]$	$[T^{-4}]$
	Models						
	Smagorinsky Eq. (2)	WALE Eq. (3)	Vreman's Eq. (3)	R_S-based Ref. [1]	σ-model Ref. [2]		
Wall-behavior	$\mathcal{O}(y^0)$	$\mathcal{O}(y^3)$	$\mathcal{O}(y^1)$	$\mathcal{O}(y^1)$	$\mathcal{O}(y^3)$		

2 Building New Proper Invariants for LES Models

At this point, it is interesting to observe that new models can be derived by imposing restrictions on the differential operators they are based on. For instance, let us consider models that are based on the invariants of the tensor GG^T

$$\nu_e = (C_M \Delta)^2 P^p_{GG^T} Q^q_{GG^T} R^r_{GG^T}, \quad \text{where} \quad -6r - 4q - 2p = -1; \quad 6r + 2q = s, \tag{4}$$

and $P_{GG^T} = 2(Q_\Omega - Q_S)$, $Q_{GG^T} = Q_G^2 + 4Z^2$ and $R_{GG^T} = R_G^2$, respectively. This tensor is proportional to the gradient model given by the leading term of the Taylor series expansion of the subgrid stress tensor $\tau(\overline{u}) = (\Delta^2/12)GG^T + \mathcal{O}(\Delta^4)$. The above-defined restrictions on the exponents follow by imposing the $[T^{-1}]$ units of the differential operator and the slope, s, for the asymptotic near-wall behavior (see Table 1), i.e. $\mathcal{O}(y^s)$. Solutions for $q(p,s) = (1-s)/2 - p$ and $r(p,s) = (2s-1)/6 + p/3$ are displayed in Fig. 1. The Vreman's model given in Eq. (3) corresponds to the solution with $s = 1$ (see Table 1) and $r = 0$. However, it seems more appropriate to look for solutions with the proper near-wall behavior, i.e. $s = 3$ (solid lines in Fig. 1). Restricting ourselves to solutions involving only two invariants of GG^T we find three new models (also represented in Fig. 1),

$$\nu_e^{S3QP} = (C_{s3qp}\Delta)^2 P^{-5/2}_{GG^T} Q^{3/2}_{GG^T}, \tag{5}$$

$$\nu_e^{S3RP} = (C_{s3rp}\Delta)^2 P^{-1}_{GG^T} R^{1/2}_{GG^T}, \tag{6}$$

$$\nu_e^{S3RQ} = (C_{s3rq}\Delta)^2 Q^{-1}_{GG^T} R^{5/6}_{GG^T}, \tag{7}$$

where the model constants, C_{s3xx}, can be related with the Vreman's constant, C_{Vr}, with the following inequality (see details in [6])

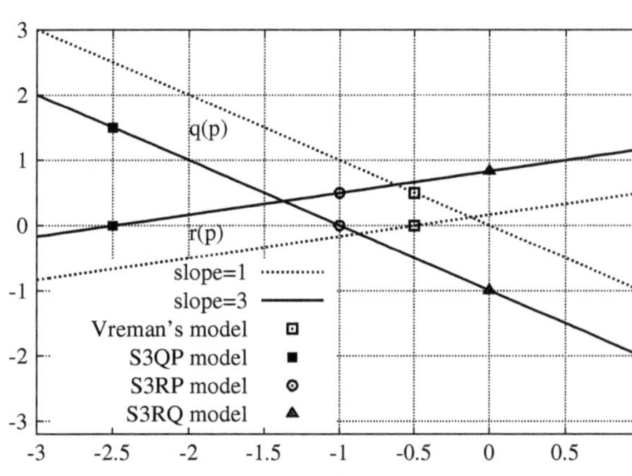

Fig. 1 Solutions for the linear system of Eq. (4) for $s = 1$ (*dashed line*) and $s = 3$ (*solid line*). Each (r, q, p) solution represents an eddy-viscosity model of the form given in Eq. (4)

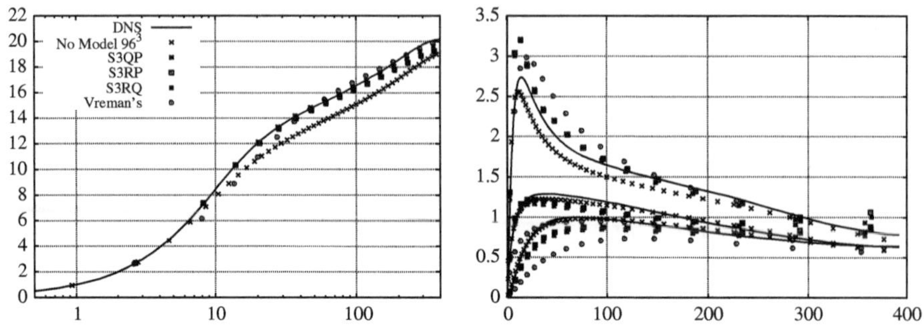

Fig. 2 Results for a turbulent channel flow at $Re_\tau = 395$ obtained with a 32^3 mesh for LES and a 96^3 mesh without model, i.e. $\nu_e = 0$. *Solid line* correspond to the DNS results by Moser et al. [7]

$$0 \leq \frac{(C_{Vr})^2}{(C_{s3xx})^2} \frac{\nu_e^{S3xx}}{\nu_e^{Vr}} \leq \frac{1}{3}. \tag{8}$$

Hence, imposing $C_{s3qp} = C_{s3rp} = C_{s3rq} = \sqrt{3} C_{Vr}$ guarantees both numerical stability and that the models have less or equal dissipation than Vreman's model, i.e. $0 \leq \nu_e^{S3xx} \leq \nu_e^{Vr}$. Figure 2 shows the performance of the proposed models for a turbulent channel flow in conjunction with the discretization methods for eddy-viscosity models proposed in [8]. The code is based on a fourth-order symmetry-preserving discretization finite volume discretization. Compared with the Vreman's model, they improve the results near the wall.

Acknowledgments This work has been financially supported by the *Ministerio de Economía y Competitividad*, Spain (ENE2014-60577-R) and a *Ramón y Cajal* postdoctoral contract (RYC-2012-11996). Calculations have been performed on the IBM MareNostrum supercomputer at the Barcelona Supercomputing Center. The authors thankfully acknowledge these institutions.

References

1. Verstappen, R.: When does eddy viscosity damp subfilter scales sufficiently? J. Sci. Comput. **49**(1), 94–110 (2011)
2. Nicoud, F., Toda, H.B., Cabrit, O., Bose, S., Lee, J.: Using singular values to build a subgrid-scale model for large eddy simulations. Phys. Fluids **23**(8), 085106 (2011)
3. Smagorinsky, J.: General circulation experiments with the primitive equations. Mon. Weather Rev. **91**, 99–164 (1963)
4. Nicoud, F., Ducros, F.: Subgrid-scale stress modelling based on the square of the velocity gradient tensor. Flow Turbul. Combust. **62**(3), 183–200 (1999)
5. Vreman, A.W.: An eddy-viscosity subgrid-scale model for turbulent shear flow: algebraic theory and applications. Phys. Fluids **16**(10), 3670–3681 (2004)
6. Trias, F.X., Folch, D., Gorobets, A., Oliva, A.: Building proper invariants for eddy-viscosity subgrid-scale models. Phys. Fluids **27**(6), 065103 (2015)
7. Moser, R.D., Kim, J., Mansour, N.N.: Direct numerical simulation of turbulent channel flow up to $Re_\tau = 590$. Phys. Fluids **11**, 943–945 (1999)
8. Trias, F.X., Gorobets, A., Oliva, A.: A simple approach to discretize the viscous term with spatially varying (eddy-)viscosity. J. Comput. Phys. **253**, 405–417 (2013)

Analysis of a Hybrid RANS/LES Model Using RANS Reconstruction

M. Nini, A. Abbà, M. Germano and M. Restelli

Abstract The Hybrid RANS/LES method proposed is based on the hybrid filter approach introduced by Germano in 2004. In this work, instead of using two explicit models respectively for LES and RANS, we use only a model for LES and we reconstruct the RANS field exploiting the properties of hybrid filter. The reconstruction is obtained from the resolved velocity and the LES subgrid stress tensor. The model has been implemented using a variational approach in a DG-FEM framework and tested for the turbulent channel flow test case. Different configurations for hybrid terms have been compared with DNS data and with pure LES results.

1 Model and Methodology

In the hybrid RANS/LES methods, RANS and LES are combined in order to achieve a detailed description of turbulent flow without incurring in unfeasible computational effort. A detailed review about hybrid RANS/LES method can be found in [1]. Starting from *Detached Eddy Simulation* [2], several methods were presented and successfully tested. Nevertheless, problems, especially in transition between RANS and LES, still remain.

The approach followed in this paper uses the hybrid filter proposed by Germano [3]. The main strength of this filter approach relies on the possibility of overcome the lack of energy and momentum transfer between RANS and LES area, through additional terms related to the characteristic of the hybrid filter. Therefore, no further artificial terms are needed. Rajamani and Kim [4] and Sanchez-Rocha and Menon

M. Nini (✉) · A. Abbà
Dip. di Scienze e Tecnologie Aerospaziali, Politecnico di Milano, Milan, Italy
e-mail: michele.nini@polimi.it

M. Germano
Department of Civil and Environmental Engineering, Duke University, Durham, NC, USA

M. Restelli
Numerische Methoden in der Plasmaphysik, Max-Planck-Institut für Plasmaphysik, Munich, Germany

J. Peinke et al. (eds.), *Progress in Turbulence VI*,
Springer Proceedings in Physics 165, DOI 10.1007/978-3-319-29130-7_15

[5, 6] studied and tested this approach, respectively for incompressible and compressible flow, showing promising results.

The hybrid filter is given by

$$\mathscr{H} = k\mathscr{F} + (k-1)\mathscr{E}, \tag{1}$$

where k is a blending factor and $\mathscr{F}$ and $\mathscr{E}$ are respectively the LES filter and the statistical RANS operator. The blending factor k, can be imposed a priori or computed dynamically [7], and in principle could vary from 1 (pure LES), where the grid resolution is high enough to resolve an adequate amount of turbulent scales, to 0 (pure RANS) where the grid become coarser. As shown by [5], the application of the hybrid filter to the compressible Navier-Stokes equations leads to several terms in the subgrid stress tensor $\tau^{\mathscr{H}}$. Nevertheless, in this work we will consider only low Mach number. Therefore, in the hypothesis of near incompressible flow, we assume that: $\tau^{\mathscr{H}} \approx \langle\rho\rangle_{\mathscr{H}}\tau_{inc}^{\mathscr{H}}$, where $\tau_{inc}^{\mathscr{H}}$ represents the subgrid stress tensor in the incompressible case. According to that, in the following, we will refer to the incompressible case; then $\tau^{\mathscr{H}}$ for hybrid RANS/LES equations can be written as

$$\begin{aligned}\tau^{\mathscr{H}}(u_i, u_j) &= k\tau^{\mathscr{F}}(u_i, u_j) + (1-k)\tau^{\mathscr{E}}(u_i, u_j) \\ &+ \frac{1-k}{k}(\langle u_i\rangle_{\mathscr{H}} - \langle u_i\rangle_{\mathscr{E}})(\langle u_j\rangle_{\mathscr{H}} - \langle u_j\rangle_{\mathscr{E}}),\end{aligned} \tag{2}$$

where $\tau^{\mathscr{E}}$ is the Reynolds stress tensor, $\tau^{\mathscr{F}}$ is the LES subgrid stress tensor and $\langle\rangle$ are used to represent quantities filtered $\mathscr{F}$, averaged $\mathscr{E}$ or *hybrid*-filtered $\mathscr{H}$.

Our work is focused on the Reynolds stress tensor $\tau^{\mathscr{E}}$. There are different possibilities to determine this term, for example using any explicit RANS model or reconstructing it by means of DNS data, experimental results or previous simulations. Here $\tau^{\mathscr{E}}$ is reconstructed from the LES model and from the resolved velocity field. We start from the definition of $\tau^{\mathscr{E}}$,

$$\begin{aligned}\tau^{\mathscr{E}}(u_i, u_j) &= \langle u_i u_j\rangle_{\mathscr{E}} - \langle u_i\rangle_{\mathscr{E}}\langle u_j\rangle_{\mathscr{E}} \\ &= \langle\tau^{\mathscr{H}}(u_i, u_j)\rangle_{\mathscr{E}} + \tau^{\mathscr{E}}(\langle u_i\rangle_{\mathscr{H}}, \langle u_j\rangle_{\mathscr{H}}).\end{aligned} \tag{3}$$

The velocity can be split into average and fluctuating part, $\langle\mathbf{u}\rangle_{\mathscr{H}} = \langle\langle\mathbf{u}\rangle_{\mathscr{H}}\rangle_{\mathscr{E}} + \langle\mathbf{u}\rangle'_{\mathscr{H}}$, so the latter term becomes:

$$\tau^{\mathscr{E}}(\langle u_i\rangle_{\mathscr{H}}, \langle u_j\rangle_{\mathscr{H}}) = \langle(\langle u_i\rangle_{\mathscr{H}} - \langle u_i\rangle_{\mathscr{E}})(\langle u_j\rangle_{\mathscr{H}} - \langle u_j\rangle_{\mathscr{E}})\rangle_{\mathscr{E}}. \tag{4}$$

Inserting the hybrid stress tensor definition (2) in (3)

$$\begin{aligned}\tau^{\mathscr{E}}(u_i, u_j) &= k\langle\tau^{\mathscr{F}}(u_i, u_j)\rangle_{\mathscr{E}} + (1-k)\tau^{\mathscr{E}}(u_i, u_j) \\ &+ \frac{1-k}{k}\langle(\langle u_i\rangle_{\mathscr{H}} - \langle u_i\rangle_{\mathscr{E}})(\langle u_j\rangle_{\mathscr{H}} - \langle u_j\rangle_{\mathscr{E}})\rangle_{\mathscr{E}} + \tau^{\mathscr{E}}(\langle u_i\rangle_{\mathscr{H}}, \langle u_j\rangle_{\mathscr{H}})\end{aligned} \tag{5}$$

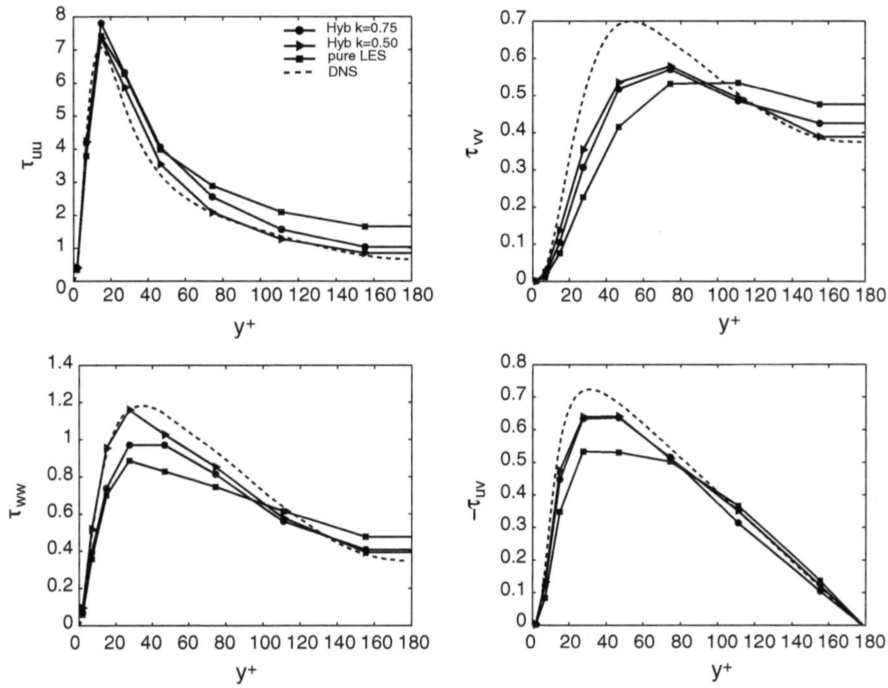

Fig. 1 Stress profiles, from *left to right* streamwise component τ_{uu}, normal component τ_{vv}, *right* spanwise component τ_{ww} and shear stress $-\tau_{uv}$

and using relation (4), Reynolds stress tensor becomes:

$$\tau^{\mathscr{E}}(u_i, u_j) = \langle \tau^{\mathscr{F}}(u_i, u_j)\rangle_{\mathscr{E}} + \frac{1}{k^2}\tau^{\mathscr{E}}(\langle u_i\rangle_{\mathscr{H}}, \langle u_j\rangle_{\mathscr{H}}). \tag{6}$$

Finally, substituting the relation (6) in (2), we obtain $\tau^{\mathscr{H}}(u_i, u_j)$

$$\begin{aligned}\tau^{\mathscr{H}}(u_i, u_j) = k\tau^{\mathscr{F}}(u_i, u_j) + (1-k)\langle \tau^{\mathscr{F}}(u_i, u_j)\rangle_{\mathscr{E}} + \frac{1-k}{k^2}\tau^{\mathscr{E}}(\langle u_i\rangle_{\mathscr{H}}, \langle u_j\rangle_{\mathscr{H}}) \\ + \frac{1-k}{k}(\langle u_i\rangle_{\mathscr{H}} - \langle u_i\rangle_{\mathscr{E}})(\langle u_j\rangle_{\mathscr{H}} - \langle u_j\rangle_{\mathscr{E}}). \end{aligned} \tag{7}$$

Therefore, we need only a model for LES. One of the most relevant drawback for this approach is that RANS reconstruction introduces a term $1/k^2$. The presence of this term limits the lower value for the blending factor. In fact, although setting $k = 0$ would not have been possible also using an explicit RANS model, as shown by (2), with RANS reconstruction the minimum k allowed increases, according to our tests the value should be $\sim 0.3 \div 0.4$. As a consequence we have to adjust the grid to compute LES with the adequate resolution. Nevertheless, in terms of numerical cost, this problem is balanced by the absence of additional equations related to the RANS model.

2 Results

Considering that the main purpose of the work is to evaluate applicability and perspectives RANS reconstruction procedure, we have tested only the constant blending factor case, in which no extra terms related to non-commutativity between hybrid filter and derivatives incur. Simulations refer to a turbulent channel flow at $Ma = 0.2$. The grid resolution in wall unit is: $\sim 23x^+$ for streamwise direction, $\sim 10y^+$ for spanwise direction for and $\sim 0.65/7.9y^+$ in the normal direction.

Results for stress profiles Fig. 1 show a good agreement with respect to incompressible DNS data obtained by Moser et al. [8]. Moreover, hybrid method, in particular for $k = 0.50$ where the hybrid terms are more relevant, improve the pure LES results, obtained with the same dynamic anisotropic model used as LES model for hybrid method.

3 Conclusion

A new methodology for hybrid filter RANS/LES method has been studied. The novelty introduced is represented by the reconstruction of RANS term, which allows to avoid the usage of explicit RANS models. The results, for constant value of blending factor and low Mach number, are promising and show that the hybrid method can integrate LES computing where the grid resolution is not enough to solve an appropriate number of turbulent scale.

Acknowledgments We acknowledge that the results of this research have been achieved using the PRACE Research Infrastructure resource provided by NIIF (Hungary), within the project DECI-11 HyDiG.

References

1. Fröhlich, J., von Terzi, D.: Hybrid les/rans methods for the simulation of turbulent flows. Prog. Aerosp. Sci. **44**(5), 349–377 (2008)
2. Spalart, P.R., Jou, W.H., Strelets, M., Allmaras, S.R.: Comments on the feasibility of les for wings, and on a hybrid rans/les approach. Adv. DNS/LES **1**, 4–8 (1997)
3. Germano, M.: Properties of the hybrid rans/les filter. Theor. Comput. Fluid Dyn. **17**(4), 225–231 (2004)
4. Kim, J., Rajamani, B.: A hybrid-filter approach to turbulence simulation. Flow Turbul. Combust. **85**, 421–441 (2010)
5. Menon, S., Sanchez-Rocha, M.: The compressible hybrid rans/les formulation using an additive operator. J. Comput. Phys. **228**(6), 2037–2062 (2009)
6. Menon, S., Sànchez-Rocha, M.: An order-of-magnitude approximation for the hybrid terms in the compressible hybrid RANS/LES governing equations. J. Turbul. **12**, 1–22 (2011)
7. Walters, D.K., Bhushan, S.: A dynamic hybrid reynolds-averaged navier stokes-large eddy simulation modeling framework. Phys. Fluids **24**, 1–7 (2012)
8. Moser, R.D., Kim, J., Mansour, N.N.: Direct numerical simulation of turbulent channel flow up to $re_\tau = 590$. Phys. Fluids **11**, 943–945 (1999)

On the Coupling of Direct Numerical Simulation and Resolvent Analysis

F. Gómez, H.M. Blackburn, M. Rudman, A.S. Sharma and B.J. McKeon

Abstract The present contribution explores the relationship between response and forcing via amplification mechanisms in the Navier–Stokes equations applied to a turbulent pipe flow. A novel numerical method coupling direct numerical simulation with the resolvent model [*J. Fluid Mech.* 658, 336-382 (2010)] is developed in order to reveal the exact distribution of the nonlinear forcing terms, originally unknown in the model. The obtained results highlight the major role of the nonlinear terms in the energy spectra.

1 Introduction

The recent theory of coherent structures in wall turbulence developed by McKeon & Sharma [1, 2] permits the prediction of different organized motion in turbulent flows. Modelizations of the interaction of streaks associated with near-wall cycle, hairpin vortices, hairpin packets, and very-large scale motions (VLMs) can be all directly extracted from the Navier–Stokes equations, under the assumption/restriction of a turbulent mean flow, via a singular value decomposition (SVD) of the resolvent operator that relates velocity fluctuations with nonlinear forcing terms, responsible for sustaining the turbulence [1, 2]. The core of this model lays in the high selectivity shown by the SVD of the resolvent operator, indicating that certain resolvent modes are more energetically and/or dynamically relevant than others. However, the distribution of the nonlinear forcing is unknown in this model and assumptions must

F. Gómez (✉) · H.M. Blackburn · M. Rudman
Department of Mechanical and Aerospace Engineering, Monash University, 3800 Victoria, Australia
e-mail: francisco.gomez-carrasco@monash.edu

A.S. Sharma
Faculty of Engineering and the Environment, University of Southampton, SO17 1BJ, Southampton, UK

B.J. McKeon
Graduate Aerospace Laboratories, California Institute of Technology, 91125 Pasadena, USA

J. Peinke et al. (eds.), *Progress in Turbulence VI*,
Springer Proceedings in Physics 165, DOI 10.1007/978-3-319-29130-7_16

be taken with respect to this. Examples of this are the work of Sharma and McKeon [2], who successfully assigned weightings to the modes according to their relative importance based on observation in the existing literature in order to model hairpin packets, and the work of Luhar et al. [3], in which a broad unit forcing is assumed across all wavenumbers for the prediction of the effectiveness of opposition control in pipe flow. Additional efforts to unveil the distribution of nonlinear forcing have been carried out by Moarref et al. [4], who applied optimization methods to weight resolvent modes in order to reconstruct a given turbulent spectra, and by Gómez et al. [5] who compared the most amplified resolvent modes with the most energetic modes arising from a dynamical mode decomposition (DMD) of direct numerical simulation (DNS) data. Here, we derive a new methodology to extract the exact contribution of the nonlinear terms.

2 Numerical Method

A spectral element–Fourier numerical discretization in cylindrical coordinates [6] is employed to solve the non-dimensional incompressible Navier–Stokes equations

$$\nabla \cdot \hat{\mathbf{u}} = 0\,, \qquad \frac{\partial \hat{\mathbf{u}}}{\partial t} + \hat{\mathbf{u}} \cdot \nabla \hat{\mathbf{u}} = -\nabla p + Re^{-1} \nabla^2 \hat{\mathbf{u}}, \tag{1}$$

where Re is the Reynolds number based on the bulk velocity and pipe diameter $D = 2R$, $\hat{\mathbf{u}} = (u, v, w)$ is the velocity vector expressed in cylindrical coordinates (x, r, θ) and p is the modified pressure. A (x, r)-domain of length $L = 2\pi D$ is employed and a Fourier discretization is applied in the periodic azimuthal direction. This permits writing the velocity obtained from the DNS as a sum of azimuthal Fourier modes $\hat{\mathbf{v}}_n(x, r, t)$

$$\hat{\mathbf{u}}(x, r, \theta, t) = \sum_{\pm n} \hat{\mathbf{v}}_n(x, r, t) e^{in\theta}. \tag{2}$$

A manipulation of the Navier–Stokes equation (1) in which both fluctuating velocity $\mathbf{u} = \hat{\mathbf{u}} - \mathbf{u}_0$ and nonlinear terms $\mathbf{f}$ are Fourier-decomposed in azimuthal wavenumbers n and frequency ω yields the linear relation

$$\mathbf{u}_{n,\omega}(x, r) = \mathscr{H}_{n,\omega} \mathbf{f}_{n,\omega}(x, r), \tag{3}$$

for each (n, ω) combination. The resolvent operator $\mathscr{H}_{n,\omega}$ acts as a transfer function between the fluctuating velocity and the nonlinear terms. A singular value decomposition (SVD) of this resolvent operator

$$\mathscr{H}_{n,\omega} = \sum_{m} \boldsymbol{\psi}_{n,\omega,m} \sigma_{n,\omega,m} \boldsymbol{\phi}^*_{n,\omega,m}, \tag{4}$$

provides an amplification relationship between two sets of orthonormal singular vectors $\boldsymbol{\psi}_{n,\omega,m}$ and $\boldsymbol{\phi}_{n,\omega,m}$ via the magnitude of the corresponding singular value $\sigma_{n,\omega,m}$, with m denoting the SVD order. Next, the nonlinear terms are projected onto the singular vectors $\boldsymbol{\phi}_{n,\omega,m}$

$$\mathbf{f}_{n,\omega} = \sum_{m} \chi_{n,\omega,m} \boldsymbol{\phi}_{n,\omega,m}, \tag{5}$$

where the unknown forcing coefficients $\chi_{n,\omega,n}$ represent the nonlinear interactions which act as the forcing that drives the turbulent fluctuations. The fluctuating velocity field is then reconstructed as a weighted sum of singular response modes

$$\mathbf{u}(x, r, \theta, t) = \sum_{\omega} \sum_{n} \sum_{m} a_{n,\omega,m} \boldsymbol{\psi}_{n,\omega,m}(x, r) e^{\mathrm{i}(n\theta - \omega t)}, \tag{6}$$

in which the unknown complex amplitudes $a_{n,\omega,m}$

$$a_{n,\omega,m} e^{-\mathrm{i}\omega t} = \int_{\omega - \Delta\omega/2}^{\omega + \Delta\omega/2} \chi_{n,\omega',m} \sigma_{n,\omega',m} e^{-\mathrm{i}\omega' t} \mathrm{d}\omega', \tag{7}$$

are defined as the product of the non-linear forcing and the resolvent amplifications integrated in frequency over a frequency bin $\Delta\omega$ with the purpose of discretizing the integral in frequency. Equivalently, these amplitudes can be approximated by

$$a_{n,\omega,m} \simeq \chi_{n,\omega,m} \sigma_{n,\omega,m} \Delta\omega. \tag{8}$$

The complex amplitudes $a_{n,\omega,m}$ are obtained by coupling the DNS solution in (2) and the resolvent model in (6), which reads

$$\hat{\mathbf{v}}_n(x, r, t) = \sum_{\pm\omega} \sum_{m} a_{n,\omega,m} \boldsymbol{\psi}_{n,\omega,m}(x, r) e^{-\mathrm{i}\omega t}, \tag{9}$$

for each azimuthal wavenumber n. Equation (9) is premultiplied by the complex conjugate of the mode corresponding to frequency η and SVD index q and integrated over the (x, r)-domain Ω

$$\int_{\Omega} \boldsymbol{\psi}^*_{n,\eta,q}(x, r) \cdot \hat{\mathbf{v}}_n(x, r, t) \mathrm{d}\Omega = \int_{\Omega} \sum_{\omega} \sum_{m} a_{n,\omega,m} \boldsymbol{\psi}^*_{n,\eta,q}(x, r) \cdot \boldsymbol{\psi}_{n,\omega,m}(x, r) e^{-\mathrm{i}\omega t} \mathrm{d}\Omega, \tag{10}$$

where $\mathrm{d}\Omega = r\mathrm{d}r\mathrm{d}x$. Making use of the orthonormality of the resolvent modes in Ω, the following scalar equation is obtained

$$P_{n,\eta,q}(t) = a_{n,\eta,q} e^{-\mathrm{i}\eta t} + \int_{\Omega} \sum_{\omega \neq \eta} \sum_{m \neq q} a_{n,\omega,m} \boldsymbol{\psi}^*_{n,\eta,q}(x, r) \cdot \boldsymbol{\psi}_{n,\omega,m}(x, r) e^{-\mathrm{i}\omega t} \mathrm{d}\Omega \tag{11}$$

where $P_{n,\omega,m}(t)$ is denoted as the temporal projection coefficients and correspond to the evolution in time of the projection of response modes onto DNS snapshots. The amplitude coefficients $a_{n,\omega,m}$ are obtained by estimating the power spectral density (PSD) of the temporal projection coefficients $P_{n,\omega,w}(t)$ at the corresponding frequency bin $\Delta\omega$. Welch's method is employed for this purpose, which in a simplified manner reads

$$a^*_{n,\omega,m}a_{n,\omega,m} \approx \int_{\omega-\Delta\omega/2}^{\omega+\Delta\omega/2} PSD\left(P_{n,\omega,w}(t)\right) \mathrm{d}\omega. \tag{12}$$

3 Results and Discussion

Figure 1 shows the distribution in frequency of the amplitude per frequency $|a_{5,\omega,1}|\ \Delta\omega$ obtained via projections, the amplification $\sigma_{5,\omega,1}$ emerging from the SVD of the resolvent operator, and the nonlinear forcing $|\chi_{5,\omega,1}|$ of the first singular value $m = 1$ corresponding to the azimuthal wavenumber $n = 5$. A characteristic sparsity is observed in all three distributions, as consequences of the finite length domain employed in the DNS [5]. We observe that the frequencies of the resolvent modes with higher amplitude (blue line) matches the most energetic frequencies obtained via DMD analysis of DNS data, indicated by bars in Fig. 1, implying the importance of the first singular vector. In addition, for this particular case at $n = 5$ we observe a significant shift of the nonlinear terms (green line) with respect to the amplification (red line), which is translated into a displacement of the amplitude peaks locations (blue line) with respect to the amplification peaks. This effect highlights the major role of the nonlinear terms in the location of the energy peaks in frequency. Understanding the distribution of nonlinear forcing remains as a future challenge.

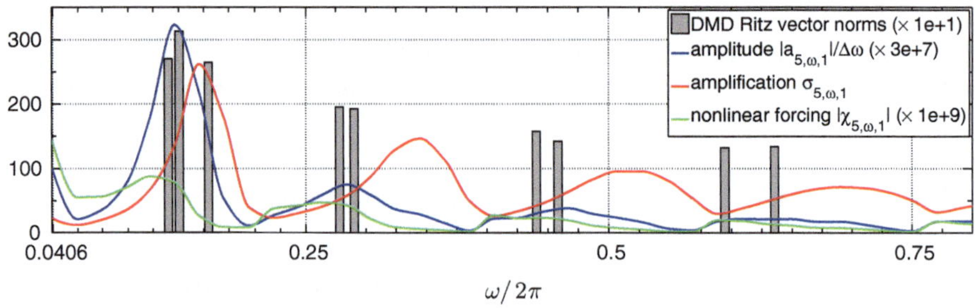

Fig. 1 Distribution of amplitude $a_{5,\omega,1}$ (*blue*), amplification $\sigma_{5,\omega,1}$ (*red*) and nonlinear forcing $\chi_{5,\omega,1}$ (*green*) in frequency of the first singular value $m = 1$ corresponding to the azimuthal wavenumber $n = 5$. Bars indicate most energetic frequencies computed via DMD of DNS data

Acknowledgments The authors acknowledge financial support from the Australian Research Council through the ARC Discovery Project DP130103103, and from Australia's National Computational Infrastructure via Merit Allocation Scheme Grant D77.

References

1. McKeon, B.J., Sharma, A.S.: J. Fluid Mech. **658**, 336 (2010)
2. Sharma, A.S., McKeon, B.J.: J. Fluid Mech. **728**, 196 (2013)
3. Luhar, M., Sharma, A.S., McKeon, B.J.: J. Fluid Mech. **749**, 597 (2014)
4. Moarref, R., Jovanović, M., Tropp, J., Sharma, A., McKeon, B.: Phys. Fluid **26**, 051701 (2014)
5. Gomez, F., Blackburn, M., Rudman, M., McKeon, B., Luhar, M., Moarref, R., Sharma, A.: Phys. Fluids **26**, 074103 (2014)
6. Blackburn, H.M., Sherwin, S.: J. Comput. Phys. **197**(2), 759 (2004)

Modeling Helicity Dissipation-Rate Equation

Nobumitsu Yokoi

Abstract Transport equation of the dissipation rate of turbulent helicity is derived with the aid of a statistical analytical closure theory of inhomogeneous turbulence. It is shown that an assumption on the helicity scaling with an algebraic relationship between the helicity and its dissipation rate leads to the transport equation of the turbulent helicity dissipation rate without resorting to a heuristic modeling.

1 Introduction

The helicity defined by $\int_V \mathbf{u} \cdot \boldsymbol{\omega} \, dV$, as well as the kinetic energy, is an inviscid invariant of the Navier–Stokes equation [V: fluid volume, $\mathbf{u}$: velocity, $\boldsymbol{\omega}(= \nabla \times \mathbf{u})$: vorticity]. Unlike the local turbulent energy density $\langle \mathbf{u}'^2 \rangle / 2$, the local turbulent helicity density $\langle \mathbf{u}' \cdot \boldsymbol{\omega}' \rangle$ is non-positive-definite and a pseudoscalar that changes its sign under the inversion or reflection ($\mathbf{u}'$: velocity fluctuation, $\boldsymbol{\omega}'$: vorticity fluctuation). Since any pseudoscalar should vanish in a mirror symmetric system, a non-zero pseudoscalar represents the breakage of mirror symmetry. In non-mirror symmetric turbulence, a finite helicity density is spatially distributed to affect the local transport. The dynamic evolution of turbulent helicity is subject to the balance among the helicity production (from the large-scale inhomogeneities), its dissipation and transport rates.

Effects of helicity (hereafter we drop "density") have been examined in the context of turbulent transports. In the dynamos, the turbulent helicity is directly connected to the so-called α effect, and plays an important role in magnetic field generation [3, 5]. Also in the non-mirrorsymmetric hydrodynamic turbulence such as a turbulent swirling flow, the turbulent helicity is expected to counterbalance the eddy viscosity [6]. The evaluation of the helicity dissipation rate is of crucial importance in determining the magnitude of effective transport.

N. Yokoi (✉)
Institute of Industrial Science, University of Tokyo, 4-6-1, Komaba, Meguro, Tokyo 153 8505, Japan
e-mail: nobyokoi@iis.u-tokyo.ac.jp

J. Peinke et al. (eds.), *Progress in Turbulence VI*,
Springer Proceedings in Physics 165, DOI 10.1007/978-3-319-29130-7_17

2 Helicity in Inhomogeneous Turbulence

The turbulent helicity $H \equiv \langle \mathbf{u}' \cdot \boldsymbol{\omega}' \rangle$ obeys an exact transport equation:

$$DH/Dt \equiv (\partial/\partial t + \mathbf{U} \cdot \nabla) H = P_H - \varepsilon_H + \nabla \cdot T_H \tag{1}$$

($\mathbf{U}$: mean velocity). Here, P_H and T_H are the production and transport rates of H, whose expressions are suppressed. The helicity dissipation rate ε_H is defined by

$$\varepsilon_H \equiv 2\nu \left\langle \frac{\partial u'_b}{\partial x_a} \frac{\partial \omega'_b}{\partial x_a} \right\rangle. \tag{2}$$

Evaluation of ε_H is of crucial importance to estimate the turbulent helicity evolution.

In order to derive the dynamic equation of ε_H, we have to express H in inhomogeneous turbulence. We follow the formal procedure of the two-scale direct-interaction approximation (TSDIA) [5–7], a combination of the multiple-scale analysis with a propagator renormalization closure theory of turbulence. In the TSDIA framework, the lowest-order velocity field is equivalent to the homogeneous isotropic turbulence, and the effects of the mean-field inhomogeneities, rotation, etc. are taken into account in a perturbation manner in the first- and higher-order velocity fields. If we introduce the Green's function of the lowest-order velocity field, $G'_{\mathrm{B}\alpha\beta}(\mathbf{k}; \tau, \tau')$, the first-order velocity field $\mathbf{u}'_1$ can be expressed in terms of this Green's function. For the lowest order velocity field $\mathbf{u}'_0$ and the Green's function, we assume the generic form for the homogeneous isotropic turbulence as

$$\frac{\langle u'_{0\alpha}(\mathbf{k};\tau)\, u'_{0\beta}(\mathbf{k}';\tau') \rangle}{\delta(\mathbf{k}+\mathbf{k}')} = D_{\alpha\beta}(\mathbf{k})\, Q_{\mathrm{B}}(k;\tau,\tau') + \frac{i}{2}\frac{k_a}{k^2}\varepsilon_{\alpha\beta a} H_{\mathrm{B}}(k;\tau,\tau'), \tag{3}$$

$$\langle G'_{\mathrm{B}\alpha\beta}(\mathbf{k};\tau,\tau') \rangle = D_{\alpha\beta}(\mathbf{k})\, G_{\mathrm{B}}(k;\tau,\tau'), \tag{4}$$

where $D_{\alpha\beta}(\mathbf{k}) = \delta_{\alpha\beta} - k_\alpha k_\beta / k^2$ is the projection operator. Here, Q_{B} and H_{B} are the spectral functions of the kinetic energy and helicity of the lowest-order fields, respectively. The second term in Eq. (3) represents the non-mirror symmetry of turbulence with H being a pseudoscalar.

The turbulent helicity is expanded as

$$H(\mathbf{x};t) = \langle \mathbf{u}' \cdot \boldsymbol{\omega}' \rangle = \langle \mathbf{u}'_0 \cdot \boldsymbol{\omega}'_0 \rangle + \langle \mathbf{u}'_0 \cdot \boldsymbol{\omega}'_1 \rangle + \langle \mathbf{u}'_1 \cdot \boldsymbol{\omega}'_0 \rangle + \cdots. \tag{5}$$

Substituting $\mathbf{u}'_0$ and $\mathbf{u}'_1$ ($\boldsymbol{\omega}'_n = \nabla \times \mathbf{u}'_n$ with $n = 0, 1$) into Eq. (5), with a renormalization procedure ($Q_{\mathrm{B}} \to Q$, $H_{\mathrm{B}} \to H$, $G_{\mathrm{B}} \to G$), we obtain

$$H(\mathbf{x};t) = I_0\{H\} - \frac{1}{3} I_0 \left\{ G, \frac{DH}{Dt} \right\} + \frac{8}{27} (\boldsymbol{\Omega} + 2\boldsymbol{\omega}_{\mathrm{F}}) \cdot I_0 \{G, \nabla Q\}, \tag{6}$$

where $\boldsymbol{\Omega}\,(= \nabla \times \mathbf{U})$ is the mean vorticity, $\boldsymbol{\omega}_\mathrm{F}$ the angular velocity, and the abbreviated forms of integral are defined by $I_n\{A\} = \int k^{2n} A(k, \mathbf{x}; \tau, \tau, t) d\mathbf{k}$ and $I_n\{A, B\} = \int k^{2n} d\mathbf{k} \int_{-\infty}^{\tau} d\tau_1 A(k, \mathbf{x}; \tau, \tau_1, t) B(k, \mathbf{x}; \tau, \tau_1, t)$.

In order to evaluate integrals in Eq. (6), we assume the propagators (correlation and response functions) in the inertial range such as $Q(k, \mathbf{x}; \tau, \tau', t) = \sigma_K(k, \mathbf{x}; t) \times \exp\left[-\omega_K(k, \mathbf{x}; t)|\tau - \tau'|\right]$, etc., where the spectra in the inertial range are assumed as

$$\sigma_K(k, \mathbf{x}; t) = \sigma_{K0} \varepsilon^{2/3} k^{-11/3}, \quad \sigma_H(k, \mathbf{x}; t) = \sigma_{H0} \varepsilon^{-1/3} \varepsilon_H(\mathbf{x}; t) k^{-11/3}, \tag{7}$$

with the time scales $\omega_K(k, \mathbf{x}; t) = \omega_{K0} \varepsilon^{1/3} k^{2/3} = \tau_K^{-1}, \omega_H(k, \mathbf{x}; t) = \omega_{H0} \varepsilon_H^{1/3} k^{2/3} = \tau_H^{-1}, \omega(k, \mathbf{x}; t) = \omega_0 \varepsilon^{1/3} k^{2/3} = \tau^{-1}$. The helicity spectrum σ_H in Eq. (7) arises from the assumption that the spectrum of the helicity is determined by the scale (k), energy and helicity transfer rates (ε and ε_H). This has been confirmed by DNS's [1, 2].

Using Eq. (7), H [Eq. (6)] can be estimated up to the first-order as

$$H(\mathbf{x}; t) = 3 \cdot (2\pi)^{1/3} \sigma_{H0} \varepsilon^{-1/3} \varepsilon_H \ell_\mathrm{C}^{2/3} + \frac{1}{6 \cdot (2\pi)^{1/3}} \frac{\sigma_{H0}}{\omega_0 + \omega_{H0}} \left(\ell_\mathrm{C} \varepsilon^{-1/3}\right)^2 \varepsilon_H \\ \times \left[\left(1 + \frac{\omega_{H0}}{\omega_0 + \omega_{H0}}\right) \frac{1}{\varepsilon} \frac{D\varepsilon}{Dt} - \frac{1}{\varepsilon_H} \frac{D\varepsilon_H}{Dt} - \left(11 + 2\frac{\omega_{H0}}{\omega_0 + \omega_{H0}}\right) \frac{1}{\ell_\mathrm{C}} \frac{D\ell_\mathrm{C}}{Dt}\right], \tag{8}$$

where ℓ_C is the size of the largest energy-containing eddies.

3 Modeling the Helicity Dissipation-Rate Equation

In constructing a system of model equations, we can choose any three of four turbulence statistical quantities $(H, \varepsilon, \varepsilon_H, \ell_\mathrm{C})$. In order that any choice among the four quantities should be equivalent (*model transferability*), some algebraic relation should be held among them [4, 8].

We solve Eq. (8) concerning ℓ_C in a perturbation manner. Up to the lowest-order analysis, we have

$$\ell_\mathrm{C} = 3^{-3/2} (2\pi)^{-1/2} \sigma_{H0}^{-3/2} \varepsilon^{1/2} \varepsilon_H^{-3/2} H^{3/2}, \tag{9}$$

or equivalently,

$$\varepsilon_H = C_H H / \tau, \quad \tau = \ell_\mathrm{C}^{2/3} \varepsilon^{-1/3}, \quad C_H = 1/[3(2\pi)^{1/3} \sigma_{H0}]. \tag{10}$$

As the lowest-order analysis of the turbulent helicity expression, we obtained an algebraic model for the turbulent dissipation rate as the first of Eq. (10) with the usual eddy turn-over time scale [the second of Eq. (10)]. This corresponds to the estimate of the turbulent helicity dissipation rate in homogeneous isotropic turbulence.

If we proceed to the first-order analysis, under the requirement of model transferability, the second term of Eq. (8) should be balanced in itself. Using Eq. (9), we change expression given in Eq. (8) based on ε, ε_H, and ℓ_C into the one based on ε, ε_H, and H. As this result, we have

$$\frac{D\varepsilon_H}{Dt} - C_{H1}(\omega_0, \omega_{H0})\frac{\varepsilon_H}{\varepsilon}\frac{D\varepsilon}{Dt} + C_{H2}(\omega_0, \omega_{H0})\frac{\varepsilon_H}{H}\frac{DH}{Dt}, \tag{11}$$

where $C_{H1}(\omega_0, \omega_{H0})$ and $C_{H2}(\omega_0, \omega_{H0})$ are coefficients determined by the time scales of turbulence. If we assume $\tau \simeq \tau_K \simeq \tau_H$ ($\omega_0 \simeq \omega_{K0} \simeq \omega_{H0}$), we have $C_{H1}(\omega_0, \omega_{H0}) \simeq 0.26$ and $C_{H2}(\omega_0, \omega_{H0}) \simeq 1.1$. Finally, we obtain

$$\frac{D\varepsilon_H}{Dt} = C_{\varepsilon_H 1}\frac{\varepsilon_H}{K}P_K - C_{\varepsilon_H 2}\frac{\varepsilon_H}{K}\varepsilon + C_{\varepsilon_H 3}\frac{\varepsilon_H}{H}P_H - C_{\varepsilon_H 4}\frac{\varepsilon_H}{H}\varepsilon_H, \tag{12}$$

where the model constants are theoretically estimated as

$$C_{\varepsilon_H 1} = 0.36,\ C_{\varepsilon_H 2} = 0.49,\ C_{\varepsilon_H 3} = C_{\varepsilon_H 4} = 1.1. \tag{13}$$

Conclusion. From the lowest-order analysis, the helicity dissipation rate is estimated by an algebraic form [Eq. (10)]. Up to the first-order analysis, the ε_H equation is derived as Eq. (12) with the theoretically-determined model constants. Reflecting the spectral form [Eq. (7)], it depends on both the energy and helicity equations.

Acknowledgments Basic calculations of this work were performed during NY's stay at the Rudolf Peierls Centre for Theoretical Physics, The University of Oxford (January 2013) and at the Consorzio RFX in Padova (February 2013) as a visiting researcher. Part of this work is supported by the Japan Society for the Promotion of Science (JSPS) Core-to-Core Program (No. 22001) Institutional Program for Young Researcher Overseas Visits and also by the JSPS Grants-in-Aid for Scientific Research (No. 24540228).

References

1. Baerenzung, J., Politano, H., Ponty, Y., Pouquet, A.: Spectral modeling of turbulent flows and the role of helicity. Phys. Rev. E **77**(046303), 1–15 (2008)
2. Lessinnes, T., Plunian, F., Stepanov, R., Carati, D.: Dissipation scales of kinetic helicities in turbulence. Phys. Fluids **23**(035108), 1–6 (2011)
3. Parker, E.N.: Hydromagnetic dynamo models. Astrophys. J. **122**, 293–314 (1955)
4. Yokoi, N.: Modeling the turbulent cross-helicity evolution: production, dissipation, and transport rates. J. Turb. **12**(27), 1–33 (2011)
5. Yokoi, N.: Cross helicity and related dynamo. Geophys. Astrophys. Fluid Dyn. **107**, 114–184 (2013)
6. Yokoi, N., Yoshizawa, A.: Statistical analysis of the effects of helicity in inhomogeneous turbulence. Phys. Fluids A **5**, 464–477 (1993)
7. Yoshizawa, A.: Statistical analysis of the deviation of the Reynolds stress from its eddy-viscosity representation. Phys. Fluids **27**, 1377–1387 (1984)
8. Yoshizawa, A.: Statistical modeling of a transport equation for the kinetic energy dissipation rate. Phys. Fluids **30**, 628–631 (1987)

Analysis of the Bivariate EMD Behavior for Separating Coherent Structures from Interference Fluctuations in Isotropic Homogeneous Turbulence

Mehdi Sadeghi, Fabrice Foucher, Karim Abed-Meraim and Christine Mounaïm-Rousselle

1 Introduction

In general, turbulent flow is undergoing the influence of coherent structures. The existence of these structures was recognized many years ago (Brown and Roshko) [1], but their exact definition is still open to debate. Hussain and Reynolds [5] introduced the so-called triple decomposition such that a physical variable of the turbulent flow is a superimposition of a coherent (i.e., phase averaged), a mean (i.e., time-averaged) and a random fluctuation. Fourier transform can be used to separate large-scale structures from small-scales but the choice of cut-off frequency is a great drawback of this approach. Lumley [7] proposed proper orthogonal decomposition (POD) to extract coherent structures from turbulent flows. This approach is associated with the statistical description of flow that restricts its interest to non-stationary dynamics. Using orthogonal wavelet analysis, Farge et al. [6] separated large scale from turbulent flow; however, the performance of the analysis is closely tied to how well the morphology of the signals under study is represented by the selected basis function.

For the analysis of nonlinear, non-stationary and noisy data, Huang et al. [3] introduced the so-called Empirical Mode Decomposition (EMD). This method is intuitive, direct and adaptive, with an posteriori-defined decomposition basis driven entirely from the data themselves. Huang et al. [4] investigated the intermittency of fully developed turbulent flows by means of EMD and Hilbert spectral analysis. The present work seeks to assess the efficiency of Bivariate EMD to extract coherent structures from synthetic interference fluctuations in isotropic homogeneous turbulence.

M. Sadeghi (✉) · F. Foucher · K. Abed-Meraim · C. Mounaïm-Rousselle
Laboratoire PRISME, Université D'Orléans, 8 Rue Léonard de Vinci,
45072 Orléans, France
e-mail: mehdi.sadeghi@etu.univ-orleans.fr

J. Peinke et al. (eds.), *Progress in Turbulence VI*,
Springer Proceedings in Physics 165, DOI 10.1007/978-3-319-29130-7_18

2 State of the Art

2.1 Empirical Mode Decomposition

EMD represents an arbitrary signal via a number of Intrinsic Mode Functions (IMF), and the residual. By design, an IMF is a function for which the number of extrema and the number of zero crossings are either equal or differ at most by one, together with the mean value of two envelopes associated with the local maxima and minima being zero.Generally, the finest component of the shortest period at each instant will be identified and decomposed into the first IMF. The components of longer periods will be identified and decomposed into the following IMFs in sequence [3]. The basic idea of bivariate EMD considers a bivariate signal, complex valued time series, as a fast rotation superimposed on slower ones [9]. In order to separate the more rapidly rotating component from slower ones, the idea is to define the slowly rotating component as the mean of some 3-dimensional tube,envelope, that tightly encloses the signal. Since the presentation of Bivariate EMD and its algorithm are not subjects of this paper, the reader is refered to [9] for more information.

2.2 Methodology

An experimental homogeneous isotropic turbulent flow field representing the turbulence field inside an engine was used as the original signal $u_i(t)$. It was generated by six fans in a spherical combustion chamber with a mean velocity inside the chamber equal to zero. The turbulence field was characterized by using high rate particle image velocimetry and Integral length-scale, Integral time-scale and energy spectra were determined Galmiche et al. [2]. This flow field is perturbed by a mono-tone signal as a synthetic fluctuation $p(t) = Asin(2\pi f.t)$, that mimics coherent structures. The perturbed signal under analysis is therefore $u_p(t) = u_i(t) + p_i(t)$. The EMD sorts the different IMFs of a velocity signal from high frequency to the lowest one. The first IMF is the representation of the high-frequency part (small scale) of the flow and depending on the decomposition level the IMF is increasingly representative of the tendencies of the velocity (large scale). The EMD analysis of a reference signal and a perturbed signal at one point is illustrated in Fig. 1.

To discriminate between the perturbation and original signal Mazellier et al. [8] introduced the resemblance criterion based on the statistical dependence between the reconstructed signals resulting from two successive steps.

$$if\ 1 < n \leq N: \quad R(n) = \frac{\langle u_n(t) u_{n-1}(t) \rangle}{\sqrt{\langle u_n(t)^2 \rangle} \sqrt{\langle u_{n-1}(t)^2 \rangle}}, if\ n = 1: \quad R(n) = 0 \quad (1)$$

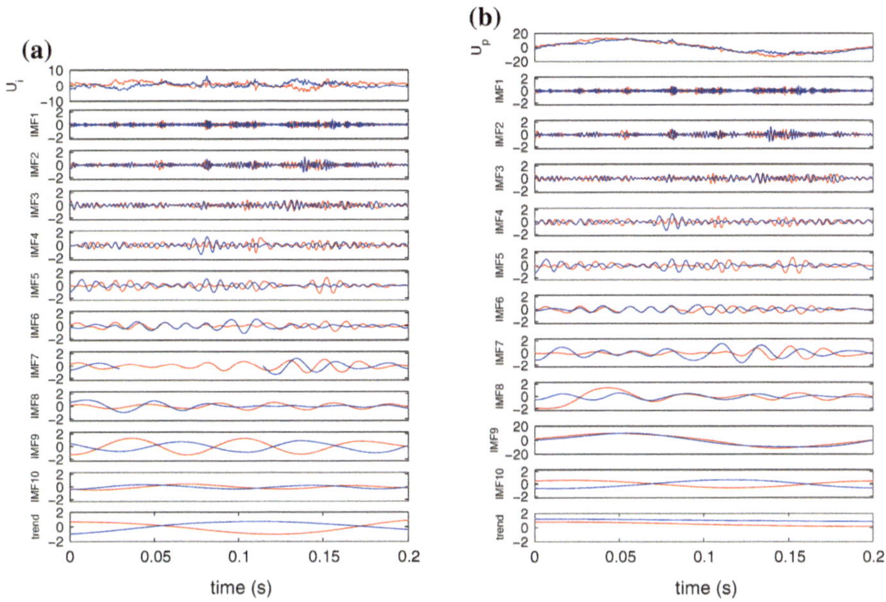

Fig. 1 **a** Initial signal, **b** Perturbated signal

where the reconstructed signal $u_n(t)$ defined as follows:

$$if\ n \leq N: \quad u_n(t) = \sum_{k=1}^{n} IMF_k(t), if\ n = N: \quad u_n(t) = \sum_{k=1}^{n} IMF_k(t) + r_N \tag{2}$$

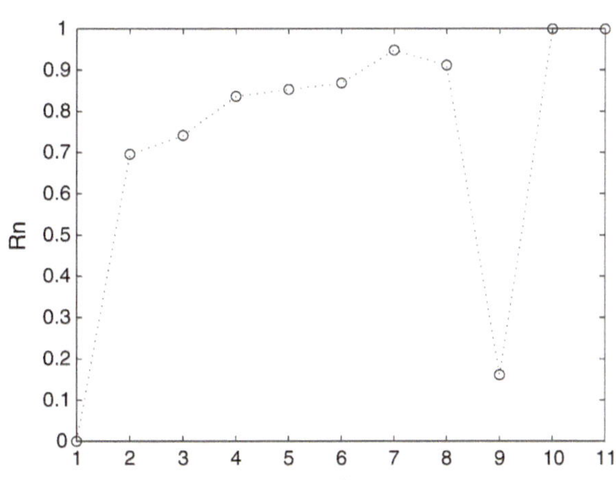

Fig. 2 Evaluation of the resemblance criterion R(n) as a function of reconstruction number n

The evolution of the resemblance criterion R(n) as a function of the reconstruction number n for the perturbation signal is shown in Fig. 2, sudden drop in IMF 9 indicates that this IMF represents a perturbation that has to be removed from the signal.

3 Results

To assess the quality of the recovering algorithm, the mean of the correlation coefficient between the recovered signal and the initial one in the case of different perturbations as a function of coupled (a, f), amplitude and frequency, is plotted in Fig. 3, One can see that the recovering procedure is only weakly sensitive to the amplitude of the perturbation but it decreases significantly when the perturbation signal frequency is increased. Also in each case of perturbation, the power spectral density of both recovered and initial flow field have been computed and the averaged values of the similarity measure as in Eq. (1) between them is illustrated in Fig. 4. This illustration also demonstrates in low frequency perturbations, an impressive agreement between the Fourier spectrum of the recovered turbulent velocity field and the initial one. To compare the large-scale features of the recovered turbulent flow field and the initial one, the integral time and length-scales of two fields are computed. As an example, the mean autocorrelation coefficient and the mean two-point correlation coefficient, in the case where the maximum perturbation amplitude is 10 m/s with different

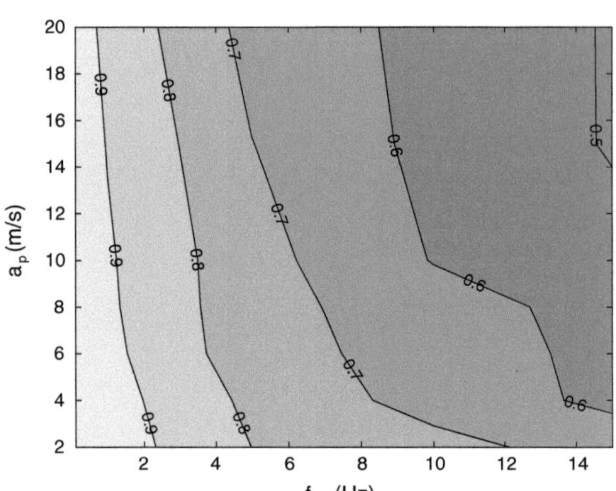

Fig. 3 Contour plot of the recovering quality indicator with respect to the parameters of the signal tone perturbation

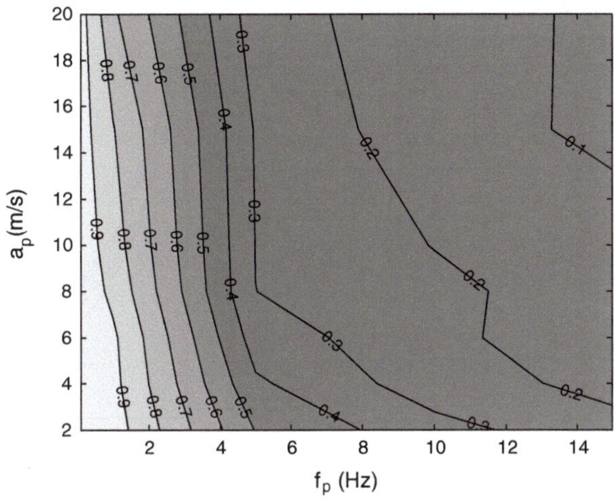

Fig. 4 Contour plot of the similarity between the mean power spectral density of recovered and initial flow field with respect to the parameters of the signal tone perturbation

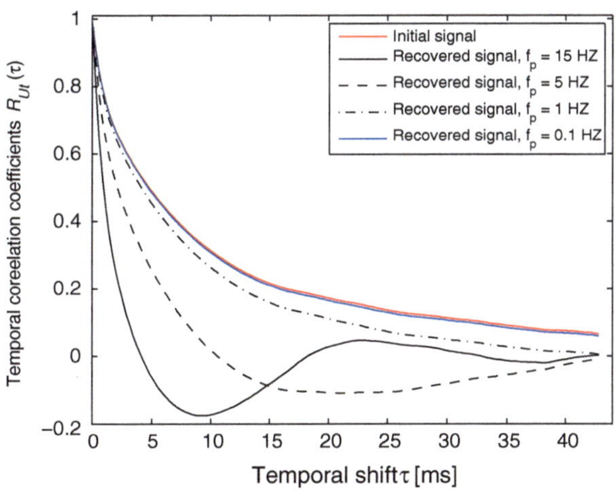

Fig. 5 Temporal correlation as cfunction of different recovered signal

perturbation frequencies, are illustrated in Figs. 5 and 6. As shown in these figures in the base frequency perturbation there is a good agreement between the large-scale behavior in the recovered velocity field and the initial one.

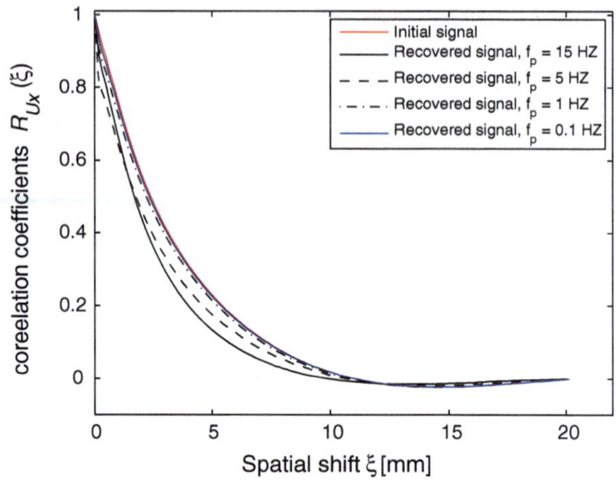

Fig. 6 Typical evolution of correlation coefficients

4 Conclusion

The bivariate EMD for the first time was applied on perturbed 2D turbulent field to recover the reference signal. The influence of amplitude and frequency of perturbation on the recovering procedure have been studied. This study justified in the case the perturbation frequency is established in a much lower range than the integral scale frequency, the original signal can be perfectly recovered.

References

1. Brown, G.L., Roshko, A.: On density effects and large structure in turbulent mixing layer. J. Fluid Mech. **54**, 775–816 (1974)
2. Galmiche, B., Mazellier, N., Halter, F., Foucher, F.: Turbulence characterization of a high-pressure high-temperature fan-stirred combustion vessel using LDV. PIV TR-PIV meas. Exp. Fluids. (2014). doi:10.1007/s00348-013-1636-x
3. Huang, N.E., Shen, Z., Long, S.R., W.u, W.C., Shih, H.H., Zheng, Q., Yen, N.C., Tung, C.C., Liu, H.H.: The empirical mode decomposition and the hilbert spectrum for nonlinear and non-stationary time series analysis. Proc R Soc Lond A 454, 903–995 (1998)
4. Huang, Y.X., Schmitt, F.G., Lu, Z.M., Liu, Y.L.: An amplitudefrequency study of turbulent scaling intermittency using empirical mode decomposition and hilbert spectral analysis. J. EPL. EuroPhys. Lett. (2008). doi:10.1209/0295-5075/84/40010
5. Hussain, A.K.M.F., Reynolds, W.C.: The mechanics of an organized wave in turbulent shear flow. J. Fluid Mech. **41**, 241–258 (1970)
6. Farge, M., Pellegrino, G., Schneider, K.: Coherent vortex extraction in 3D turbulent flows using orthogonal wavelets. Phys. Rev. Lett. **87**(5), 054501 (2001)
7. Lumley, J.L.: The structure of inhomogenous turbulence. In: Yalgom, A.M., Tatarski, V.I. (eds.) Atmospheric turbulence and wave propagation, pp. 166–178. Nauka, Moscow (1967)

8. Mazellier, N., Foucher, F.: Separation between coherent and turbulent fluctuations: what can we learn from the empirical mode decomposition? Exp. Fluids (2011). doi:10.1007/s00348-011-1069-3
9. Rilling, G., Flandrin, P., Goncalves, P., Lilly, J.M.: Bivariate empirical mode decomposition. IEEE Signal Process. Lett. **14**, 936–939 (2007)

Turbulent Shear Flows Described by the Algebraic Difference-Quotient Turbulence Model

Peter W. Egolf and Kolumban Hutter

Abstract It is shown that Newton's shear law for laminar flow and the Difference-Quotient Turbulence Model (DQTM) for turbulent flow are the analog constitutive laws describing the relations between shear strain rate and shear stress. Whereas the laminar case is fully linear and local, the turbulent counterpart is nonlinear and nonlocal. In this brief article the capacity of the new model is outlined by quoting references and related articles which contain results of convincing simplicity and accuracy and a presentation and discussion of the resulting analytical solutions of plane turbulent Couette flow. Newton's linear velocity profile between sheared plates for laminar flow is embedded in this solution as a special case. The solution of the corresponding nonlinear differential equation also reveals a cooperative phenomenon.

1 Introduction

Since the early zero-equation turbulence models appeared (see e.g. Taylor [1], Prandtl [2], etc.), it became clear that a strong analogy exists between momentum exchange by molecules in the laminar case and the turbulent momentum exchange by eddies in turbulent flow. However, the molecules with small diameters compared to their mean free path length lead to Newton's linear gradient shear law, whereas an analog law developed for turbulent shear flows, describing multi-size eddies (which even may interpenetrate), cannot be linear. Thus, the basic law of turbulent flow calculations, based on a too close analogy with Newton's shear law, which is at present still the well-accepted constitutive equation of Boussinesq [3], is only a crude approximation to describe the occurring phenomena in turbulent flows.

Different authors already dedicated their efforts to develop new constitutive equations as a replacement of Boussinesq's equation. Pope [4] proposed a more general

P.W. Egolf (✉)
University of Applied Science of Western Switzerland, Yverdon, Switzerland
e-mail: Peter.egolf@heig-vd.ch

K. Hutter
Swiss Federal Institute of Technology, ETH, Zurich, Switzerland
e-mail: hutter@vaw.baug.ethz.ch

J. Peinke et al. (eds.), *Progress in Turbulence VI*,
Springer Proceedings in Physics 165, DOI 10.1007/978-3-319-29130-7_19

effective viscosity hypothesis; Bernard and Handler [5] investigated the fluid mechanisms related to non-locality by numerical simulations. Important work has been published by Huang [6], who proposed a fading memory effect as main ingredient to improve the situation. Also Hamba [7] proposed a nonlocal model of the Reynolds stress in turbulent shear flows. Note that memory and non-locality effects are analog features in time and space. An alternative new algebraic closure model, namely the Difference-Quotient Turbulence Model (DQTM), was developed and published in 1991 [8]. Applying this closure, analytical solutions of numerous fundamental flow problems have been derived [9–13].

2 Deformation of a Laminar and Turbulent Flowing Fluid

The mean deformation of a turbulent flowing fluid with mean shear is analogous to that of laminar flows [14]. However, still some important deviations shall be outlined. From Fig. 1 it is concluded that

$$\frac{d\gamma_{T,21}}{dt} = -\frac{d\delta_{T,21}}{dt} = \lim_{\Delta t \to 0} \frac{\arctan\left(\frac{\overline{\Delta x_1}}{\overline{\Delta x_2}}\right)}{\Delta t}. \qquad \text{(1a,b)}$$

At first, because the fluid fluctuates, also the deformation is a fluctuating quantity. Therefore, time-averaged increments are considered and the angles are denoted by the index T (turbulent). Furthermore, the velocities are also replaced by their mean values and the vertical displacement $\overline{\delta x_2}$ is ignored.

Admitting that a linear gradient law in the x_2-direction is wrong, the limit $\overline{\Delta x_2} \rightarrow 0$ is not taken. This is the main idea and subtlety that opens the door for the derivation of a new description of the shearing processes in turbulent flows. Now, the time derivative of the mean complementary angle takes the form

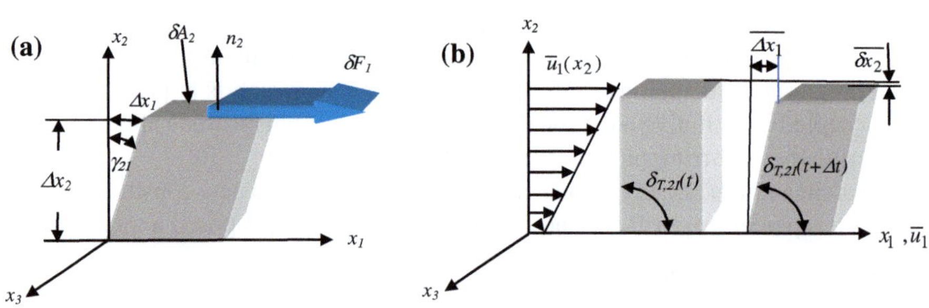

Fig. 1 The static deformation of a solid body is shown in (**a**) and the dynamic mean deformation of a fluid element of a turbulent flowing fluid between the times t and $t+\Delta t$ in (**b**)

$$\frac{d\delta_{T,21}}{dt} = -\lim_{\Delta t\to 0}\frac{\arctan\left(\frac{\left[\bar{u}_1(x_2+\overline{\Delta x_2})-\bar{u}_1(x_2)\right]\Delta t}{\overline{\Delta x_2}}\right)}{\Delta t} = -\frac{\bar{u}_1(x_2+\overline{\Delta x_2})-\bar{u}_1(x_2)}{\overline{\Delta x_2}} = -\frac{\Delta\bar{u}_1}{\overline{\Delta x_2}}. \tag{2a-c}$$

However, here, different from the laminar case, the spatial difference $\overline{\Delta x_2}$ and the velocity difference $\Delta\bar{u}_1$ are large-scale quantities! The limit $\Delta t \to 0$ causes $\overline{\Delta x_1}$ also to be small even for large $\Delta\bar{u}_1$. It is demanded that this velocity difference is bounded, which is indeed the case. Because an effect of large-scale eddies in turbulent flows, which by scaling represents the entire cascade of eddies of different sizes, is a well-established fact, it is now decided to let $\overline{\Delta x_2}$ tend to its maximum value—rather than toward its minimum (zero). In the studied example of shear flow this position is at the top boundary, denoted by $x_{2\,\max}$, where the maximum velocity, $\bar{u}_{1\,\max}$, occurs. These considerations imply

$$\frac{d\delta_{T,21}}{dt} = -\frac{\bar{u}_1(x_{2_{\max}})-\bar{u}_1(x_2)}{x_{2\,\max}-x_2} = -\frac{\bar{u}_{1\,\max}-\bar{u}_1(x_2)}{x_{2\,\max}-x_2}, \tag{3a-b}$$

It can be seen that $x_{2\,\max}$ is defined as the position where $\bar{u}_1$ is at its maximum $\bar{u}_{1\,\max}$. The turbulent dynamic viscosity is the ratio of the turbulent shear stress to the rate of turbulent shear strain,

$$\mu_T = \frac{\tau_{T,21}}{\frac{d\gamma_{T,21}}{dt}} = -\frac{\tau_{T,21}}{\frac{d\delta_{T,21}}{dt}} = \frac{\tau_{T,21}}{\frac{\bar{u}_{1\,\max}-\bar{u}_1(x_2)}{x_{2\,\max}-x_2}}, \quad \nu_T = \frac{\mu_T}{\rho}, \tag{4a-d}$$

where ν_T denotes the effective (turbulent) kinematic viscosity. Now, it follows that

$$\frac{\tau_{T,21}}{\rho} = -\overline{u_2'u_1'} = \sigma\,(x_{2\,\max}-x_2)\,(\bar{u}_1-\bar{u}_{1\,\min})\,\frac{\bar{u}_{1\,\max}-\bar{u}_1}{x_{2\,\max}-x_2}, \quad \sigma = \frac{const}{3}, \tag{5a-c}$$

where in analogy to statistical mechanics for molecular dynamics an expression for the turbulent viscosity has been derived (see also Table 1). Instead of molecules of a single size, this modeling takes into consideration a cascade of eddies of different sizes. Equation (5a-c) define the algebraic DQTM, which has been derived in this article in a different manner than in earlier work, where, for example, a nonlocal integral presentation with a kernel consisting of Heaviside distributions was introduced [11].

The full solution of turbulent Couette flow, calculated with the DQTM, is presented in [10]. Table 1 shows results in [10] and of this article, in comparison with analog quantities of the laminar case.

Table 1 Comparison of numerous quantities of laminar and turbulent flows based on the DQTM

	Laminar flow	Turbulent flow
Characteristic length	λ	$\Lambda_{T,2} = x_{2\,\max} - x_2$
Characteristic velocity	u_{mol}	$\bar{u}_1 - \bar{u}_{1\,\min}$
Shear stress	$\tau_{21} = \mu \frac{du_1}{dx_2}$	$\tau_{T,21} = \mu_T \frac{\bar{u}_{1\,\max} - \bar{u}_1(x_2)}{x_{2\,\max} - x_2}$
(Effective) dynamic viscosity	$\mu = \sigma \rho \lambda u_{mol}$	$\mu_T = \sigma \rho \Lambda_{T,2} (\bar{u}_1 - \bar{u}_{1\,\min})$
Dimensionless variable	$\eta = x_2 / x_{2\,\max}$	$\eta = x_2 / x_{2\,\max}$
Dimensionless function	$g_1 = u_1(\eta)/U$	$g_1 = \bar{u}_1(\eta)/U$
Differential equation	$g_1'' = 0$	$2Re_c/Re \cdot g_1' + 4\mathrm{X} \cdot g_1 (1 - g_1) - 1 = 0$
Order parameter (see Fig. 2)	$\mathrm{X}\,(\mathrm{Re} < \mathrm{Re}_c) = 0^{a}$	$Re_c/Re(\mathrm{X}) = \sqrt{\mathrm{X}(1-\mathrm{X})} / \arctan \left[\sqrt{\mathrm{X}/(1-\mathrm{X})}\right]$
Boundary conditions	$g_1(-1) = 0, g_1(1) = 1$	$g_1(-1) = 0, g_1(1) = 1$
Mean downstream velocity	$g_1 = 1/2\,(1 + \eta)$	$g_1 = 1/2 \left\{1 + \sqrt{(1-\mathrm{X})/\mathrm{X}} \cdot \tan\left[\arctan\left(\sqrt{\mathrm{X}/(1-\mathrm{X})}\right)\eta\right]\right\}$
Dimensionless mean shear stress	$g_{21} = g_1' = 1/2$	$g_{21} = -4\mathrm{X} g_1 (1 - g_1)$

[a] X is related to the mean shear S (see Ref. [10], Eq. (16)). With $S = -2\bar{\omega}_3 = -2\bar{\omega}$ X is directly related to the mean vorticity $\bar{\omega}$

3 Turbulence, a Cooperative Phenomenon in Analogy to Magnetism

The DQTM reveals a critical phenomenon. Such also occur in systems of other physical domains, e.g. in structural phase transitions of solid crystals, solid/fluid and fluid/gas systems, fluid mixtures (e.g. water/phenol), magnetism, superconductivity, super-fluidity, etc. In an open system a (constant) flux of energy into and out of it, by self-organization, may also create two phases with different entropy densities. In turbulence, vortices are occurring in well-defined patches or spots (low entropy phase) separated from laminar streaks (belonging to the high entropy phase). In a magnetic system the phase regions with oriented spins (low entropy) are called Weiss domains (see white patches in Fig. 2, first panel on the left), and it is the magnetization M (order parameter) as a function of temperature T(stress parameter), that describes how large these patterns are (second panel). In analogy, we call the white vorticity-rich domains turbulent domains (third panel) and notice that our analytical theory also delivers the order parameter X(Re) [10] (fourth panel), as e.g. a spin-wave theory does for the magnetic system. The critical value of the magnetic system is its Curie temperature, T_c, and of plane turbulent Couette flow the critical Reynolds number, Re_c. By setting the derivatives (zero entropies) of these curves with respect to the

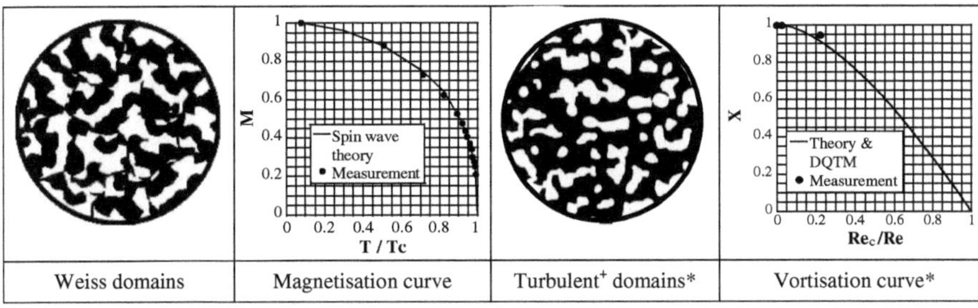

Fig. 2 Analogy between the magnetic and the turbulent system is shown with a picture of two phases each and figures showing magnetisation (Ref. [13] and ref. therein) and vortisation*. * Denotes preliminary new name, $^{+}$ In analogy to 'Weiss' a scientists name could eventually be more accurate

stress parameters equal to zero, these follows in both cases the validity of Nernst's theorem for (non-degenerate) systems.

References

1. Taylor, G.I.: Eddy motion in the atmosphere. Phil. Trans. R. Soc. Lond. **215**, 523–537 (1915)
2. Prandtl, L.: Bemerkungen zur Theorie der freien Turbulenz. ZAMM **5**(22), 241–243 (1942)
3. Boussinesq, J.: Mém. Prés. Div. Savant Acad. Sci. **23**, 46 (1872)
4. Pope, S.B.: A more general effective-viscosity hypothesis. J. Fluid Mech. **72**(2), 331–340 (1975)
5. Bernard, P.S., Handler, R.A.: Reynolds stress and the physics of turbulent momentum transport. J. Fluid Mech. **220**, 90–124 (1990)
6. Huang, Y.-N.: On modeling the Reynolds stress in the context of continuum mechanics, communications in nonlinear. Sci. Num. Sim. **9**, 543–559 (2004)
7. Hamba, F.: Nonlocal analysis of the Reynolds stress in turbulent shear flow. Phys. Fluids **17**, 115102 (2005)
8. Egolf, P.W.: A new model on turbulent shear flows. Helv. Phys. Acta **64**, 944–945 (1991)
9. Egolf, P.W.: Difference-quotient turbulence model: a generalization of Prandtl's mixing-length theory. Phys. Rev. E **49**(2), 1260–1268 (1994)
10. Egolf, P.W., Weiss, D.A.: Model for turbulent plane Couette flow. Phys. Rev. Lett. **75**(16), 2956–2959 (1995)
11. Egolf, P.W.: Difference-quotient turbulence model: the axi-symmetric isothermal jet. Phys. Rev. E **58**(1), 459–470 (1998)
12. Egolf P.W., Weiss D.A.: Difference-quotient turbulence model: Analytical solutions for the core region of plane Poiseuille flow, Phys. Rev E **62**/1-A, 553–563 (2000)
13. Egolf, P.W.: Lévy statistics and beta model: a new solution of "wall" turbulence with a critical phenomenon. Int. J. Refrig. **32**, 1815–1836 (2009)
14. Reif, F.: Fundamentals of Statistical and Thermal Physics. MacGraw Hill Company, New York (1965)

Lyapunov Stability Criteria for Reacting Ionic Fluid Flows

Martina Costa Reis and Adalberto Bono Maurizio Sacchi Bassi

Abstract In this work, the stability of reacting ionic fluid flows is studied by using the Lyapunov direct method. For this purpose, one considers the flow of a reacting ionic fluid inside a bounded domain under the action of external sources. The Lyapunov candidate function characteristic for this kind of fluid depends on the barycentric velocity of the fluid, mass concentration, temperature, pressure, and electromagnetic fields. Nevertheless, whenever more strict physical conditions are imposed to the reacting ionic fluid flow, the obtained Lyapunov candidate function is identical to a well-known thermodynamic potential of classical thermodynamics.

1 Introduction

Among the variational methods reported in the literature, the Lyapunov direct method is of great relevance because its wide scope allows to study the stability of linear and non-linear autonomous systems even without knowing their equilibrium states [1]. Thus, in this work, the Lyapunov direct method is employed to investigate the stability of reacting ionic fluid flows.

Section 2 shows some basic ideas about the Lyapunov stability criteria. In turn, in Sect. 3 a proper Lyapunov candidate function is constructed to a reacting ionic fluid flow, and the thermodynamic conditions that characterize the stability of equilibrium states of this fluid are investigated. The aim of this work is to discuss how the Lyapunov direct method can be used to predict the evolution of a reacting ionic fluid flow toward the stability.

M.C. Reis (✉) · A.B.M.S. Bassi
Department of Physical Chemistry, Institute of Chemistry,
University of Campinas-UNICAMP, Campinas-sp, Brazil
e-mail: marreis@iqm.unicamp.br

A.B.M.S. Bassi
e-mail: bassi@iqm.unicamp.br

J. Peinke et al. (eds.), *Progress in Turbulence VI*,
Springer Proceedings in Physics 165, DOI 10.1007/978-3-319-29130-7_20

2 Basic Notions of Lyapunov Stability Criteria

Consider a continuous body $\mathscr{S}$. A thermodynamic state $\chi(t) = (\Gamma_1(t), \ldots, \Gamma_n(t))$ of a continuous body is a n-tuple of time-dependent functions locally defined over $\mathscr{S}$. Moreover, a thermodynamic equilibrium state $\chi(t^*) = \chi^*$ of $\mathscr{S}$ is a state where the continuous body is at rest in relation to an inertial reference frame κ, and the functions $\Gamma(t)$ are uniform and time-independent fields over $\mathscr{S}$.

It is worth mentioning that thermodynamic equilibrium states are never rigorously reached. From the practical point of view, in continuum thermodynamics one says that a system reaches the equilibrium state when changes in macroscopic parameters can no longer be observed by the available measuring devices.

Definition 1 Suppose that a continuous body possesses an equilibrium state χ^*. A scalar-valued function $\mathscr{L}(t)$ is a Lyapunov candidate function, if for every attainable process it holds the following properties:

1. $\mathscr{L}(t)$ is a non-increasing function of time in a neighborhood of χ^*;
2. $\mathscr{L}(t)$ is a continuous function at χ^*;
3. $\mathscr{L}(t)$ has a strong local minimum at χ^*. In other terms, χ^* is a strong local minimum of $\mathscr{L}(t)$, if

$$\mathscr{L}\left(t^*\right) < \mathscr{L}(t) \quad \forall \chi(t) \in \mathscr{N}\left(\chi\left(t^*\right), \delta\right),$$

where $\mathscr{N}(\chi(t^*), \delta)$ is the set of feasible points contained within the δ-neighborhood of χ^*.[1]

As a consequence of the above properties, in a neighborhood of χ^* the Lyapunov candidate functions obey the inequality,

$$\frac{d}{dt}\mathscr{L}(t) \leq 0. \tag{1}$$

In fact, Lyapunov candidate functions are of great relevance to stability problems because from $\mathscr{L}(t)$ one can test the stability of an equilibrium state by using the Lyapunov theorem [1]. Furthermore, Lyapunov candidate functions may be used to exclude those constitutive functions that are not coherent with the physical conditions that characterize the thermodynamic equilibrium states.

According to the Lyapunov theorem, an equilibrium state is said to be stable, if the time derivative of $\mathscr{L}(t)$ is negative semi-definite. On the other hand, an equilibrium state is said to be asymptotically stable, if the time derivative of $\mathscr{L}(t)$ is negative definite. Of course, in view of this condition, an asymptotically equilibrium state is

[1] The δ-neighborhood of the state χ^* is defined as the region that contains the current states of the continuous body $\mathscr{S}$ as soon as the initial external conditions of $\mathscr{S}$ are restored.

also a stable equilibrium state. Example of a stable χ^* is the equilibrium state of an idealized gravity pendulum subject to no air friction and drag. In turn, the equilibrium state of a real gravity pendulum subject to air friction and drag can be considered an example of an asymptotically stable χ^*.

3 An Example of Lyapunov Stability Criteria Application: Reacting Ionic fluid Flows

Consider a flow of a reacting ionic fluid inside a bounded domain under action of external sources. Besides, assume that this fluid is governed by the same equations of a heat conducting mixture of fluids with diffusion and chemical reactions subject to the influence of electromagnetic fields [2]. By supposing that **(i)** the fluid is a non-polar continuum, **(ii)** the mixture is so diluted that its dynamics is given by the solvent dynamics, **(iii)** the motion of the fluid is incompressible, and **(iv)** the entropy production is a non-negative quantity for any physically admissible process of the fluid (second law of thermodynamics), then one obtains from the balance laws in their integral forms the following expression

$$\frac{d}{dt}\int_{\mathscr{V}} \rho\left(\varepsilon + \frac{1}{2}\mathbf{v}\cdot\mathbf{v} - \theta\eta + \sum_{a=1}^{n-1}\xi_a\mu_a + \frac{p}{\rho} + \frac{\mathbf{D}}{\rho}\cdot\mathbf{E}\right)dv - \int_{\mathscr{V}}\sum_{r=1}^{R}\Omega_r\Upsilon_r dv \leq 0, \tag{2}$$

where $\mathscr{V}$ is a part of the material domain of the whole continuous mixture, dv and da are the volume and surface area elements of the mixture, respectively, ρ is the mixture mass density, $\rho\varepsilon$ is the mixture internal energy density, $\mathbf{v}$ is the mixture barycentric velocity, θ is the absolute temperature, $\rho\eta$ is the mixture entropy density, ξ_a is the mass concentration of the ath mixture constituent, μ_a is the chemical potential of the ath constituent, p is the hydrostatic pressure, $\mathbf{D}$ is the electric displacement vector, $\mathbf{E}$ is the electric field vector, Ω_r is symmetric to the chemical affinity of a chemical reaction r, and Υ_r is the rate of conversion of r by supposing R independent chemical reactions.

In obtaining Eq. (2), it was considered that the entropy source is a linear combination of the linear momentum and energy sources, and the ionic fluid is a linear dielectric. Furthermore, it was considered that the entropy flux vector of the whole continuous mixture includes contributions from the heat flux vector $\mathbf{h}$, temperature, diffusive flux $\mathbf{j}_a$, and chemical potential, *viz.* $\boldsymbol{\Phi} = \mathbf{h}/\theta - \sum_{a=1}^{n-1}\mu_a\mathbf{j}_a/\theta$. This expression, whose deduction can be found in reference [3], is widely used in the constitutive modeling of multiconstituent continuous bodies.

The discussion about the stability of a reacting ionic fluid flow proceeds by placing the first physical restriction: although chemical reactions continue to occur in the continuous mixture, the chemical equilibrium of each reaction r has been reached. Hence, one has

$$\frac{d}{dt}\mathscr{L}_1(t) \leq 0 \quad \mathscr{L}_1(t) = \int_{\mathscr{V}} \rho \left(\varepsilon + \frac{1}{2}\mathbf{v}\cdot\mathbf{v} - \theta\eta + \sum_{a=1}^{n-1} \xi_a \mu_a + \frac{p}{\rho} + \frac{\mathbf{D}}{\rho}\cdot\mathbf{E} \right) dv, \tag{3}$$

where $\mathscr{L}_1(t)$ is the Lyapunov candidate function for a reacting ionic fluid flow whose chemical reactions are at chemical equilibrium. In this case, $\mathscr{L}_1(t)$ includes contributions arising from the barycentric velocity of the mixture, mass concentration, temperature, pressure, and electromagnetic fields. Moreover, this Lyapunov candidate function is also valid for the case of a heat conducting mixture of fluids with diffusion, but non-reactive.

If one imposes more strict physical constrains to the fluid, namely absence of diffusive fluxes, chemical equilibrium, and the fluid is at mechanical rest in relation to an inertial reference frame, then it follows that

$$\frac{d}{dt}\mathscr{L}_2(t) \leq 0 \qquad \mathscr{L}_2(t) = \int_{\mathscr{V}} \rho \left(\varepsilon - \theta\eta + \frac{p}{\rho} + \frac{\mathbf{D}}{\rho}\cdot\mathbf{E} \right) dv. \tag{4}$$

In fact, the restrictions imposed to the fluid are the very familiar physical and chemical conditions that classical thermodynamics uses to study the stability of chemical systems. Briefly, in this case the Lyapunov candidate function is the Gibbs energy, also referred to as free enthalpy, including electromagnetic fields.

4 Final Remarks

In this work, the basic notions about Lyapunov stability criteria are provided. As an example of the Lyapunov direct method exploitation, the stability of a reacting ionic fluid flow is considered. For this purpose, the balance laws of mass concentration, linear momentum, energy, and entropy of a reacting mixture subject to the influence of electromagnetic fields were used to construct a Lyapunov candidate function $\mathscr{L}(t)$, a scalar-valued function that characterizes the changes of a proper system energy toward the stable equilibrium state. From this function, it was shown that as the reacting ionic fluid flow approaches to its stable equilibrium state, $\mathscr{L}(t)$ cannot increase.

Acknowledgments The first author acknowledges financial support from the São Paulo Research Foundation (grant 2013/20872-2).

References

1. La Salle, J., Lefschetz, S.: Stability by Liapunov's Direct Method with Applications. Academic Press, London (1961)
2. Reis, M.C., Bassi, A.B.M.S.: Reactive continuum mixtures under the influence of electromagnetic fields. In: Talamelli, A., Peinke, J., Oberlack, M. (eds.) Progress in Turbulence V, pp. 195–200. Springer, Berlin (2014)
3. Reis, M.C., Wang, Y., Bassi, A.B.M.S.: Toward the thermodynamic modeling of reacting ionic mixtures. Contin. Mech. Thermodyn. **26**, 753–769 (2014)

DNS and LES of Viscoplastic-Type Non-Newtonian Fluid Flows

A. Carmona, O. Lehmkuhl, C.D. Pérez-Segarra and A. Oliva

Abstract The aim of this work is to delve into the *turbulent flow of viscoplastic-type non-Newtonian fluids* through specific and essential numerical aspects for them. Specifically, the numerical problems introduced by the *transpose diffusive term* which are associated with the *velocity field discontinuity* and the *artificial viscous diffusion* will be addressed from the beginning. The objective is to gain insight into the underlying physics of this class of non-Newtonian fluid flows.

1 Introduction

In the specialised literature on the topic of viscoplastic-type non-Newtonian fluid flows many researchers use the generalised Newtonian model together with a specific rheological law in order to analyse non-Newtonian fluid flows which exhibit viscoplastic stresses (see, for instant, Refs. [1–4]). Nevertheless, from our point of view, specific numerical techniques for this class of non-Newtonian fluid flows are not being taken into account in the corresponding numerical simulations.

2 Mathematical Formulation

The *transport equations* describing the non-Newtonian behaviour for an incompressible fluid are:

$$\nabla \cdot \mathbf{v} = 0 \tag{1}$$

A. Carmona (✉) · O. Lehmkuhl · C.D. Pérez-Segarra · A. Oliva
Heat and Mass Transfer Technological Centre, Technical University of Catalonia, ETSEIAT, C/Colom 11, 08222 Terrassa, Spain
e-mail: angel.cfd@gmail.com

O. Lehmkuhl
Termo Fluids, S.L., Avda. Jaquard, 97 1-E, 08222 Terrassa, Spain

J. Peinke et al. (eds.), *Progress in Turbulence VI*,
Springer Proceedings in Physics 165, DOI 10.1007/978-3-319-29130-7_21

$$\rho \frac{\partial \boldsymbol{v}}{\partial t} + \rho \nabla \cdot (\boldsymbol{v}\boldsymbol{v}) = -\nabla p + \nabla \cdot (\eta \nabla \boldsymbol{v}) + \nabla \cdot \left(\eta (\nabla \boldsymbol{v})^T\right) + \rho \boldsymbol{g} \tag{2}$$

The *finite volume discretisation* and the *collocated variable arrangement* have been used to obtain the *discrete form* of these equations:

$$M \boldsymbol{v}_f = 0 \tag{3}$$

$$\rho_c \frac{\partial \boldsymbol{v}_c}{\partial t} + C \boldsymbol{v}_c = -G p_c + D \boldsymbol{v}_c + \nabla_h \cdot \left(\eta (\nabla_h \boldsymbol{v})^T\right) + B_g \tag{4}$$

In the numerical analysis of viscoplastic-type non-Newtonian fluid flows certain key aspects must be taken into account from the beginning to overcome the numerical problems introduced by the *transpose diffusive term* and associated with the velocity field discontinuity, the artificial viscous diffusion and the transpose viscous coupling. For these reason, the transpose diffusive term, symbolically written here, will be numerically analysed in more detail.

3 Numerical Analysis

Concerning the *velocity field discontinuity*, the discretised transpose diffusive term should be composed of contiguous values of the collocated discrete variable in order to reproduce faithfully the non-Newtonian behaviour. Consequently, an expression for the transpose diffusive term based on the *divergence theorem* and an arithmetic mean (this one for approaching the corresponding staggered velocity gradients) is proposed:

$$\begin{aligned} \nabla_h \cdot (\eta (\nabla_h \boldsymbol{v})^T) &= \frac{1}{\Omega_P} \left(\frac{1}{2\,\Omega_P} \sum_{PF} \mathbf{A}_{PF} \frac{\boldsymbol{v}_F}{2} \right) \cdot \sum_f \eta_f \mathbf{A}_f \\ &+ \frac{1}{\Omega_P} \sum_f \left(\frac{\eta_f}{2\,\Omega_F} \sum_{F2F} \mathbf{A}_{F2F} \frac{\boldsymbol{v}_{2F}}{2} \right) \cdot \mathbf{A}_f \quad \Rightarrow \quad \exists\, \mathbf{v}_P \end{aligned}$$

However, *artificial viscous diffusion* can be introduced by this staggered discrete operator:

$$\mathrm{AVD}^{AM} = \frac{\eta}{\Omega_P} \sum_f \left(\frac{1}{2\,\Omega_F} \sum_{F2F} \frac{\mathbf{v}_{2F}}{2} \cdot \mathbf{A}_{F2F} \right) \mathbf{A}_f \quad \Rightarrow \quad \neq 0$$

As for the *artificial viscous dissipation*, the aforementioned term should be cancelled when the non-Newtonian viscosity takes the value of the Newtonian viscosity under the hypothesis of incompressible fluid, this time with the objective of reproducing accurately the Newtonian behaviour. Accordingly, another expression for the transpose diffusive term based on the *Green's first identity* is proposed:

$$\mathrm{AVD}^{CO} = \frac{1}{\Omega_P}\left[\left(\frac{1}{\Omega_P}\sum_{PF}\mathbf{A}_{PF}\frac{\mathbf{v}_F}{2}\right)\cdot\left(\eta\sum_{f}\mathbf{A}_f\right)^{0}\right] \Rightarrow\ = 0$$

Although, a velocity field discontinuity can be allowed by this collocated discrete operator:

$$\nabla_h\cdot(\eta(\nabla_h \boldsymbol{v})^T) = \frac{1}{\Omega_P}\left[\left(\frac{1}{\Omega_P}\sum_{PF}\boldsymbol{A}_{PF}\frac{\boldsymbol{v}_F}{2}\right)\cdot\left(\sum_{f}\eta_f\,\boldsymbol{A}_f\right)\right] \Rightarrow\ \nexists\,\mathbf{v}_P$$

4 Numerical Simulations

Taylor-Green vortex [G. I. Taylor and A. E. Green (1937)]

$$\psi = \frac{1}{k^2}sin(kx - x_0)sin(kz - z_0)$$

Constricted rigid pipe (CRP, see Refs. [1–4])

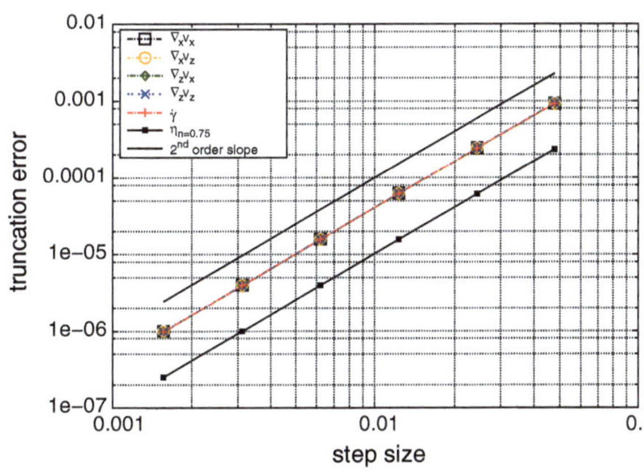

Fig. 1 Truncation error of the velocity gradient components, the shear rate and the non-Newtonian viscosity. The divergence theorem for the volume integrals and the midpoint rule for the surface integrals have been used

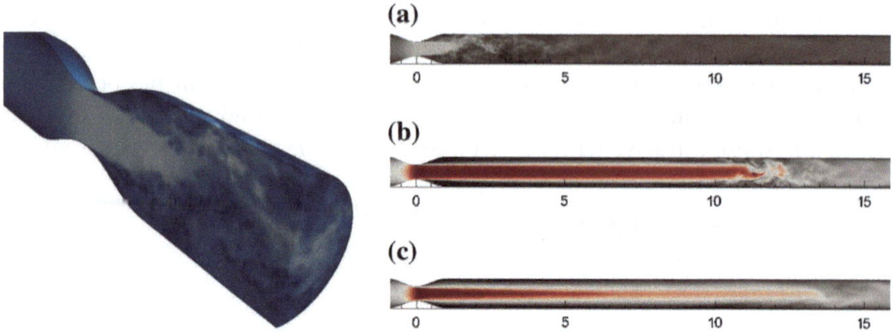

Fig. 2 Preliminary LES instantaneous snapshots of the velocity field, Re = 1000 **a** Pseudoplastic fluid (n = 0.75). **b** Newtonian fluid (n = 1.00). **c** Dilatant fluid (n = 1.50)

5 Conclusions

The numerical problems introduced by the transpose diffusive term which are associated with the velocity field discontinuity and the artificial viscous diffusion have been highlighted (Figs. 1 and 2). A *specific evaluation* for the *non-Newtonian viscosity at the control face*, in order to avoid the velocity field discontinuity in the collocated discrete operator, and a *unique definition* for the *velocity at the control face*, in order to eliminate the artificial viscous diffusion in the staggered discrete operator, are being analysed. Thus, specific numerical techniques for viscoplastic-type non-Newtonian fluid flows must be taken into account in the corresponding numerical simulations with the objective of reproducing faithfully and accurately the non-Newtonian behaviour.

Acknowledgments This work has been financially supported by the Spanish *Ministerio de Ciencia e Innovación* (Ref. ENE2010-17801).

References

1. Rudman, M., Blackburn, H.M., Graham, L.J.W., Pullum, L.: Turbulent pipe flow of shear-thinning fluids. J. Non-Newtonian Fluid Mech. **118**, 33–48 (2004)
2. Rudman, M., Blackburn, H.M.: Direct numerical simulation of turbulent non-newtonian flow using a spectral element method. Appl. Math. Model. **30**, 1229–1248 (2006)
3. Molla, M.M., Hossain, A., Wang, B.-C., Kuhn, D.C.S.: Large-eddy simulation of pulsatile non-newtonian flow in a constricted channel. Prog. Comput. Fluid Dyn. **12**, 231–242 (2012)
4. Ohta, Takashi, Miyashita, Masahito: DNS and LES with an extended smagorinsky model for wall turbulence in non-newtonian viscous fluids. J. Non-Newtonian Fluid Mech. **206**, 29–39 (2014)

Part II
Wall Bounded Flows

Flow Features in Three-Dimensional Turbulent Duct Flows with Different Aspect Ratios

Ricardo Vinuesa, Philipp Schlatter and Hassan M. Nagib

Abstract Direct numerical simulations of turbulent duct flows with width-to-height ratios 1, 3, 5, 7 and 10, at a friction Reynolds number $Re_{\tau,c} \simeq 180$, are carried out with the spectral element code Nek5000. The aim of these simulations is to gain insight into the kinematics and dynamics of Prandtl's secondary flow of second kind, and its impact on the flow physics of wall-bounded turbulence. The secondary flow is characterized in terms of the cross-plane mean kinetic energy $K = \left(V^2 + W^2\right)/2$, and its variation in the spanwise direction of the flow. Our results show that averaging times of at least 3,000 time units are required to reach a converged state of the secondary flow, which extends up to $z^* \simeq 5\,h$ from the side walls. We also show that if the duct is not wide enough to accommodate the whole extent of the secondary flow, then its structure is modified by means of a different spanwise distribution of energy. Future proposed work includes coherent structure eduction, quadrant analysis at the corner, and comparisons with spanwise-periodic channels at comparable Reynolds numbers.

1 Introduction

Turbulent duct flows of different aspect ratios AR (defined as the duct width W divided by its total height H) are of great importance for a number of technological applications. The most remarkable characteristic of ducts is the formation of the so-called Prandtl's secondary flow of second kind at the duct corners, which arises from Reynolds stress difference $\overline{v^2} - \overline{w^2}$ and the deviatoric Reynolds shear stress $\overline{vw}$. Due to the Reynolds-stress-induced nature of this kind of secondary flow, currently available Reynolds Averaged Navier Stokes (RANS) models widely used in industry in general fail to predict its effect on the flow.

R. Vinuesa (✉) · P. Schlatter
Linné FLOW Centre, KTH Mechanics, Osquars Backe 18, Stockholm, Sweden
e-mail: rvinuesa@mech.kth.se

H.M. Nagib
MMAE Department, Illinois Institute of Technology, 60616 Chicago, USA

J. Peinke et al. (eds.), *Progress in Turbulence VI*,
Springer Proceedings in Physics 165, DOI 10.1007/978-3-319-29130-7_22

The studies by Huser and Biringen [1] and Pinelli et al. [2] have shown the potential of using numerical simulations to study turbulent duct flows. They performed direct numerical simulations (DNSs) of square ducts at low Reynolds numbers below $Re_\tau = 300$. The velocity and length scales are the friction velocity $u_\tau = \sqrt{\tau_w/\rho}$ (where τ_w is the mean shear stress at the wall and ρ is the fluid density) and the duct half-height h respectively, whereas Re_τ is the friction Reynolds number. On the other hand, the impact of the secondary flow on the mean velocity profile and the turbulent fluctuations for increasing aspect ratios has not been characterized in detail before, and this is the aim of the present study.

2 Numerical Simulations

Turbulent duct flows with aspect ratios 1, 3, 5, 7 and 10 at $Re_{\tau,c} \simeq 180$ (where $u_{\tau,c}$ is the friction velocity at the duct centerplane in the spanwise direction) are simulated by means of DNS. The database was initially presented by Vinuesa et al. [3], and has been extended as a part of the present study. The simulations were performed with the code Nek5000, developed by Fischer et al. [4], and based on the spectral element method (SEM). In the SEM the computational domain is decomposed into elements, and the solution is expanded in terms of Lagrange polynomials of order N inside those elements. The location of the nodes within elements is fixed, and follows the Guass–Lobatto–Legendre distribution, whereas there are no restrictions regarding the position of the elements in the domain. In the present study we considered the $\mathbb{P}_N - \mathbb{P}_{N-2}$ formulation with $N = 11$.

With respect to the flow setup, periodicity is assumed in the streamwise direction x in all the cases, and no-slip conditions are imposed at the walls along the vertical (y) and spanwise (z) directions. All the ducts have a streamwise length of $L_x = 25h$, and the same Reynolds number $Re_{b,c} \simeq 2{,}800$ is maintained the centerplane in all the aspect ratio cases ($Re_{b,c}$ is based on centerplane bulk velocity and h). All the simulations are started from a laminar solution, and transition to turbulence is tripped by means of a localized volume force acting in y. Its parameters are designed to create strong, instationary streaks that lead to rapid turbulent breakdown as described by Schlatter and Örlü [5].

Figure 1 shows an instantaneous streamwise velocity field obtained from the duct with AR $= 10$. Near-wall streaks can be easily identified in this figure, with the well documented spacing in the spanwise direction $\Delta z^+ \simeq 100$. It is interesting to observe how near-wall streaks are also formed on the side walls, also with an approximate spacing of $\Delta y^+ \simeq 100$, and how at the corner the effect of the two walls inhibits the formation of such structures. Instead, the flow field in Fig. 1 shows how the bursting events from horizontal and vertical walls interact at the corner. These interactions are reported by Huser and Biringen [1] to result in a redistribution of energy from $\overline{v^2}$ to $\overline{w^2}$ in square ducts, which eventually leads to the formation of the mean secondary flow.

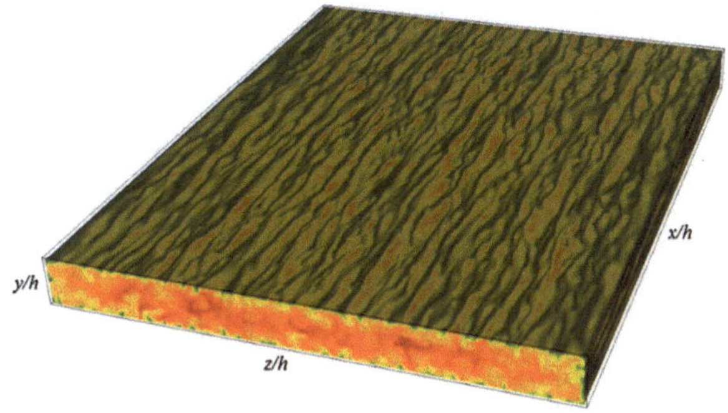

Fig. 1 Instantaneous streamwise velocity from the aspect ratio 10 case at 2,000 convective time units from the beginning of the simulation. *Green* and *orange* represent minimum and maximum velocities in the field respectively. Flow is from *left* to *right*, and walls have been removed for clarity

3 Results and Conclusions

The secondary flow is characterized in terms of its kinetic energy $K = (V^2 + W^2)/2$ at different spanwise locations z. Convergence analyses show that the spatially averaged kinetic energy $\langle K \rangle_{yz}$ at the duct corner, i.e., averaged from $y/h = -1$ to 1 and from the corner up to a spanwise distance of h, converges to an asymptotic value for averaging times of around $T_A \simeq 3{,}000$ convective time units (where bulk velocity and h are used to nondimensionalize the time). Based on this, K can be evaluated by analyzing velocity fields averaged in the homogeneous streamwise direction, and in time for at least 3,000 time units.

The spanwise variation of the kinetic energy integrated in the wall-normal direction $\langle K \rangle_y$ is shown in Fig. 2 for all the ducts under consideration in this study. This figure shows that the secondary flow is stronger close to the corner, and becomes gradually attenuated as the core of the duct is approached. It is also interesting to note that although all the aspect ratios show a similar structure, with two local maxima and one local minimum in between, their actual values are strongly influenced by the aspect ratio. Thus, AR determines the structure of the secondary flow, where for instance it is clear that the square duct exhibits a higher concentration of energy close to the corner and has a steeper decay of kinetic energy as the centerplane is approached. It is also interesting to observe that the AR $= 3$ and 5 cases show larger peak values at distances $z^* \simeq 0.4h$ from the corner than the cases with AR $= 7$ and 10. This is due to the fact that the secondary flow extends up to $z^* \simeq 5h$ from the corner, and therefore if the duct is not wide enough to accommodate the whole extent of the secondary flow, then its structure is modified by means of a different spanwise distribution of energy.

The presented database will be further analyzed to characterize the kinematics and dynamics of the secondary flow, as well as its interaction with the mean flow. Results with aspects ratios of up to 10 indicate that only the long-time average statistics of the duct are in agreement with those obtained from the spanwise-periodic channels. However, the dynamics of the turbulence and short-time averages of the vortical flow have revealed differences in the core region between the channel and ducts with aspect ratios up to 10 [6]. This suggests that experimental facilities intended to

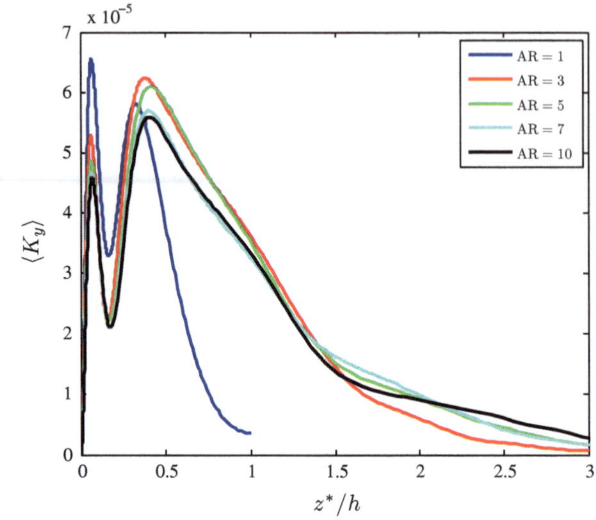

Fig. 2 Spanwise variation of the wall-normal integrated kinetic energy of the secondary flow $\langle K \rangle_y$, for all the duct cases under consideration. Results obtained from the streamwise-averaged fields, with averaging times of at least 3,000 time units. Note that z^* is measured from the corner to allow easier comparison between the various cases

represent the canonical channel flow and to compare the turbulence structures and dynamics and vortical flow details may require aspect ratios larger than facilities with highest ratios used so far of 12 and may need to be as large as 18, as in [7].

References

1. Huser, A., Biringen, S.: Direct numerical simulation of turbulent flow in a square duct. J. Fluid Mech. **257**, 65–95 (1993)
2. Pinelli, A., Uhlmann, M., Sekimoto, A., Kawahara, G.: Reynolds number dependence of mean flow structure in square duct turbulence. J. Fluid Mech. **644**, 107–122 (2009)
3. Vinuesa, R., Noorani, A., Lozano-Durán, A., El Khoury, G.K., Schlatter, P., Fischer, P.F., Nagib, H.M.: Aspect ratio effects in turbulent duct flows studied through direct numerical simulation. J. Turbul. **15**, 677–706 (2014)
4. Fischer, P. F., Lottes, J. W., Kerkemeier, S. G.: nek5000 Web page, http://nek5000.mcs.anl.gov (2008)
5. Schlatter, P., Örlü, R.: Turbulent boundary layers at moderate Reynolds numbers. Inflow length and tripping effects. J. Fluid Mech. **710**, 5–34 (2012)
6. Vinuesa, R., Schlatter, P., Nagib, H.M.: On minimum aspect ratio for duct flow facilities and the role of side walls in generating secondary flows. J. Turbul. **16**, 588–606 (2015)
7. Vinuesa, R., Bartrons, E., Chiu, D., Dressler, K.M., Rüedi, J.-D., Suzuki, Y., Nagib, H.: M: New insight into flow development and two dimensionality of turbulent channel flows. Exp. Fluids. **55**, 1759 (2014)

DNS of the Turbulent Flow Evolving in a Plane Channel from the Entry to the Fully Developed State

M. Capuano, A. Cadiou, M. Buffat and L. Le Penven

Abstract Direct numerical simulation of the turbulent flow spatially developing in a plane channel is performed. The channel is long enough for the turbulence to achieve the statistically invariant state in space and time. Spatial evolution of statistical quantities and Reynolds stress budgets are calculated. Results are compared to the canonical flows consisting of the zero-pressure gradient turbulent boundary layer and of the periodical channel flow.

1 Introduction

Developing entrance flows are of interest in a large number of application areas. For sufficiently large Reynolds numbers, from the entrance section, the flow evolves towards a turbulent state which is statistically steady and invariant in the downstream direction. Many aspects of these flows have not been completely described, for example, the entrance length L_e after which the flow reaches a developed state, the lowest Reynolds number corresponding to the fully developed turbulent flow, and characteritics of the turbulence in its spatial development. The laminar entry length is a linear function of the Reynolds number Re_h (based on the bulk velocity U_0 and the channel half-width h) [2] for Re_h values larger than 100, thus locating the asymptotic state at a distance hundreds of time the channel height from the entry. In the turbulent régime, L_e appears to be substantially smaller than in the laminar case, but with a weaker dependance with the Reynolds number. The Cole–Fernholtz scaling of the boundary layer thickness yields to $L_e/h \sim (\ln Re_h)^2$ [5], whereas $L_e/h \sim Re_h^{1/4}$ [12] is obtained with a $\frac{1}{7}$−law of the mean velocity. Recent experimental study [7] however did not report any particular trend for L_e but proposed instead $L_e/h = 260$ as a reasonable estimation of the development length. There exists an extensive literature about the fully-developed turbulent flow in a periodic plane channel (since [4, 10]), disregarding the aspects of the flow evolution. The behaviour of the turbu-

M. Capuano · A. Cadiou (✉) · M. Buffat · L. Le Penven
Laboratoire de Mécanique des Fluides et d'Acoustique, École Centrale de Lyon, INSA de Lyon, CNRS, Université de Lyon, 36 av. guy de collongue, Écully, France
e-mail: anne.cadiou@ec-lyon.fr

J. Peinke et al. (eds.), *Progress in Turbulence VI*,
Springer Proceedings in Physics 165, DOI 10.1007/978-3-319-29130-7_23

lence in this canonical flow has been studied since then (see e.g. [3]) for increasing Reynolds numbers. The analysis of zero-pressure-gradient (ZPG) turbulent boundary layers (TBL) has followed the same evolution since [11] with recent simulations at larger Reynolds number (see e.g. [6, 13]). Their characteritics differ in confinement and pressure-gradient effect in the channel flow and because of intermittency in the boundary layer flow. In near-wall region, statistics of both canonical flows are known to be nearly identical with respect to the inner scale. Differences have been identified in the large-scale structures away from the wall [9]. In order to get more insight into thc spatial evolution of the turbulent flow that enters a plane channel and reaches the statistical fully developed state, direct numerical simulations are performed and Reynolds stress budgets are calculated.

2 Flow Configuration and Numerical Method

DNS of spatially developing channel flows are performed with a spectral Galerkin method [1, 8] for several Reynolds numbers Re_h ranging from 1250 to 20,000. The computational domain starts at a small distance downstream from the origin of the boundary layers (at a location corresponding to a fixed $Re_{x_0} = 40{,}000$) and extends hundreds of time the channel half-width. Laminar-turbulent transition is triggered near the channel entrance, by a perturbation imposed at the inlet of the computational domain. As the flow evolves from a fixed Re_{x_0}, which corresponds to an increasing x_0/h, with decaying Reynolds number, the turbulent length evolves as given in Fig. 1. It seems to follow a $L_e/h \sim Re_h^{2/5}$ power law in this Reynolds number range. The correlations are fitted on the entrance length obtained for the largest Reynolds number, multiplying the constant of [12] by a factor 2. We will focus here on the $Re_h = 10{,}000$ case, where the boundary layers are well separated in the entry section

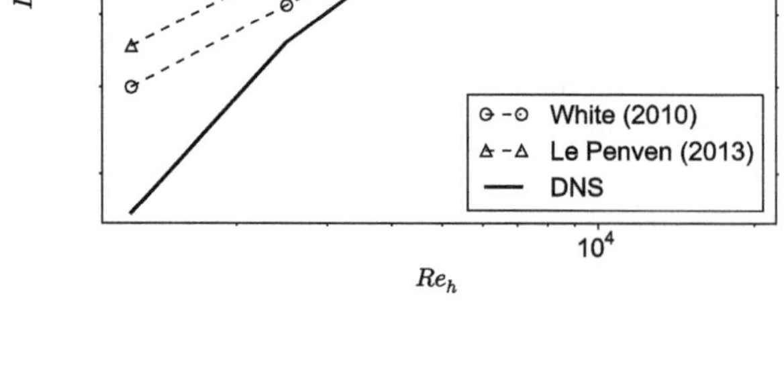

Fig. 1 Turbulence entrance length

(at $x/h = 4$). The laminar entrance length extends to $L_e/h \sim 1800$ and is consistent with [2]. In the DNS, the turbulence entrance length is $L_e/h \sim 110$.

The computational domain ($280h \times 2h \times 6.4$) has about 1.5 billions of modes ($23040 \times 128 \times 512$). The transition occurs in the boundary layers (at $x/h \sim 10$) whose thickness grow, accelerating the flow until turbulence occupies all the channel width (at $x/h = 50$). After $x/h = 110$, the turbulent intensity in the channel mid-plane becomes constant as well as the friction velocity. In the region located here between $x/h = 50$ and $x/h = 110$, the flow evolves from a boundary-layer type flow (with weak favorable pressure gradient, corresponding to a Clauser parameter of $\beta = -0.007$) to a periodic-channel type flow (with $Re_\tau = 567$).

3 Spatial Evolution of Statistical Quantities and Budgets

The flow evolution and the region described in the previous section can be seen on the skin friction evolution (Fig. 2). Between $x/h = 10$ and $x/h = 50$ the turbulent boundary layers grow from $Re_\theta = 633$ to $Re_\theta = 1045$. As the pressure gradient is very weak, the statistics compare well to recent simulations of zero-pressure gradient boundary layer on a flat plate [6] (between $Re_\theta = 670$ and $Re_\theta = 1{,}000$). After $x/h = 110$, the skin friction reaches a constant value, which corresponds to the periodic channel flow of $Re_\tau = 567$. Results are in close agreement to the of DNS of the periodic channel flow of $Re_\tau = 590$ [10].

The turbulence is found to be in equilibrium once the boundary layers have merged (after $x/h = 110$), as indicated by the ratio of the wall-normal integrated turbulent kinetic energy production P and dissipation ε within the boundary layers, which is quasi-constant (and equal to 1). Wall turbulence statistics are quite similar in all sections, as shown by the Reynolds stress components (Fig. 3) for $y^+ < 30$. Inner

Fig. 2 Skin friction coefficient

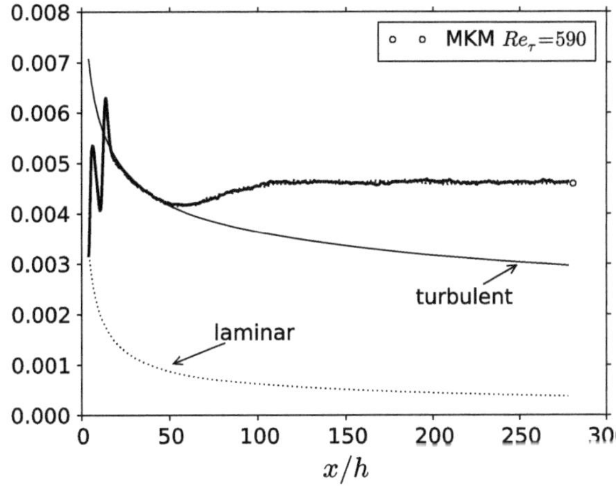

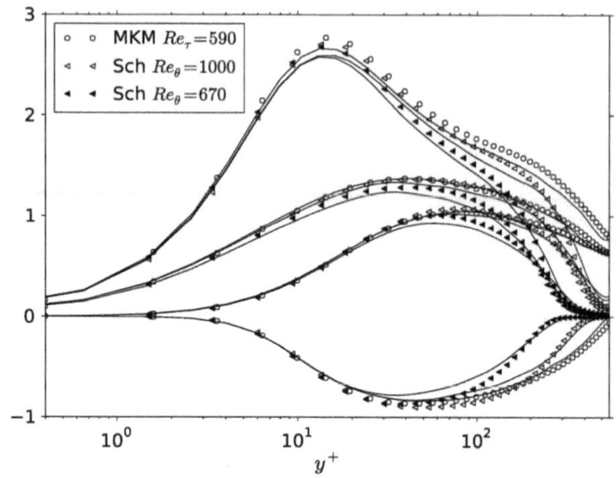

Fig. 3 Reynolds stresses

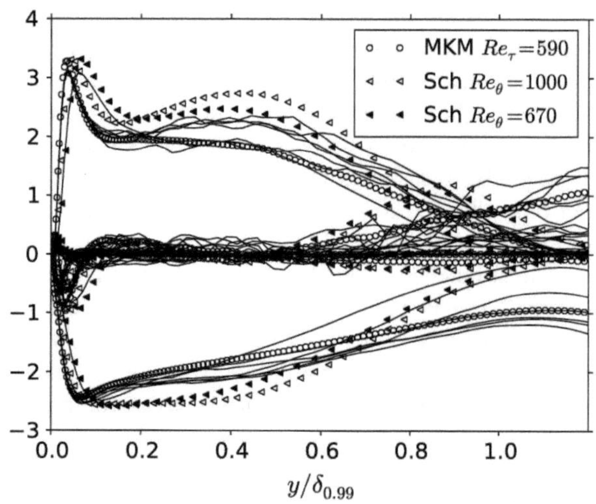

Fig. 4 Kinetic energy budgets

scaled kinetic energy budgets are identical (Fig. 4). As expected, differences occurs on the edge of the boundary layers. To enhance the balance in the boundary layer, the budget terms are multiplied with $y/\delta_{0.99}$ and scaled in outer units $u_\tau^3/\delta_{0.99}$ [6]. Convergence of the high order correlations (turbulent diffusion, velocity-pressure correlation) should be improved by accumulating more samples in order to evaluate scaling laws. The shear stress follows a linear evolution in the middle of the channel showing that the fully developed turbulent flow is reached.

4 Conclusion

DNS of the flow evolving in a plane channel from the entry region to the fully developed state is discussed. Results are consistent with the literature, the turbulence evolving gradually from a slighly accelerated boundary layer to a fully periodic channel flow [6, 10]. Turbulence is found in equilibrium once the turbulent entry length is reached, which is supposed to be located where the time-averaged velocity in the channel mid-plane becomes constant. The asymptotic state, where the turbulence could be considered as statistical invariant in longitudinal space direction and time, as in a periodical channel flow, is reached further downstream. This corresponds to sections where the Reynolds shear stress becomes linear in the middle of the channel.

Acknowledgments The simulations were performed on BlueGene/Q platforms thanks to GENCI resources. The authors would also like to thank the computer center P2CHPD at Université Claude Bernard Lyon 1, member of the Fédération Lyonnaise de Modélisation et Sciences Numériques, for providing computer facilities.

References

1. Buffat, M., Le Penven, L., Cadiou, A.: An efficient spectral method based on an orthogonal decomposition of the velocity for transition analysis in wall bounded flow. Comput. Fluids **42**, 62–72 (2011)
2. Durst, F., Ray, S., Ünsal, B., Bayoumi, O.A.: The development lengths of laminar pipe and channel flows. J. Fluids Eng. **127**, 1154–1160 (2005)
3. Hoyas, S., Jiménez, J.: Scaling of the velocity fluctuations in turbulent channels up to $Re_\tau =$ 2003. Phys. Fluids **18**, 011702 (2006)
4. Kim, J., Moin, P., Moser, R.: Turbulence statistics in fully developed channel flow at low Reynolds number. J. Fluid Mech. **177**, 133–166 (1987)
5. Le Penven, L.: Une décomposition orthogonale des champs de vecteurs appliqué á létude de la transition turbulente en canal plan. HDR report, Université de Lyon (2013)
6. Li, Q., Brethouwer, G., Johansson, A.V., Henningson, D.S., Schlatter, P.: Simulations of spatially evolving turbulent boundary layers up to $Re_\theta = 4300$. Int. J. Heat Fluid Flow **31**, 251–261 (2010)
7. Lien, K., Monty, J.P., Chong, M.S., Ooi, A.: The entrance length for fully developed turbulent channel flow. In: 5th Australasian Fluid Mechanics Conference (2004)
8. Montagnier, J., Cadiou, A., Buffat, M., Le Penven, L.: Towards petascale simulation for transition analysis in wall bounded flow. Int. J. Numer. Methods. Fluids **72**(7), 709–723 (2013)
9. Monty, J.P., Hutchins, N., Ng, H.C.H., Marusic, I., Chong, M.S.: A comparison of turbulent pipe, channel and boundary layer flows. J. Fluid Mech. **632**, 431–442 (2009)
10. Moser, R., Kim, J., Mansour, N.: Direct numerical simulation of turbulent channel flow up to $Re_\tau = 590$. Phys. Fluids **11**(4), 943–945 (1999)
11. Spalart, Ph.R.: Direct simulation of a turbulent boundary layer up to $R_\theta = 1410$. J. Fluid Mech. **187**, 61–98 (1988)
12. White, F.M.: Fluid Mechanics, 7th edn. McGraw-Hill, New York (2010)
13. Wu, X., Moin, P.: Direct numerical simulation of turbulence in a nominally zero-pressure-gradient flat-plate boundary layer. J. Fluid Mech. **630**(1), 5–41 (2009)

Downstream Evolution of Perturbations in a Zero Pressure Gradient Turbulent Boundary Layer

E. Rodriguez-Lopez, P.J.K. Bruce and O.R.H. Buxton

Abstract This paper examines the evolution of perturbations generated by various trips in a zero pressure gradient turbulent boundary layer. Measurements taken using hot-wire anemometry show that the evolution of the boundary layer towards the natural state is strongly dependent on the trip geometry. In particular the mechanisms creating the boundary layer appear to depend primarily on the wall normal distribution of blockage ratio, recovering the natural properties more rapidly for a uniform distribution of blockage (wall normal cylinders) than for non-uniform blockage (saw-tooth fence). The relative size of the trip with respect to the boundary layer is shown to be a second order effect. Standard behaviour (characterized by the skin friction coefficient, C_f, the wake component, Π, and the shape factor, H) is recovered successfully $500D \sim 75h$ downstream, presenting 175 % higher momentum thickness, θ, than the natural case for the same downstream distance.

1 Introduction

Upstream perturbations modify the development of a boundary layer, making it fully turbulent [3], or thicker than its natural size [2]. These perturbations will evolve with downstream distance (x) towards the natural state of the turbulent boundary layer (TBL), described in [1]. Obtaining a thick TBL in a short distance is of primary importance to the study of high Reynolds number flows in short wind tunnels *only if* asymptotic properties are fully recovered and the trip effects have disappeared. The present objective is to study the influence of various trips (divided into uniform and non-uniform wall normal distributions of blockage) located at different downstream positions, on the TBL properties (thickness, log-law, friction coefficient, wake component [1], turbulence intensity, shape factor and spectra) and the effect of the TBL formation mechanism on the recovery distance of natural properties.

E. Rodriguez-Lopez (✉) · P.J.K. Bruce · O.R.H. Buxton
Imperial College London, London, UK
e-mail: eduardo.rodriguez-lopez12@imperial.ac.uk

J. Peinke et al. (eds.), *Progress in Turbulence VI*,
Springer Proceedings in Physics 165, DOI 10.1007/978-3-319-29130-7_24

2 Experimental Set-Up

Experiments are conducted at Imperial College London in a wind tunnel of $0.91 \times 0.91\,\text{m}^2$ section and 4.8 m length with $U_e = 10\,\text{ms}^{-1}$. A wooden flat plate provided with a flap is mounted vertically spanning the whole height of the tunnel ensuring zero pressure gradient. Measurements are taken using constant temperature anemometry. Sampling at 100 kHz for 30 s and low-pass filtering at 30 kHz allow resolution of $0.2 < t^+ = tu_\tau^2/\nu < 0.5$. The different trips described in Table 1 are located at $x = \{160,\ 890\}$ mm where $\delta_{99} = \{2.3,\ 15.7\}$ mm, $\theta = \{0.35,\ 2.11\}$ mm and $\delta_\gamma = \{2.4,\ 20.4\}$ mm corresponding to $h/\delta_\gamma \gg 1$ and $h/\delta_\gamma \lesssim 1$ respectively; δ_γ is defined as the first point without turbulence intermittency and for fully turbulent boundary layers it is approximately $\delta_\gamma \sim 1.3\delta_{99}$. Velocity profiles are taken for $0.9\,\text{m} < x < 3.9\,\text{m} \Leftrightarrow 0.57 \times 10^6 < Re_x = U_e x/\nu < 2.42 \times 10^6$ (far field), and $\{2, 4, 6, 10, 20, 40\}$ diameters downstream of the trips (near field). The wall shear stress and the wall location are extrapolated from the velocity profile [4] with accuracies $\Delta u_\tau < \pm 1\,\%$, $\Delta y^+ = \pm 0.5$. Where $y^+ = yu_\tau/\nu$ and $u^+ = u/u_\tau$.

3 Results

Experiments measuring the far field created by all the trips from Table 1 located such that $h/\delta_\gamma \gg 1$ show that: (i) Different arrays of cylinders generate a thicker TBL than the natural case (Fig. 1a) and with the same properties (Fig. 1b, c). (ii) The

Table 1 Trips description. $Z = 10$ mm for cylinders [2] and $Z = 20$ mm for Saw. $h = max\{H1, H2\}$

	1row10	*1row20*	*2row10*	*10and20*	*20and10*	*2row20*	*2stagger*	*Saw Inv*	*Saw*	*Natural*
Type	Cyl(*)							Saw(**)		-
Symbol	△	∗	▷	□	◇	★	+	◁	▽	○
H1(mm)	10	20	10	10	20	20	20	15.3	15.3	0
H2(mm)	0	0	10	20	10	20	20	-	-	0
D(mm)	3	3	3	3	3	3	3	-12.8	12.8	0

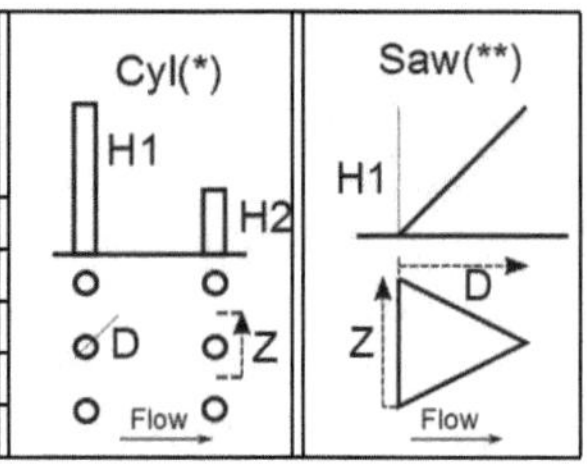

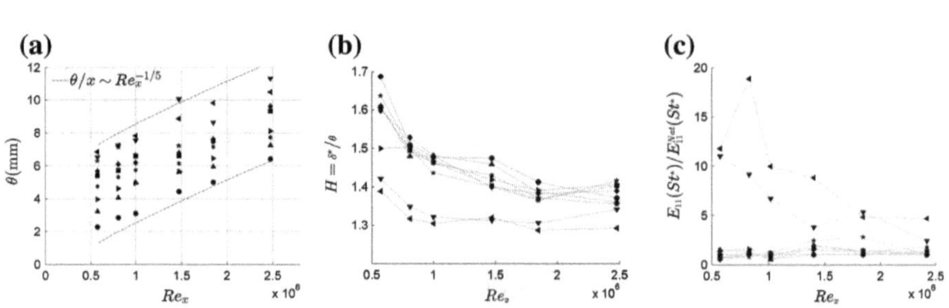

Fig. 1 **a** Momentum thickness (θ) evolution. **b** Shape factor (H) evolution. **c** Relative magnitude of spectra at $y/\theta = 19$ ($y \gg \delta_\gamma$) for $St^\star$. Symbols are given in Table 1

thickness of the TBL created is a function of the height of the highest cylinder, not of the number of rows nor their order. (iii) The sawtooth fences generate an even thicker TBL (Fig. 1a) but with a different shape factor (Fig. 1b). (iv) It also generates an acoustic perturbation in the freestream ($y/\theta = 19$) up to 20 times stronger than the natural case at a fixed Strouhal number $St^\star = f^\star \theta / U_e \sim 0.019$ which persists for more than 2 m (Fig. 1c). These observations suggest different formation mechanisms for the TBL using cylinders or sawtooth. Both are exhaustively studied henceforth.

As seen in Fig. 3a, b cylinders generate a spanwise periodic combination of jets and wakes which becomes uniform after $40D$. This low momentum, highly turbulent flow is seen as the outer flow by the TBL growing underneath and it is entrained increasing TBL thickness by augmenting its wake region. This generates a development region (until $500D \sim 75h$) where the wake component (Π) [1], friction coefficient, (C_f) and shape factor (H), present non standard behaviour (Fig. 2). It also shows that the effect of h/δ_γ is secondary compared with the distance from the trips.

On the other hand, the sawtooth, due to its non uniform blockage ratio in the wall normal direction, generates a velocity profile downstream of the gaps ($z = 0$) which resembles a boundary layer but does not reflect its properties (Fig. 3e). Downstream of the obstacle ($z = Z/2$) a recirculation and a strong tip vortex are seen. Although these two spanwise periodic regions become uniform after $8h$, the turbulence intensity is artificially high until $70h$, where the inner peak starts to appear (Fig. 3f). Spectral data (not included for brevity) shows a shedding from the cylinders at $St = fD/U_e = 0.19$ disappearing before $20D$. The broadband turbulence generated by the recirculation and the tip eddies in the sawtooth covers $0.1 < St = fh/U_e < 0.5$ and persists for more than $40h$. Downstream of the location where the inner peak (at $y^+ = 15$) is developed no extraordinary spectral features are seen. Note that the turbulence intensity in the case of cylinders increases with x; whereas in the sawtooth case it initially decreases from a high value near the trips, before increasing again with x after a longer development region.

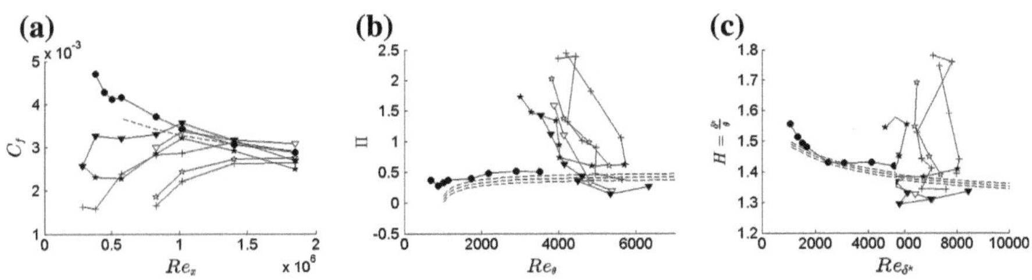

Fig. 2 **a** Friction coefficient (C_f) evolution. *Dashed line* is 1/5th correlation. **b** Wake component (Π) evolution. *Dashed line* by [1]. **c** Shape factor (H) evolution. *Dashed line* by [1]. Symbols are given in Table 1. *Filled symbols* are $h/\delta_\gamma \gg 1$ and *white symbols* are $h/\delta_\gamma \lesssim 1$

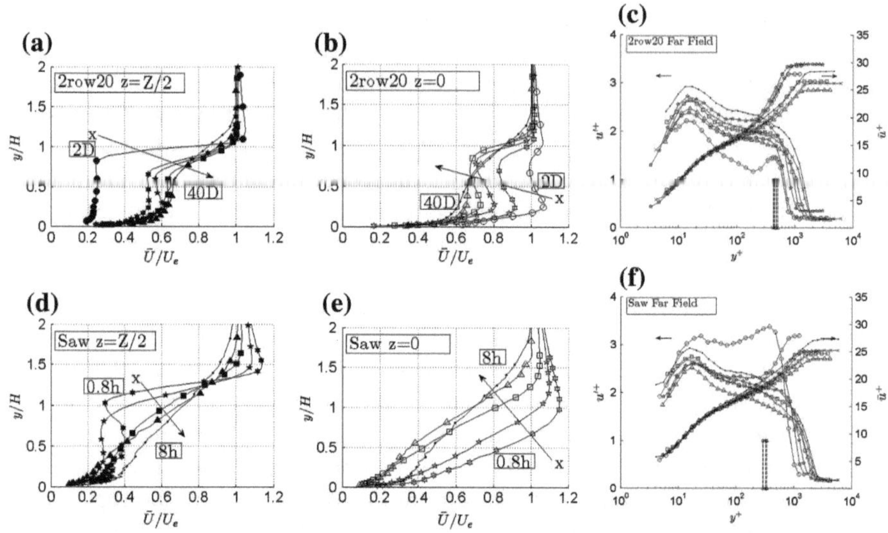

Fig. 3 Velocity profiles for Near ($z = \{0, Z/2\}$) and Far Field using *2row20* and *Saw* trips. **a**, **b**→[∘ ∗ ⋆□△·]→$\hat{x}$ = [$2D, 4D, 6D, 10D, 20D, 40D$] from the trip (*2row20* Near field). **d**, **e**→[∗ ⋆ □△·]→$\hat{x}$ = [$0.8h, 1.2h, 2h, 4h, 8h$] from the trip (*Saw* Near field). **c**, **f**→[∘ ∗ ⋆□ △ × ·]→x = [0.46, 0.6, 0.9, 1.3, 1.6, 2.2, 2.9] m from leading edge (Far field). *Dashed line* marks the non-dimensionalized height of the obstacle, $h^+ = hu_\tau/\nu$

4 Conclusions

Different arrays of spanwise distributed trips generate perturbations which evolve downstream of them. This study shows that the standard properties of the TBL can be recovered at a distance downstream of the obstacle which depends on the trip geometry. In particular, the wall normal distribution of blockage has been demonstrated to play the main role while the relative size h/δ_y seems to be secondary. The recovery of standard TBL properties is shown to be faster and more effective when using arrays of cylinders than when using the sawtooth due to the different formation mechanisms. The use of these trips can increase the TBL thickness by up to 175 %; thus generating a high Reynolds number TBL in a shorter distance than the natural case while maintaining the same properties.

References

1. Chauhan, K.A., Monkewitz, P.A., Nagib, H.M.: Criteria for assessing experiments in zero pressure gradient boundary layers. Fluid Dyn. Res. **41**(2), 021404 (2009)
2. Kornilov, V.I., Boiko, A.V.: Wind-tunnel simulation of thick turbulent boundary layer. Thermophys. Aeromech. **19**(2), 247–258 (2012)

3. Preston, J.H.: The minimum Reynolds number for a turbulent boundary layer and the selection of a transition device. J. Fluid Mech. **3**(4), 373–384 (1957)
4. Rodríguez-López, E., Bruce, P., Buxton, O.: A robust post-processing method to determine skin friction in turbulent boundary layers from the velocity profile. Exp. Fluids **56**(4), 1–16 (2015)

Scale Dependent Stochastic Self-energy Model of the Energy Transfers in Turbulent Channel Flows

V. Kitsios, J.A. Sillero, J.S. Frederiksen and J. Soria

1 Introduction

The forward and inverse energy cascades in turbulent channel flow are studied by extracting flow specific subgrid coefficients. The effect that subgrid scales of motion have on the resolved scales is quantified by developing subgrid models from the statistics of the differences between a reference direct numerical simulations (DNS) and a truncated DNS using the method of [1]. This approach is inherently linked to large eddy simulation (LES), where the large eddies are explicitly resolved on a computational grid and the unresolved subgrid-scale interactions are modelled. Here LES is used as a means of validating the subgrid model coefficients, from which physical interpretations are subsequently made. This is in contrast to typical approaches where one starts with a physical hypothesis leading to a subgrid model.

V. Kitsios (✉) · J. Soria
Laboratory for Turbulence Research in Aerospace and Combustion, Department of Mechanical and Aerospace Engineering, Monash University, Clayton, Australia
e-mail: vassili.kitsios@gmail.com

J.A. Sillero
School of Aeronautics, Universidad Politécnica de Madrid, Madrid, Spain

J.S. Frederiksen
CSIRO Oceans and Atmosphere Flagship, 107-121 Station St, Aspendale, Australia

J. Soria
Department of Aeronautical Engineering, King Abdulaziz University, Jeddah, Kingdom of Saudi Arabia

J. Peinke et al. (eds.), *Progress in Turbulence VI*,
Springer Proceedings in Physics 165, DOI 10.1007/978-3-319-29130-7_25

2 Direct Numerical Simulation

The incompressible isothermal Navier–Stokes equations are solved in a channel of periodic streamwise direction x, inhomogeneous wall-normal direction y, and periodic spanwise direction z, with associated velocity components u, v and w. The system is nondimensionalised by the centreline velocity (u_0) and channel half height (h). A collocated Chebyshev method of polynomial index j is used in y, and Fourier discretisation is used in the x and z directions of respective wavenumber indices α and β [2]. The associated pseudo wavenumber vector is defined as $\mathbf{k} \equiv (\alpha, j, \beta)$. The DNS is solved over the wavenumber set

$$\begin{aligned}\mathbf{T} = \mathbf{C}(T,N) = [\, \mathbf{k}',\ \mathbf{k}'' \mid\ & 0 \le \alpha' \le T\ ,\ 0 \le j' \le N\ ,\ -T \le \beta' \le T \\ & 0 \le \alpha'' \le T\ ,\ 0 \le j'' \le N\ ,\ -T \le \beta'' \le T\,],\end{aligned} \tag{1}$$

where T is the maximum wavenumber, and N the maximum Chebyshev polynomial index. The case simulated here has a Reynolds number $Re_\tau \equiv u_\tau h/\nu = 950$, where ν is the kinematic viscosity, $u_\tau = \sqrt{\tau_{wall}/\rho}$ the friction velocity, with τ_{wall} the magnitude of the wall shear stress. The flow is simulated in a doubly periodic box of extent $L_x = \pi$, $L_y = 2$, and $L_z = \pi/2$, with $T = 127$ and $N = 385$.

3 Stochastic Self-energy Subgrid Model

We study the energy transfers by decomposing the scales of motion into a resolved wavenumber set $\mathbf{R} = \mathbf{C}(T_R, N)$, where T_R is the LES truncation wavenumber ($T_R < T$), and a subgrid wavenumber set $\mathbf{S} = \mathbf{T} - \mathbf{R}$. To facilitate the discussion we introduce the state vector $\mathbf{u}(\mathbf{k}) \equiv (u, v, w)$, of fluctuating component $\hat{\mathbf{u}}$. The time derivative of $\mathbf{u}$ is decomposed such that $\mathbf{u}_t(t) = \mathbf{u}_t^{\mathbf{R}}(t) + \mathbf{u}_t^{\mathbf{S}}(t)$. The tendency of the resolved scales is $\mathbf{u}_t^{\mathbf{R}}$, where all triadic interactions involve wavenumbers within the resolved set $\mathbf{R}$. The remainder is the subgrid tendency, $\mathbf{u}_t^{\mathbf{S}}$, which has at least one wavenumber within the subgrid set $\mathbf{S}$ involved in the triadic interactions. The latter is further decomposed into its fluctuating ($\hat{\mathbf{u}}_t^{\mathbf{S}}(t)$) and time averaged ($\overline{\mathbf{f}}$) components. The former is parameterised by the matrix equation

$$\hat{\mathbf{u}}_t^{\mathbf{S}}(\mathbf{k}, t) = -\mathbf{D_d}(\mathbf{k})\ \hat{\mathbf{u}}(\mathbf{k}, t) + \hat{\mathbf{f}}(\mathbf{k}, t). \tag{2}$$

The drain dissipation is determined by post-multiplying (2) by $\hat{\mathbf{u}}^\dagger(t_0)$, integrating over the decorrelation period τ, ensemble averaging, and rearranging to produce

$$\mathbf{D_d} = -\left\langle \int_{t_0}^{t} \hat{\mathbf{u}}_t^{\mathbf{S}}(\sigma)\hat{\mathbf{u}}^\dagger(t_0)d\sigma \right\rangle \left\langle \int_{t_0}^{t} \hat{\mathbf{u}}(\sigma)\hat{\mathbf{u}}^\dagger(t_0)d\sigma \right\rangle^{-1}, \tag{3}$$

where † denotes the Hermitian conjugate for vectors and matrices [1]. The model for $\hat{\mathbf{f}}$ is determined by calculating the noise covariance matrix $\mathscr{F}_b = \mathbf{F_b} + \mathbf{F_b}^\dagger$, where $\mathbf{F_b} = \langle \hat{\mathbf{f}}(t)\ \hat{\mathbf{u}}^\dagger(t)\rangle$. Post-multiplying both sides of (2) by $\hat{\mathbf{u}}^\dagger(t)$, and adding the conjugate transpose of (2) pre-multiplied by $\hat{\mathbf{u}}(t)$ yields the balance equation

$$\left\langle \hat{\mathbf{u}}_t^{\mathbf{S}}(t)\hat{\mathbf{u}}^\dagger(t)\right\rangle + \left\langle \hat{\mathbf{u}}(t)\hat{\mathbf{u}}_t^{\mathbf{S}\dagger}(t)\right\rangle = -\mathbf{D_d}\left\langle \hat{\mathbf{u}}(t)\hat{\mathbf{u}}^\dagger(t)\right\rangle - \left\langle \hat{\mathbf{u}}(t)\hat{\mathbf{u}}^\dagger(t)\right\rangle \mathbf{D_d}^\dagger + \mathscr{F}_b. \quad (4)$$

Given that $\mathbf{D_d}$ is known, $\mathscr{F}_b$ can now be calculated. For the implementation of the stochastic subgrid parameterisation, $\hat{\mathbf{f}}$ is represented as a white noise process such that $\langle \hat{\mathbf{f}}(t)\ \hat{\mathbf{f}}^\dagger(t')\rangle = \mathscr{F}_b\, \delta(t-t')$. One can approximate $\hat{\mathbf{u}}_t^{\mathbf{S}}$ by the deterministic model $\hat{\mathbf{u}}_t^{\mathbf{S}}(\mathbf{k},t) = -\mathbf{D_{net}}(\mathbf{k})\ \hat{\mathbf{u}}(\mathbf{k},t)$, where the net dissipation $\mathbf{D_{net}} = \mathbf{D_d} + \mathbf{D_b}$ and the backscatter dissipation $\mathbf{D_b} = -\mathbf{F_b}\left\langle \hat{\mathbf{u}}(t)\ \hat{\mathbf{u}}^\dagger(t)\right\rangle^{-1}$.

The drain, backscatter and net upper diagonal coefficients for wall normal scales of $\lambda_y^+ = \lambda_y u_\tau/\nu = 2$ are illustrated in Fig. 1a–c respectively, where we define $\lambda_y = 1 - \cos(2\pi/j)$. These coefficients are plotted in the horizontal wavenumber index half plane $(\alpha, \beta > 0)$, as they are symmetric about the α axis. The drain dissipation is positive indicating that the parameterisation deterministically sends energy from the resolved (large) to the subgrid (small) scales. The backscatter dissipation is negative indicating energy is stochastically injected into the resolved scales. The net dissipation is positive indicating that the net effect of the drain and backscatter is such that energy is being transferred out of the large scales and into the small scales representative of a forward energy cascade. The dissipations are also scale selective with the magnitude of the coefficients increasing as they approach the truncation boundaries. This implies that the small resolved vortices near the truncation boundary have more significant interactions with the subgrid than the larger resolved structures. Coefficients with wall normal scales $\lambda_y^+ > 2$ have qualitatively the same properties. The associated coefficients for wall normal scales of $\lambda_y^+ = 0.4$ are illustrated in Fig. 1d–f. Here at certain wavenumber pairs the stochastic backscatter overwhelms the drain dissipation, indicating that these small vertical and small horizontal scales receive a net injection of energy from the subgrid scales, which is representative of an inverse energy cascade. Further details on the coefficients can also be found in [3].

The subgrid coefficients are validated by comparing a LES where these coefficients have been adopted to the original DNS. The equations governing the LES are the same as for the DNS, but solved over the wavenumber set $\mathbf{R}$ instead of $\mathbf{T}$, with either the deterministic or stochastic models used to represent $\mathbf{u}_t^{\mathbf{S}}(\mathbf{k})$. Comparisons are made on the basis of the time-averaged kinetic energy spectra. The spectrum at $y^+ = yu_\tau/\nu = 133$ is plotted against the streamwise (α) and spanwise (β) wavenumbers for the no subgrid model case in Fig. 2a. This figure illustrates a clear accumulation of energy at the smallest resolved scales of motion, which distorts the entire spectrum. The subgrid interactions would have drained this excess energy out of the resolved scales and into the subgrid, to be eventually damped by viscous dissipation. The spectrum is corrected when adopting either the deterministic or stochastic subgrid models as illustrated in Fig. 2b, c.

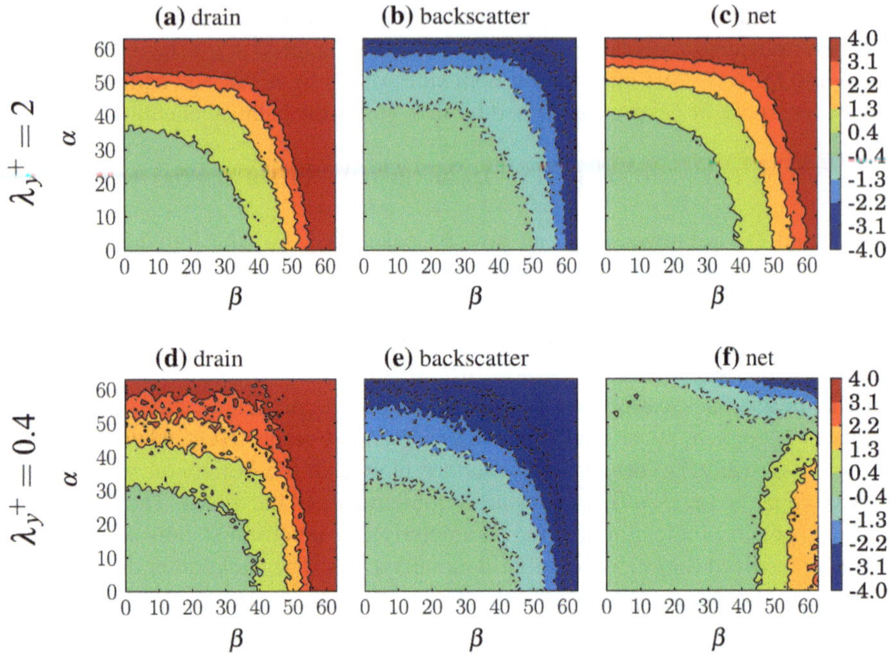

Fig. 1 The upper diagonal component of the subgrid dissipation matrices illustrated in the horizontal wavenumber index half-plane ($\alpha, \beta \geq 0$). **a–c** $\lambda_y^+ \equiv \lambda_y u_\tau/\nu = 2$; **d–f** $\lambda_y^+ = 0.4$

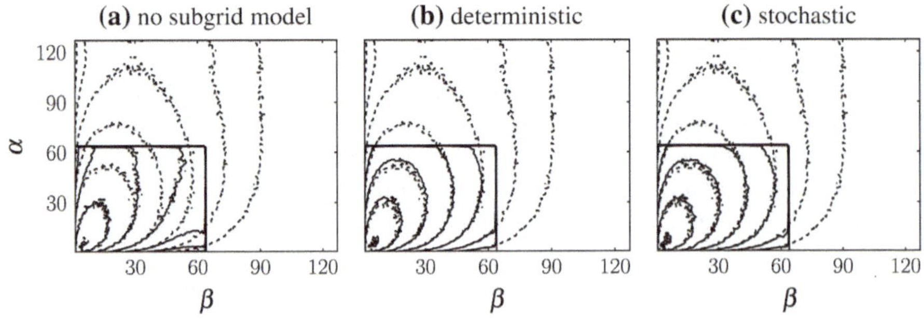

Fig. 2 Comparison of the DNS (*dotted line*) and LES (*solid line*) time-averaged kinetic energy spectra: **a** no subgrid model, where $\hat{\mathbf{u}}_l^{\mathbf{S}} = 0$; **b** deterministic, where $\hat{\mathbf{u}}_l^{\mathbf{S}} = -\mathbf{D}_{\mathbf{net}}\hat{\mathbf{u}}$; and **c** stochastic, where $\hat{\mathbf{u}}_l^{\mathbf{S}} = -\mathbf{D}_{\mathbf{d}}\hat{\mathbf{u}} + \hat{\mathbf{f}}$

4 Concluding Remarks

Deterministic and stochastic self-energy subgrid models have been calculated from the DNS statistics of turbulent channel flow. LES incorporating these subgrid models have been shown to successfully reproduce the DNS within the resolved scales. The coefficients of the subgrid models indicate the presence of both forward and inverse cascades in wall bounded turbulence.

Acknowledgments The authors acknowledge the computational resources provided by the NCI, and the funding from the European Research Council, Australian Research Council, and CSIRO. Julio Soria gratefully acknowledges the support of the Australian Research Council Discovery Outstanding Researcher Award fellowship.

References

1. Frederiksen, J.S., Kepert, S.M.: Dynamical subgrid-scale parameterizations from direct numerical simulations. J. Atmos. Sci. **63**, 3006–3019 (2006)
2. Kim, J., Moin, P., Moser, R.: Turbulence statistics in fully developed channel flow at low Reynolds number. J. Fluid Mech. **177**, 133–166 (1987)
3. Kitsios, V., Sillero, J.A., Soria, J., Frederiksen, J.S.: Stochastic self-energy subgrid model for the large eddy simulation of turbulent channel flows. J. Phys. Conf. Ser. **506**, 012001 (2014)

Effect of Irregular Surface in a Turbulent Channel

David Sassun and Paolo Orlandi

Abstract The effect of highly irregular surfaces in a turbulent channel is studied through DNS through the use of the immersed boundary method to reproduce complex surfaces. The height of the surfaces analyzed is proportional to the streamwise velocity component u or the pressure p in a wall parallel plane of a smooth channel at $y^+ = 12$. The surface with height proportional to u is drag reducing, that proportional to p is drag increasing. The relative magnitude of the friction components allows to identify the type of roughness as k-type, since the form drag is larger than the viscous drag. An additional surface has been obtained by smoothing a regular geometry made of staggered cubes. In this case, the resistance is doubled, and the form drag is much larger than the viscous drag.

1 Introduction

The effect of surface roughness on a flow is of practical interest in many disciplines. Examples include flows over cities or vegetation, physiological flows and turbomachinery. The main practical focus of the studies on surface roughness is the effect on drag and on heat transfer. From the theoretical side, the interest is to try to relate the slope of the surface to the boundary conditions at the plane of the crests and, as a consequence, to the production of turbulent kinetic energy near the wall.

Perry et al. [6] categorized the turbulent flows past rough surfaces in two classes: k- and d-type flows. In k-type, the form drag produced by the surface is large, and vortices of the same size as the roughness are shed into the main flow. On the other hand, the viscous drag is predominant in d-type roughness, and the vortices are confined within the roughness elements, and weak interaction with the overlying flow was depicted by Leonardi et al. [4]. It has been shown that 3D and bidimensional

D. Sassun (✉) · P. Orlandi
Dipartimento di Meccanica e Aerospaziale, Sapienza Universitá di Roma,
Via Eudossiana 16, Rome, Italy
e-mail: david.sassun@uniroma1.it

P. Orlandi
e-mail: orlandi@kolmogorov.ing.uniroma1.it

J. Peinke et al. (eds.), *Progress in Turbulence VI*,
Springer Proceedings in Physics 165, DOI 10.1007/978-3-319-29130-7_26

transverse roughness with large w/k are drag increasing, while longitudinal geometries, under some conditions, can be drag reducing [2]. The reduction of the drag may be related to ordered structures that are more anisotropic, while the structures that are linked to a drag increase are more isotropic. The effect of irregular rough surface was studied experimentally, providing only an incomplete picture of the flow near the wall. To have a better understanding of the complex physics the present DNS may help.

2 Simulation Details

The details of the numerics and of the simulation parameters are similar to those in Orlandi [5], therefore they are not repeated. Instead the numerical procedure to evaluate the metrics for the Immersed Boundary technique is briefly described. The height of the surface is given at every point of the domain. The metric is evaluated at the first point outside the surface with a subtraction (in y) or a linear interpolation between adjacent points (in x and z). This method allows us to have a surface defined in every point of the computational grid. Methods based on ray-tracing algorithms are slower when the number of points of the surface increases.

From the DNS field of the smooth channel, we take a plane parallel to the wall at $y^+ = 12$, where the production of turbulent kinetic energy is maximum. The values of the streamwise velocity and the pressure are used to get the height of the surface, generating a periodic surface in the streamwise and spanwise directions. The distribution of these variables can produce very narrow valleys, that could not be reproduced well by the immersed boundary technique. Therefore, a smoothing procedure, with an accurate choice of the number of iterations and of neighbouring points, leads to the surfaces with the k_a and k_{rms} given in Table 1. Although the surfaces Us and Ps have comparable k_a and k_{rms}, the surface shapes are different (Fig. 1a, b). In fact, the low speed streaks have been transformed into elongated valleys, similar to the grooves between regular bars, while in Ps the roughness is homogeneously distributed. The surface Cs, obtained from a regular 3D geometry made of staggered cubes with the smoothing procedure, leads to a wavy surface (Fig. 1c).

Table 1 Re_τ is evaluated separately for the rough (R) and smooth (S) walls; T_D, V_D and F_D are the total, viscous and form drag of the lower wall; h_R is the distance from the rough wall of the maximum of velocity, used as reference length for $Re_{\tau,R}$

Case	k^+_{max}	k_a	k_{rms}	$Re_{\tau,R}$	$Re_{\tau,S}$	T_D	V_D	F_D	h_R	$\tilde{v}^+_R$	ΔU^+
Ps	37.6	−1.106	0.030	188	165	1.090	0.287	0.807	1.042	0.721	11.5
Us	33.0	−1.107	0.041	165	172	0.954	0.343	0.611	0.992	0.585	10.3
Cs	78.8	−1.149	0.061	394	123	2.260	0.358	1.900	1.361	0.920	13.2

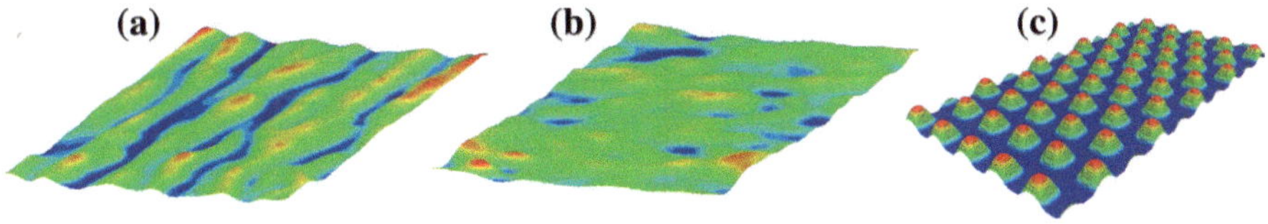

Fig. 1 Surfaces **a** Us, **b** Ps, **c** Cs. Only a quarter of the domain is shown for Cs

3 Results

The setup with the rough surfaces in one side and a smooth wall on the opposite produces an asymmetry in the channel, therefore the reference lengths to evaluate the friction Reynolds number is the distance of the plane of the crests from the point of the maximum of the velocity profile. The distance from the rough wall h_R ($h_S = 2 - h_R$), together with the values of Re_τ in the two sides, are listed in Table 1. The total drag T_D on the rough surface has been scaled with that on the smooth surface. The value of h_R for Cs, together with the high $u_{\tau,R}$, produces the highest $Re_{\tau,R}$. Consequently, the corresponding value of h_S reduces the $Re_{\tau,S}$. This surface yields also the highest drag on the rough wall. The $Re_{\tau,R}$ and drag of Ps are also increased with respect to the smooth channel. The surface of Us is the only case that allows instead to achieve a reduction of $u_{\tau,R}$ and of the drag.

It is necessary to evaluate the viscous $V_D = \nu\partial\langle U\rangle/\partial y|_R$ and the form $F_D = \langle uv\rangle_R$ drag components to understand the variation of the total drag caused by the different shape of the surfaces. These values are computed at the plane of the crests. The integration of the streamwise Navier–Stokes equation inside the roughness allows us to express the pressure drag as a function of the Reynolds stress $\langle uv\rangle$ at the plane of the crests, as done in Leonardi et al. [3]. For the geometries in Table 1, the viscous drag is comparable and the main difference to the total drag is caused by the form drag. Therefore, to increase the drag, F_D should increase, and this can be achieved with 3D roughness elements (Cs), which generate strong ejections of fluid from between the roughness elements into the overlying flow. The same occurs for Ps, Although with a smaller effect. It is difficult to achieve drag reduction because it is necessary to decrease V_D while not letting F_D increase. The surface of Us is characterized by elongated valleys, which do not differ too much from the largely studied triangular riblets [2, 5]. Both these surfaces are drag-reducing, but the components balance differently. In the riblets, V_D is large and F_D is low. Us, instead, has F_D larger than V_D. The large form drag is due to the irregularity of the surface in x that produce strong ejections. Therefore, it seems that the drag is linked to the structures elongated in the x direction.

The turbulent viscosity at the wall $\nu_{T,R}/\nu$ is defined as F_D/V_D, and its variation with the wall normal stress $\tilde{v}_R^+$ is shown in Fig. 2a. For these plots, several surface have been used, and in particular those used by Orlandi [5] to parametrize the roughness function ΔU^+ with $\tilde{v}_R^+$. $\nu_{T,R}/\nu$ allows to distinguish between flows dominated by viscous or form drag, in fact if $\nu_{T,R}/\nu$ is greater than 1, then $F_D > V_D$. Figure 2a shows

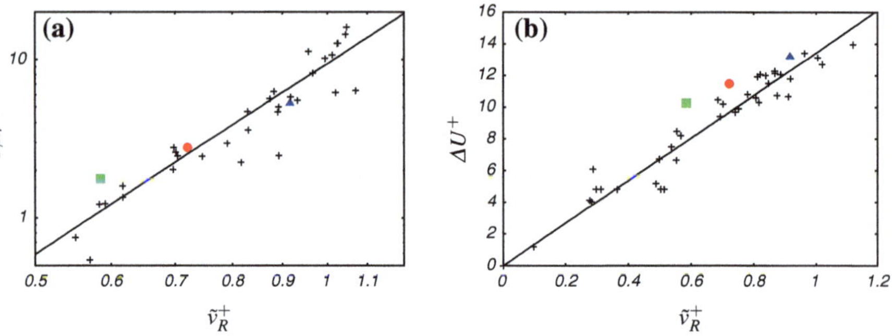

Fig. 2 ●: Ps, ■: Us, ▲ : Cs, +: [5]; **a** Normalized turbulent viscosity at the plane of the crests as a function of $\tilde{v}_R^+$; *solid line*: fit with $\nu_T/\nu = A(\tilde{v}_R^+)^4$; **b** Roughness function as a function of $\tilde{v}_R^+$; *solid line*: $\Delta U^+ = B\kappa^{-1}\tilde{v}_R^+$

that $\nu_{T,R}/\nu$ increases proportionally to $(\tilde{v}_R^+)^4$. Orlandi [5] found that $\Delta U^+ = B\kappa^{-1}\tilde{v}_R^+$ (shown in Fig. 2b). These simple relationships imply that the normal stress at the wall is an important quantity to parametrize the roughness, and the one for the turbulent viscosity might be useful for an extension of the Spalart-Allmaras model over rough walls. In fact this model uses the turbulent viscosity, and Aupoix and Spalart [1] imposed a nonzero ν_T^+ at the boundary to account for the roughness. With our results, this constant is related to a flow property. What is missing is a connection between $\tilde{v}_R^+$ and the geometrical parameters of the surface.

References

1. Aupoix, B., Spalart, P.R.: Extensions of the Spalart–Allmaras turbulence model to account for wall roughness. Int. J. Heat Fluid Flow **24**, 454–462 (2003)
2. García-Mayoral, R., Jiménez, J.: Hydrodynamic stability and breakdown of the viscous regime over riblets. J. Fluid Mech. **678**, 317–347 (2011)
3. Leonardi, S., Orlandi, P., Smalley, R.J., Djenidi, L., Antonia, R.A.: Direct numerical simulations of turbulent channel flow with transverse square bars on one wall. J. Fluid Mech. **491**, 229–238 (2003)
4. Leonardi, S., Orlandi, P., Antonia, R.A.: Properties of d- and k- type roughness in a turbulent channel flow. Phys. Fluids **19**, 12501–12506 (2007)
5. Orlandi, P.: The importance of wall-normal Reynolds stress in turbulent rough channel flows. Phys. Fluids **25**, 110813 (2013)
6. Perry, A.E., Schofield, W.H., Joubert, P.: Rough wall turbulent boundary layers. J. Fluid Mech. **37**, 383–413 (1969)

On the Effects of Surface Morphology on the Structure of Wall-Turbulence

Marco Placidi and Bharathram Ganapathisubramani

Abstract Experiments were conducted in the fully-rough regime on surfaces with large relative roughness ($h/\delta \approx 0.1$) generated by regularly distributed LEGO™ bricks of uniform height, arranged in different configurations. Measurements were made with high resolution PIV on six different frontal solidities, λ_F, at fixed plan solidity, λ_P. Results indicate that the spatial underlying structure of the turbulence across the different surface morphologies is universal in both its shape and orientation in relation to the flow velocity. Harpin packets inclination with respect of the wall is also found to be consistent not only across the different wall surfaces but also when compared to previous studies on smooth walls. Slices of two-point correlations for both streamwise and wall-normal velocity fluctuations and Reynolds shear stresses present a good collapse across the entire y/δ range for all wall morphologies.

1 Motivations and Background

Surface roughness is commonly encountered in nature and it represents a challenge in numerous engineering applications. Nevertheless, very little is understood about rough walls compared to their smooth wall counterpart [10]. Luckily, smooth and rough walls present similarities which are commonly employed in the study of boundary layers [16]. Among these, it is well documented that hairpin-like vortical structures populate smooth wall turbulent boundary layers at low and moderate Reynolds numbers and tend to align coherently to form larger-scale structures termed hairpin vortex packets [1, 7]. Therefore, given that vortex organisation in the outer region is commonly observed in instantaneous snapshots of wall turbulence, these structures are bound to leave their imprint upon the statistics of the flow [5]. The main feature ascribed to harpin packets is a series of vortices aligned in the streamwise direction forming a large-scale forward-leaning alternation of high and low momentum

M. Placidi (✉) · B. Ganapathisubramani
University of Southampton, Southampton SO17 1BJ, UK
e-mail: m.placidi@soton.ac.uk

B. Ganapathisubramani
e-mail: G.Bharath@soton.ac.uk

J. Peinke et al. (eds.), *Progress in Turbulence VI*,
Springer Proceedings in Physics 165, DOI 10.1007/978-3-319-29130-7_27

regions. Few studies have also supported the presence of similar harpin-like structures in the outer layer of rough wall boundary layers, providing that the roughness is small compared to the boundary layer thickness ($h/\delta < 0.05$). The morphology of these rough surfaces were either irregular, based on damaged turbine blades [19], or regular woven mesh [17]. However, it is still unclear if and to what extent the same vortical structure signature would persist in flows over large surface roughness. PIV measurements were therefore acquired in the (x, y) plane of fully turbulent flows over different high relative roughness morphologies ($h/\delta \approx 0.1$). The aim was to explore the effect of these severe wall conditions on the different types of structures in the outer region of turbulent boundary layers.

2 Experimental Facility and Details

Experiments were carried out in the suction wind tunnel at the University of Southampton. The tunnel has a working section of 4.5 m in length, with a 0.9 m × 0.6 m cross section. The free-stream turbulence intensity is homogenous and less than 0.3 %. The same facility has been used for previous studies on rough walls [2–4, 14, amongst others]. The streamwise, wall-normal and spanwise directions are here given along the x–y–z directions and u–v–w are the corresponding velocities. Fluctuating velocities are denoted with a $'$. Experiments were conducted in nominally zero-pressure-gradient ($K = (\nu/U_e^2)[dU_e/dx] \approx 5 \times 10^{-8}$) at 11.5 m/s. For rough surfaces, this study used a LEGO™ baseboard onto which rectangular LEGO™ bricks (or blocks), uniformly distributed in staggered array, were securely fixed. These bricks presented a uniform height ($h = 11.4$ mm). Six different patterns at fixed plan coverage were adopted to examine the effects of frontal solidity on the structure of the turbulence ($\lambda_F = 0.09, 0.12, 0.15, 0.18, 0.21, 0.24$). The different cases were designed on the basis of previous studies' predictions for the peak in drag, $D = D(\lambda_F)$ [10]. A fetch length of about 20δ was covered with brick elements to guarantee the fully rough regime [3]. Measurements were acquired using planar Particle Image Velocimetry (PIV). Some of the main parameters characterising the different surfaces are given in Table 1. The reader is referred to [13] for further details on the surface morphology and the experimental setup.

Table 1 Relevant experimental parameters

Dataset	λ_F	λ_P	δ (mm)	h/δ	U_τ (m/s)	Re_τ	δ^* (mm)	h^+	$\alpha_{u'u'}$ (0.4)	$\alpha_{u'u'}$ (0.5)
LF1	0.09	0.27	111	0.102	0.60	4600	17	463	13°	12°
LF2	0.12	0.27	122	0.090	0.66	5500	22	518	11°	10°
LF3	0.15	0.27	121	0.093	0.69	5700	22	541	13°	13°
LF4	0.18	0.27	122	0.093	0.75	6300	24	588	11°	10°
LF5	0.21	0.27	129	0.088	0.81	7200	27	635	14°	11°
LF6	0.24	0.27	127	0.090	0.80	7000	27	628	13°	11°

3 Results

As [1] and [15] pointed out, one looks at the streamwise velocity correlations to infer information about the structure of the turbulence (i.e. vortex packets). Figure 1a shows contours of the two-point correlations of the streamwise fluctuating velocity, $R_{u'u'}$. A forward-leading structure of positive correlation is shown revealing a large-scale structure coherency consistent with imprint of packets [1]. The correlations are inclined toward the flow direction. Here an example is shown for the LF4 case with the correlation centred at $y_{ref} = 0.4\delta$. However, this is representative of all the other cases and similar conclusions can be drawn when alternative wall-normal locations are considered. The inclination angle of these structures can be inferred, following [6], by a least-square fit procedure along the points further away from the auto-correlation peak at $y = y_{ref}$, along different contour lines. The results of this procedure are reported in Table 1. Inclination angles were found in the range of 10–13° and a slight sensitivity to the change in surface morphology was noticed, although a clear trend is difficult to infer. The current results are in line with previous findings on both smooth and rough walls investigations, which suggested similar values for

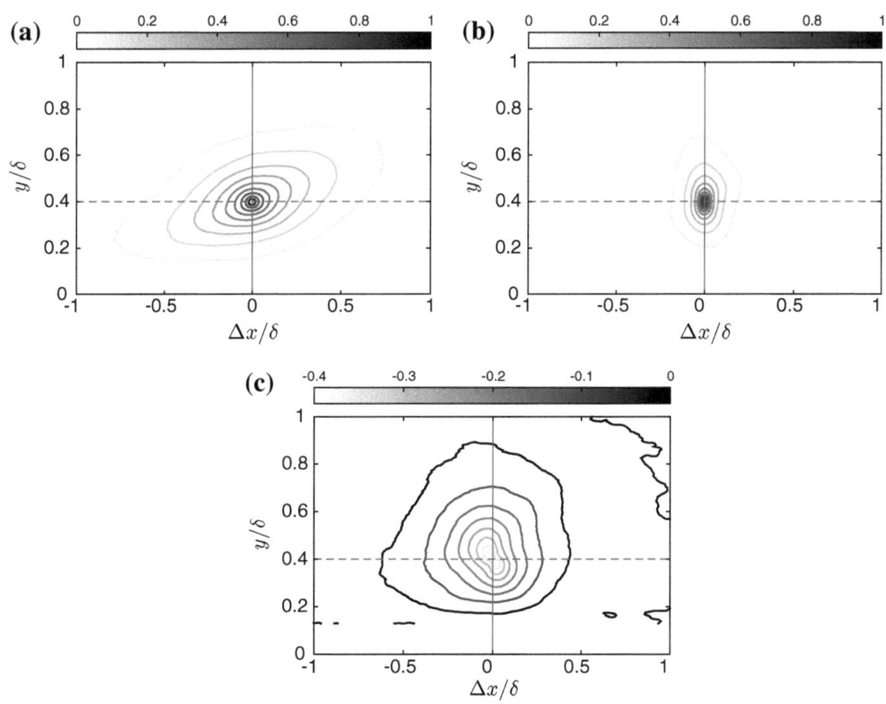

Fig. 1 Two-point correlation of **a** streamwise, **b** wall-normal and **c** Reynolds shear stress fluctuations. Colorbar represents the normalised correlation coefficient, $R_{u'u'}$, $R_{v'v'}$ and $R_{u'v'}$. Flow is left to right. Example for LF4

the characteristic inclination of the packets [1, 5, 6, 8, 9, 12, 15, 17, 18]. On the contrary, the current results do not seem to suggest that the surface morphology can have a significant influence on the vortex packets inclination, certainly not to the extend some researchers have previously documented [11].

Also reported in Fig. 1b are correlations of the wall-normal fluctuations, $R_{v'v'}$. These correlation structures are found to be compact in both streamwise and wall-normal directions with an extent that is much lower than for $R_{u'u'}$. This is consistent with previous findings [8, 17, 18, amongstothers]. Finally, two-point correlations

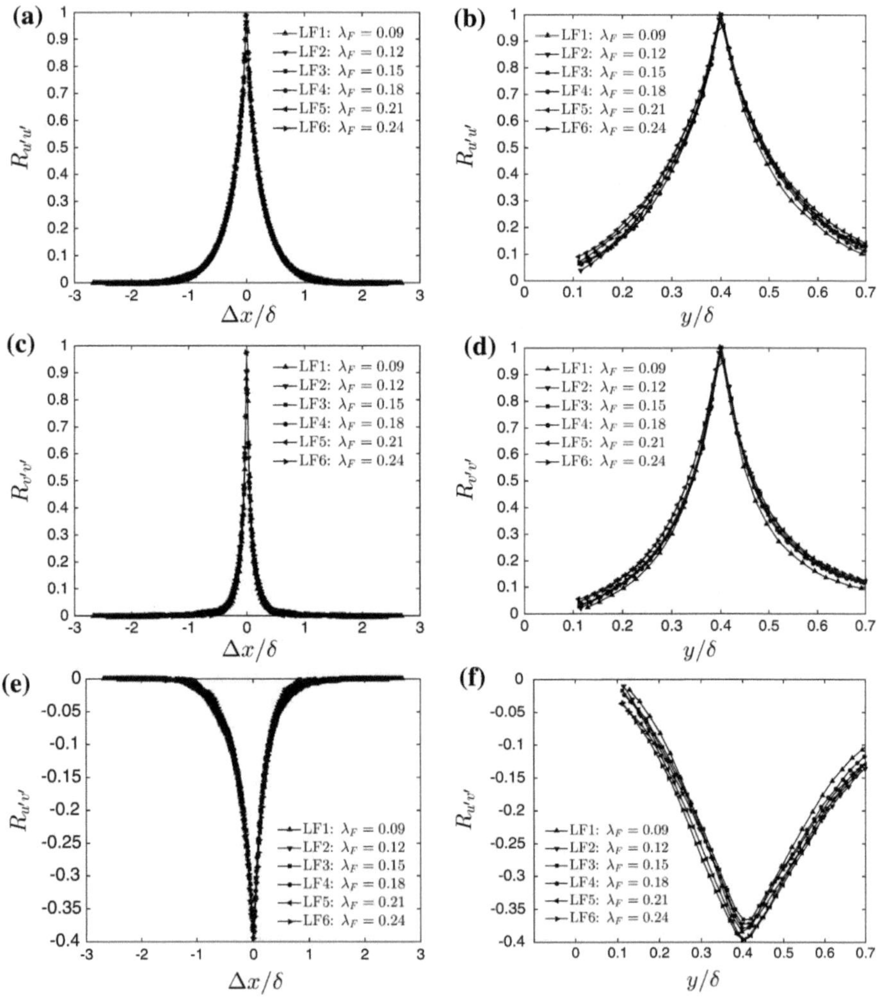

Fig. 2 (*Left*) Streamwise and (*right*) wall-normal slices through auto-correlation points of (*top*) $R_{u'u'}$, (*centre*) $R_{v'v'}$ and (*bottom*) $R_{u'v'}$ contours as a function of λ_F ($\lambda_P = const = 0.27$). *Solid lines* represent the real data resolution whist markers are spaced every five vectors for clarity.

of the Reynolds shear fluctuations, $R_{u'v'}$, are presented in Fig. 1c. These are characterised by a backward leaning structure of strong negative correlation, the extent of which is larger than the wall-normal correlation but considerably smaller than the streamwise coherency, as previously reported [17].

The effect of the considered surface morphologies on the turbulent structure can be further explored taking a slice through the auto-correlation points of $R_{u'u'}$ $R_{v'v'}$ and $R_{u'v'}$ in both streamwise and wall-normal directions, as highlighted by the dashed and solid lines in Fig. 1 respectively. These are presented in Fig. 2a–e. A very good collapse in both the streamwise and wall-normal cuts is found across all the different cases although this becomes poorer in $R_{u'v'}$ due to the higher experimental uncertainty in the determination of this quantity. To conclude, both the consistency in the inclination of the vortex packets and the collapse of all the main correlation cuts across different roughness topologies seem to offer a clear indication that the underlying spatial structures of the turbulence (i.e. vortex packets and their characteristics) are largely unaffected by changes in surface morphology—hence universal. This universality also extend to the smooth wall cases.

References

1. Adrian, R.J., Meinhart, C.D., Tomkins, C.D.: Vortex organization in the outer region of the turbulent boundary layer. J. Fluid Mech. **442**, 1–54 (2000)
2. Amir, M., Castro, I.P.: Turbulence in rough-wall boundary layers: universality issues. Exp. Fluids **51**, 313–326 (2011)
3. Castro, I.P.: Rough-wall boundary layers: mean flow universality. J. Fluid Mech. **585**, 469–485 (2007)
4. Cheng, H., Castro, I.P.: Near wall flow over urban-like roughness. Bound. Layer Meteorol. **105**, 411–432 (2002)
5. Christensen, K.T., Adrian, R.J.: Statistical evidence of hairpin vortex packets in wall turbulence. J. Fluid Mechan. **431**, 433–443 (2001)
6. Christensen, K .T., Wu, Y.: Characteristics of vortex organization in the outer layer of wall turbulence. In: Fourth International Symposium on Turbulence and Shear Flow Phenomena, Williamsburg, Virginia, pp. 1025–1030. January (2005)
7. Ganapathisubramani, B., Longmire, E.K., Marusic, I.: Characteristics of vortex packets in turbulent boundary layers. J. Fluid Mech. **478**, 35–46 (2003)
8. Ganapathisubramani, B., Schultz, M. P.: Turbulent boundary layer structure over sparsely distributed roughness. In: 7th International Symposium on Turbulence and Shear Flow Phenomena, Ottawa, Canada, 28–31 July (2011)
9. Head, M.R., Bandyopadhyay, P.: New aspects of turbulent boundary-layer structure. J. Fluid Mech. **107**, 297–338 (1981)
10. Jimenez, J.: Turbulent flows over rough walls. Annu. Rev. Fluid Mech. **36**, 173–196 (2004)
11. Krogstad, P.A., Antonia, R.A.: Structure of turbulent boundary layers on smooth and rough walls. J. Fluid Mech. **227**, 1–21 (1994)
12. Nakagawa, S., Na, Y., Hanratty, T.J.: Influence of a wavy boundary on turbulence. Exp. Fluids **35**(5), 422–436 (2003)
13. Placidi, M.: On the effect of surface morphology on wall turbulence. Ph.D. thesis, University of Southampton. Engineering and the Environment, April (2015)
14. Reynolds, R.T., Castro, I.P.: Measurements in an urban-type boundary layer. Exp. Fluids **45**, 141–156 (2008)

15. Tomkins, C.D., Adrian, R.J.: Spanwise structure and scale growth in turbulent boundary layers. J. Fluid Mech. **490**, 37–74 (2003)
16. Townsend, A.A.: The Structure of Turbulent Shear Flow. Cambridge University Press, Cambridge (1976)
17. Volino, R.J., Schultz, M.P., Flack, K.A.: Turbulence structure in rough- and smooth-wall boundary layers. J. Fluid Mech **592**, 263–293 (2007)
18. Volino, R.J., Schultz, M.P., Flack, K.A.: Turbulence structure in a boundary layer with two-dimensional roughness. J. Fluid Mech. **635**, 75–101 (2009)
19. Wu, Y., Christensen, K.T.: Outer-layer similarity in the presence of a practical rough-wall topography. Phys. Fluids **19**(8), 085108 (2007)

Contradictions in the Large-Wavelength Approximation of Turbulent Flow Past a Wavy Bottom

Paolo Luchini

1 A Classical Problem: The Formation of Ripples and Dunes

As described, for instance, by Charru et al. [3], the steady or oscillatory motion of a liquid above a granular bed leads to the formation of ripples. The sand ripples one observes on a beach at low tide are an example: these ripples were formed by the oscillations induced by the surface waves when the beach was covered with shallow water. The mechanism of their formation, related to fluid inertia, sits in the relative phase of bottom shear stress oscillations with respect to oscillations of the bottom itself, with positive phase advance of the shear stress dragging the particles toward crests during each half-period (Fig. 1).

Classically the response of the bottom shear stress to a small wall undulation has been studied in limiting cases. The short-wave limit, corresponding to a boundary-layer approximation, was introduced by Benjamin [2] for laminar flow and by Jackson and Hunt [4] for turbulent flow. The long-wave limit, corresponding to a shallow-water approximation, was adopted by Benjamin [1] and Yih [11] for laminar flow and by Shkadov [10] and many others for turbulent flow. Recently, Luchini and Charru [6, 7] introduced a consistent viscous form of the shallow-water approximation and applied it to this problem.

All of these studies predict a negative quadrature component of the bottom shear stress (an *unstable* erodible bottom). Several experiments by Hanratty et al. in the 1980s confirm the presence of a negative quadrature component.

All of these authors use an eddy-viscosity model to represent turbulence.

P. Luchini (✉)
DIIN, Università di Salerno, Via Giovanni Paolo II 132, 84084 Fisciano, SA, Italy
e-mail: luchini@unisa.it

J. Peinke et al. (eds.), *Progress in Turbulence VI*,
Springer Proceedings in Physics 165, DOI 10.1007/978-3-319-29130-7_28

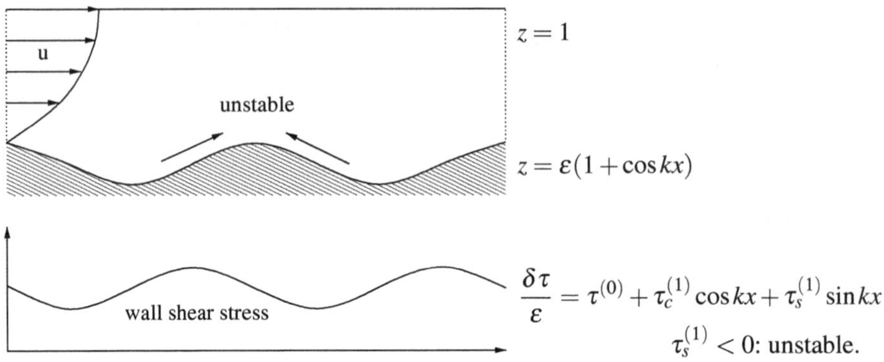

Fig. 1 Effect of the *bottom* shear stress and its phase lead on the stability of a sandy bed

2 The Failure of Eddy Viscosity

Luchini and Russo [8, 9] observed that in Luchini and Charru's long-wave expansion the steady-streaming effect of the wall becomes equivalent to an inertia-force term proportional to the x-derivative of the mean velocity, just as in the corresponding laminar-flow analysis of [1]. Therefore, an equivalent problem could be formulated in which turbulent flow in a straight channel is driven by an externally imposed, steady volume force. This problem can be readily solved by direct numerical simulation (DNS), thus leading to a rare head-to-head comparison between eddy viscosity and DNS.

The main outcome of the comparison, reported in Fig. 4 of [9], uncovered a profound disagreement between the results of eddy viscosity and DNS, to the point that the flow rate induced by a given volume-force profile is of opposite signs in one and the other. This conclusion turns out to be independent of the details of the eddy-viscosity model, as long as viscosity is positive.

3 A Contradiction

Luchini and Russo's conclusion, while on one hand it provides a severe warning against the use of eddy-viscosity models in stability calculations, on the other encounters a contradiction with all the previous results (including experimental ones) concerning a sandy bed; for, as a consequence of the sign change, the phase of the bottom shear stress is now predicted to be positive, that is, the sandy bottom is predicted to be stable and the formation of ripples not to happen.

4 Immersed-Boundary Numerical Simulations of Flow in a Wavy Channel

A new set of direct numerical simulations, which are the main object of this presentation, allows this contradiction to be reconciled. The code used is based on second-order finite differences in all directions, with a stretched coordinate along the wall-normal z coordinate only, is fully parallel, and adopts an explicit time advancement for all but the pressure-correction equation. Its immersed-boundary implementation in staggered variables uses internal (to the fluid) points only, is continuous with respect to boundary crossing and stable in iteration at all distances from the boundary. It already proved its usefulness in a previous Stokes-flow application [5]. Most attention was paid to evaluating the shear stress at the wall precisely. The common expedient of summing virtual forces is not sufficient: the contribution of pressure gradient over fractional cell boundaries is of the same order as the effect under investigation. In addition, the amplitude ε of the wall oscillation has to strike a compromise between linearity and statistical fluctuation of the time averaging. An algorithm was developed to estimate the expected value of this statistical error.

After a confirmation test on laminar flow, the code was applied to the DNS of turbulent flow in wavy channels of various wavelengths, spanning the short-wave, long-wave and intermediate regimes. Already the laminar test showed that special care must be applied in even identifying these regimes: the governing parameter is not just the wavenumber k of the wall's wavyness (made dimensionless with the channel's depth) but its product with the Reynolds number, kRe. It follows that, in order for the flow to behave in the shallow-water regime, it is not enough for k to be small compared to 1, but it must be small compared to $1/Re$ as well.

Figure 2 shows the real and imaginary part (in-phase and quadrature Fourier component) of the wall-shear-stress response to a wall undulation of wavenumber k, in the long-wave (shallow-water) range of wavenumbers. There are several surprises in this figure. First, the wavelength involved is much larger than the typical computational box adopted in DNS of a straight channel (4π long, usually deemed to be

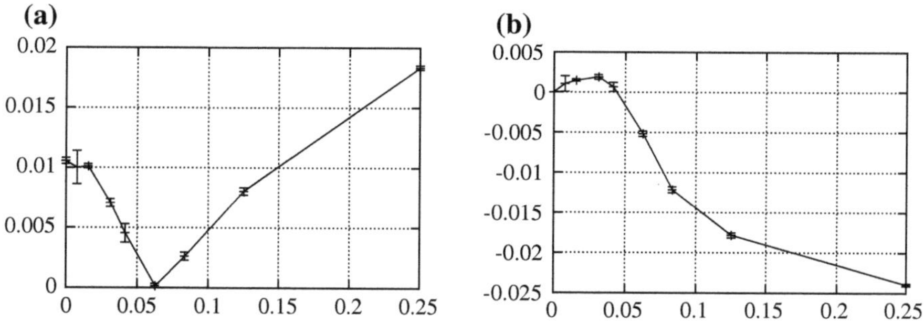

Fig. 2 *Bottom* shear stress of turbulent flow in a wavy channel versus wavenumber k, at $Re = 1450$ ($Re_\tau = 100$). **a** In-phase Fourier component; **b** quadrature Fourier component

sufficient for the longitudinal correlation of velocity fluctuations to become negligible); in order to draw this plot we had to adopt computational boxes as large as 256π. Second, the in-phase response sharply drops to near zero for a wavelength of 32π (roughly a hundred times the channel depth), in striking contrast to the laminar (and eddy-viscosity) response which increases monotonically from its value at $k = 0$, thus exposing an unforeseen (as far as we know) resonance. Third, what is most relevant to the sand-ripple problem, the quadrature component of the response changes sign for a similarly low wavenumber.

5 Conclusion

The solution to the potential contradiction of Sect. 3 is in that the mathematical long-wave limit of the quadrature shear stress, confirmed to be of the same sign as in the previous DNS of Luchini and Russo [8, 9] and corresponding to a stable sand ripple, is only achieved at a much longer wavelength than tested in all experiments. Whereas both turbulent numerical simulations, with an equivalent volume force and with an actual wavy wall, invalidate eddy-viscosity models (and all the classical calculations based on them) in the long-wave limit, for most practical wavelengths the sign of the wall shear stress corresponds to an unstable ripple, as it must since sand ripples do self-produce in reality.

Simulations at higher Reynolds number, which we do not have the space to display here, confirm this behaviour. A dip, if becoming rounded, in the in-phase response and a sign change in the quadrature response continue to be present. Both move towards higher k.

References

1. Benjamin, T.B.: Wave formation in laminar flow down an inclined plane. J. Fluid Mech. **2**, 554–574 (1957)
2. Benjamin, T.B.: Shearing flow over a wavy boundary. J. Fluid Mech. **6**, 161–205 (1959)
3. Charru, F., Andreotti, B., Claudin, P.: Sand Ripples and Dunes. Annu. Rev. Fluid Mech. **45**, 469–93 (2013)
4. Jackson, P.S., Hunt, J.C.R.: Turbulent wind flow over a low hill. Q. J. R. Meteorol. Soc. **101**, 929–955 (1975)
5. Luchini, P.: Linearized no-slip boundary conditions at a rough surface. J. Fluid Mech. **737**, 349–367 (2013)
6. Luchini, P., Charru, F.: Consistent section-averaged equations of quasi-one-dimensional laminar flow. J. Fluid Mech. **656**, 337–341 (2010)
7. Luchini, P., Charru, F.: The phase lead of shear stress in shallow-water flow over a perturbed bottom. J. Fluid Mech. **665**, 516–539 (2010)
8. Luchini, P., Russo, S.: A comparison between eddy-viscosity models and direct numerical simulation: the response of turbulent flow to a volume force. 64th Annual Meeting of the APS Division of Fluid Dynamics, Baltimore, MD 20–22, Bull. Am. Phys. Soc. 56–18 (2011)

9. Luchini, P., Russo, S.: A comparison between eddy-viscosity models and direct numerical simulation: the response of turbulent flow to volume forcing. Proceedings of XX Congresso AIMETA di Meccanica Teorica e Applicata, pp. 1–9. Bologna 12–15 Sep 2011
10. Shkadov, V.Ya.: Wave modes in the gravity flow of a thin layer of a viscous fluid. Izv. Akad. Nauk. SSSR, Mekh. Zhidk. Gaza 3: 43–51 (1967)
11. Yih, C.S.: Stability of liquid flow down an inclined plane. Phys. Fluids **6**, 321–334 (1963)

Wall Oscillation Induced Drag Reduction of Turbulent Boundary Layers

Martin Skote, Maneesh Mishra, Prabal Singh Negi, Yanhua Wu, Hsiao Mun Lee and Philipp Schlatter

Abstract Spanwise oscillation applied on the wall under a turbulent boundary layer flow is investigated using direct numerical simulation. The temporal wall-forcing produces considerable drag reduction over the region where oscillation occurs. The turbulence fluctuations downstream of the oscillations are presented for the first time. Simulations with identical oscillation parameters have been performed at different Reynolds numbers to investigate the effect on the drag reduction. One of the simulations replicates an earlier experiment to test the fidelity of the current simulations. In addition, we present the future work in this area with an integrated experimental and computational investigation to explore the possibility of applying travelling waves (oscillations in both time and space) as the mode of wall motion for active control of near-wall turbulence.

1 Introduction

Turbulent drag reduction (DR) is one of the active areas of research in fluid mechanics especially due to its global impact on sustainability challenges we are facing today.

The first observations of wall oscillation as means for DR was made by Jung et al. [2] through direct numerical simulations (DNS) of a channel flow. Since then, a lot of research efforts have been made in this direction for internal flow such in a channel or pipe flow. The boundary layer flows have only recently started to be investigated [3, 6–10]. The most commonly form of control studied is realized with temporal wall oscillations, which is specified as $w(t) = W_m \sin(\omega t)$, where $w(t)$ is the spanwise wall velocity, W_m is the amplitude and ω is the frequency of the imposed oscillations, which is related to the period (T) through $\omega = 2\pi/T$.

M. Skote (✉) · M. Mishra · P.S. Negi · Y. Wu · H.M. Lee
School of Mechanical and Aerospace Engineering, Nanyang Technological University, 50 Nanyang Avenue, Singapore 639798, Singapore
e-mail: mskote@ntu.edu.sg

P. Schlatter
Linné FLOW Centre, KTH Mechanics, Stockholm, Sweden
email: pschlatt@mech.kth.se

J. Peinke et al. (eds.), *Progress in Turbulence VI*,
Springer Proceedings in Physics 165, DOI 10.1007/978-3-319-29130-7_29

Table 1 Numerical parameters

Case	L_x	L_y	L_z	N_x	N_y	N_z	Δx^+	Δy^+_{min}	Δz^+	$Re_{\Theta\,start}$
1	600	30	34	800	201	144	16	0.04	5.1	375
2	600	30	34	800	201	144	16	0.04	5.1	505
3	2400	75	34	3200	301	144	20	0.04	5.1	1400
4	2400	75	34	3200	301	144	20	0.04	5.1	963

L and N refer to the size (measured in displacement thickness at $x = 0$) and the number of modes, respectively

The aim of the present study is to compare our results with the boundary layer experiments by Ricco and Wu [5], who obtained the DR by measuring the streamwise velocity component in the viscous sublayer, to test the fidelity of the current simulations. Furthermore, the behaviour of the turbulence downstream of the oscillating region are presented for the first time.

All four simulations presented here were performed with identical oscillation parameters of $T^+ = 67$ and $W_m^+ = 11.3$ (wall scaling with the friction velocity taken from the starting point of the oscillations). The parameters pertaining to the numerical set-up is given in Table 1, where $Re_{\Theta\,start}$ denotes the starting point of the oscillations. The spectral code (SIMSON) used for the turbulent boundary layer simulations in this work was developed at KTH, Stockholm.

2 Results

Figure 1 (left) shows the degradation of DR performance with Reynolds number (Re). If we extrapolate the DR based on the lower-Re data alone (cases 1 and 2), we would conclude that the decay of DR is severe and there would be no substantial energy saving at higher Re as shown by the solid arrow. However, in simulations performed at higher Re (case 3), we observe a DR of much higher degree which indicates a Re dependence illustrated by the dashed arrow. This brings us to the question of which route the DR takes as we go to higher Re. The answer to this question lies in performing new simulations with larger domains to examine the complete development as a function of streamwise location. The (preliminary) result is shown in Fig. 1 (right) as the red curve (case 4). Clearly, the development of the DR downstream indicates that the dashed arrow in the left figure illustrates the correct behaviour. In addition, the excellent agreement between case 3 and the experimental results (indicated with rings), at an identical Re, is revealed in Fig. 1.

In order to investigate the stresses independent of the choice of scaling, we show the ratio between the peak values from the oscillating and reference cases in Fig. 2. While u_{rms} and $\overline{u'v'}$ have reached their peak values within the zone of oscillation, v_{rms} is the only quantity which is still decreasing at the point where oscillation stops for case 3. Even after the termination of oscillation, v_{rms} continues to decrease for

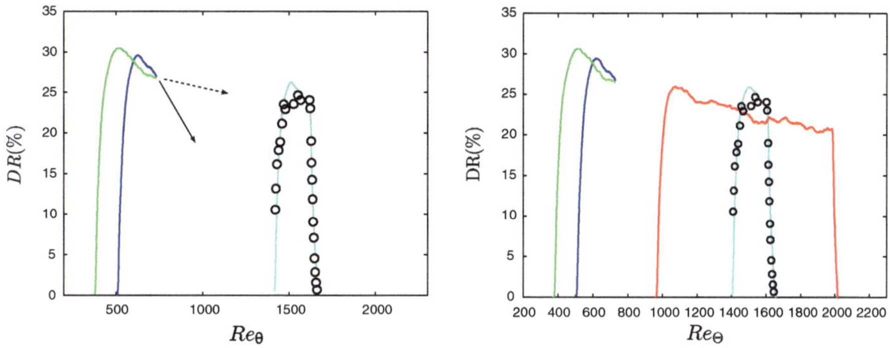

Fig. 1 *Left* DR (%) as a function of Re with temporal wall oscillation introduced at (*green*) $Re_\theta = 375$ (case 1); (*blue*) $Re_\theta = 505$ (case 2); (*cyan*) $Re_\theta = 1400$ (case 3); (∘) Experimental results from Ricco and Wu [5] at $Re_\theta = 1400$. *Arrows* indicate extrapolations based on the low-Re simulation data (see text). *Right* same legend as in (*left*) plus (*red*) $Re_\theta = 963$ (case 4)

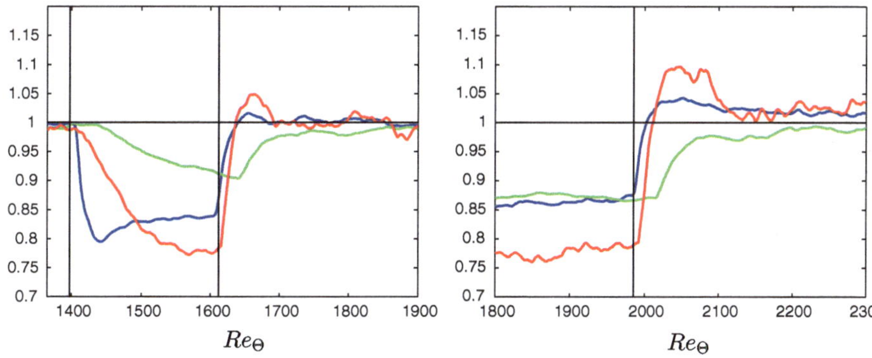

Fig. 2 Ratio of peak values from the reference case and case 3 (*left*), and case 4 (*right*). (*blue*) u_{rms}; (*green*) v_{rms}; (*red*) $\overline{u'v'}$. The *vertical lines* mark the zone of oscillation

short distance downstream. However, for case 4, the v_{rms} has reached its equilibrium but continues nevertheless to remain affected for the same extent as for case 3.

3 Future Work

Travelling waves were used by Quadrio et al. [4] as wall forcing in DNS studies of channel flow, and the DR variation with the wavenumber and angular frequency was investigated. The only experimental data to date for this type of wall forcing are provided by [1], who applied the streamwise travelling wave on the pipe flow. To our knowledge, the current work constitutes the first attempt to experimentally study wall forcing by travelling waves in the boundary layer. The wind tunnel facility

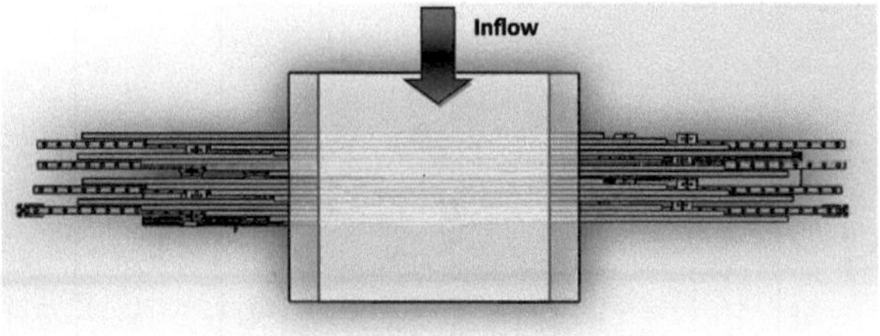

Fig. 3 *Top-view* of the experimental setup

to be used is an open-loop type in the Fluid Mechanics Lab at NTU, Singapore. After considering several points on the normalized DR map in [4] and checking the dimensional values of the system with the physical constraints, a final parameter set was determined which corresponds to a travelling wave with a frequency of 4.6 Hz and wavelength of 79.3 mm.

The design consists of four linear guides, each supporting 4 discretized segments of the travelling wave. Eight linear motors are required to control the motion of the 16 segments forming two wavelengths. A linear encoder is used for the feedback control for each of the controlling motors. The top view of the entire setup is shown in Fig. 3 with an overlapping part of the wind tunnel test section (made translucent). The wall shear stress will be directly measured using advanced oil-film-interferometry technique. High resolution volumetric three-dimensional tomographic PIV measurements will be performed to quantify in great details the evolution of the turbulent boundary layer affected by the travelling waves.

The corresponding DNS of an identical set-up will be conducted using 1024 cores at high performance computing facilities in NTU and KTH.

Acknowledgments Supported by Singapore MOE Tier-2 Grant No. MOE2012-T2-1-030.

References

1. Auteri, F., Baron, A., Belan, M., Campanardi, G., Quadrio, M.: Experimental assessment of drag reduction by traveling waves in a turbulent pipe flow. Phys. Fluids **22**(11), 115103 (2010)
2. Jung, W.J., Mangiavacchi, N., Akhavan, R.: Suppression of turbulence in wall-bounded flows by high-frequency spanwise oscillations. Phys. Fluids A **4**(8), 1605–1607 (1992)
3. Negi, P.S., Mishra, M., Skote, M.: DNS of a single low-speed streak subject to spanwise wall oscillations. Flow Turbul. Combust. **94**(4), 795–816 (2015)
4. Quadrio, M., Ricco, P., Viotti, C.: Streamwise-travelling waves of spanwise wall velocity for turbulent drag reduction. J. Fluid Mech. **627**, 161–178 (2009)
5. Ricco, P., Wu, S.: On the effects of lateral wall oscillations on a turbulent boundary layer. Exp. Therm. Fluid Sci. **29**, 41–52 (2004)
6. Skote, M.: Turbulent boundary layer flow subject to streamwise oscillation of spanwise wall-velocity. Phys. Fluids **23**(8), 081703 (2011)
7. Skote, M.: Temporal and spatial transients in turbulent boundary layer flow over an oscillating wall. Int. J. Heat Fluid Flow **38**, 1–12 (2012)

8. Skote, M.: Comparison between spatial and temporal wall oscillations in turbulent boundary layer flows. J. Fluid Mech. **730**, 273–294 (2013)
9. Skote, M.: Scaling of the velocity profile in strongly drag reduced turbulent flows over an oscillating wall. Int. J. Heat Fluid Flow **50**, 352–358 (2014)
10. Yudhistira, I., Skote, M.: Direct numerical simulation of a turbulent boundary layer over an oscillating wall. J. Turbul. **12**, 9 (2011)

Wall-Turbulence Structure with Pressure Gradient Around 2D Hump

Aiko Yakeno, Soshi Kawai, Taku Nonomura and Kozo Fujii

Abstract Direct numerical simulation around a two-dimensional hump shape is conducted at the Reynolds number $Re_h = 16,000$, based on the hump height. We investigate wall-turbulence structures around the hump in order to predict and control them to suppress separation. At this Reynolds number, specific striped wall-turbulence structure appears at the leading-edge near the wall surface. Its spanwise length-scale is close to that of the streak in a fully-developed turbulent channel flow. That is $\lambda_y = 0.08$ scaled with the hump height, which corresponds to $\lambda_y^+ = 150$ in the local viscous unit. We identify two more different spanwise-correlated scales, $\lambda_y = 0.40$ and 0.13 around the hump. Spanwise length-scale of $\lambda_y = 0.40$ is around $\lambda_y^+ = 600$. On the other hand, the other scale $\lambda_y = 0.13$ is not dependent on the local viscous scale.

1 Introduction

Flow-separation around fluidic devices is a serious problem that it decreases their performance. The separation is known to be suppressed by applying a periodic excitation control at the leading edge [1]. It is considered because the excitation promotes the two-dimensional roll vortex due to hydrodynamic instability and enhances turbulent momentum-transfer in the downstream. At the higher Reynolds number, flow characteristics such like wall-turbulence motion drastically change. In the present study, we investigate cases at $Re_h = 16,000$ based on the hump height, which the separation occurs in a turbulent flow state. At the leading edge, wall-turbulence structures appear. One of them looks like the low-speed streak that is usually observed in the turbulent channel flow or in the bypass transition of a boundary layer. This is propagated linearly with temporal energy growth due to nonmodality of the

A. Yakeno (✉) · S. Kawai · T. Nonomura · K. Fujii
Institute of Space and Astronautical Science, Japan Aerospace Exploration Agency, 3-1-1 Yoshinodai, Chuo, Sagamihara, Kanagawa 252-5210, Japan
e-mail: yakeno@flab.isas.jaxa.jp

J. Peinke et al. (eds.), *Progress in Turbulence VI*,
Springer Proceedings in Physics 165, DOI 10.1007/978-3-319-29130-7_30

linear system [2]. In the present study, we investigate turbulence structures appearing around a two-dimensional hump, in order to predict and control them to suppress separation at higher Reynolds number.

2 Computational Procedures

The hump geometry is analytically-defined semicircular column with Bessel functions, which is joined smoothly with the wall. Schematic of the hump geometry, computational domain and boundary conditions are shown in Fig. 1. All variables in the report are normalized by outflow density ρ^*_∞, velocity u^*_∞ and height of the two-dimensional hump h^*. The Reynolds number Re_h based on the outflow velocity and the hump height is set at 16, 000. Constant Blasius profile is employed at the input boundary with the thickness of $\delta_{in} = 0.25$, which is normalized by the hump height. Flow becomes turbulent around the top of the hump at the present Reynolds number. At the out flow boundary, variables are extrapolated from the adjacent nodes, and the second order filtering is applied in the streamwise direction. Computational grids are set concentrated around the hump to resolve turbulent structures near the wall. The governing equations are three-dimensional compressible Navier-Stokes equations. An in-house solver is employed. The spatial derivatives of the convective and viscous terms, symmetric conservative metrics and Jacobian [3] are evaluated by a sixth-order compact difference scheme [4]. Compact-difference-scheme-related subroutines in the code is well tuned for the K supercomputer [5]. The present computation scheme and grid are based on verification studies.

3 Wall-Turbulence Structure

The pressure coefficient C_p and its gradient in the streamwise direction $\partial C_p/\partial x$ are plotted in Fig. 2a, b, respectively. It is calculated with a static pressure $\bar{p}$ at the surface as $C_p = (\bar{p} - p_\infty)/(1/2\rho_\infty u_\infty^2)$. The favorable pressure gradient is around at the leading-edge, where the pressure gradient becomes negative as shown in Fig. 2b. The magnitude of the pressure gradient is decreasing $(\partial^2\bar{p}/\partial x^2 > 0)$ at around $x = -0.6 \sim -0.25$. This region is described as a rectangle of dotted line in the figure.

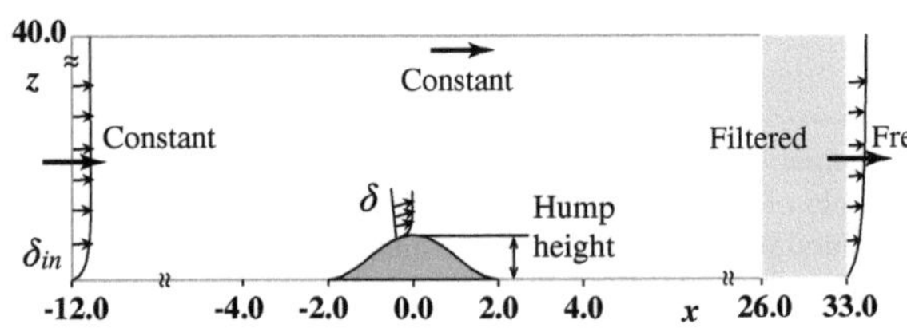

Fig. 1 Schematic of the computation domain and boundary conditions

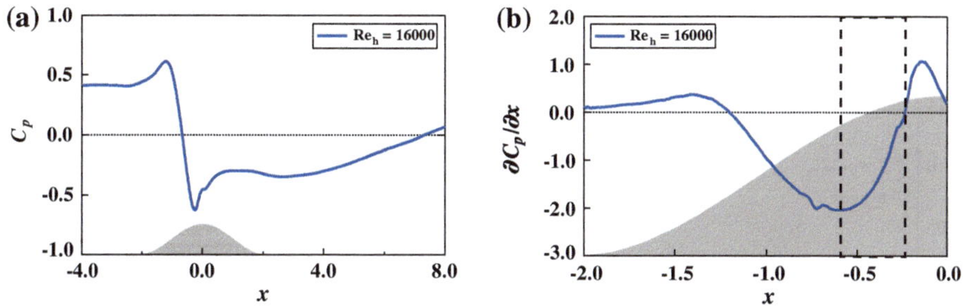

Fig. 2 **a** Pressure coefficient C_p, and **b** the gradient $\partial C_p/\partial x$ at $Re_h = 16,000$. A rectangle of *dotted line* in (**b**) represents where the magnitude of pressure gradient decreases ($\partial^2\bar{p}/\partial x^2 > 0$)

Side and top visualizations of the instantaneous flow field with iso-surfaces of the second invariant of deformation tensor $Q(=Q^*h^{*2}/u_\infty^{*2}) = 0.28$ and contours of u are shown in Fig. 3. The flow is separated at around $x = -0.07$ in a turbulent state. Near the wall, a striped turbulence-structure appears at around $x = -0.6 \sim -0.25$ as shown red-colored contours of u at $z = 0.005$ in Fig. 3b.

The maximum value of root mean square of tangential velocity $u'_{t\,rms}$ and its distance from the wall ζ_{peak} are plotted in Fig. 4a. The value of $u'_{t\,rms}$ has two peaks in the wall-normal direction. Those amplitudes exchange at around $x = -0.7$ shown as discontinuous lines in the figure. Correlations of wall-normal fluctuation w' in the spanwise direction at each ζ_{peak} are calculated as shown in Fig. 4b. We identified three spanwise correlated scales from the plot. First is the near-wall structure scale around $\lambda_y = 0.08$ as the first negative peak, I., $\lambda_y/2 = 0.04$, which is distributed at around $x = -0.50 \sim -0.30$. Spanwise width of the striped structure in Fig. 3b corresponds to this. Second is the larger structure scale around $\lambda_y = 0.40$ shown as the second negative peak, II. $\lambda_y/2 = 0.2$, which is found in the upstream at around $x = -1.0$. Third is $\lambda_y = 0.13$, which is widely distributed at around $x = -0.50 \sim 0.0$ as shown as the positive peak, III. in the figure.

Correlations are replotted in the viscous unit as shown in Fig. 5a, b. Two of spanwise structure scales $\lambda_y = 0.08, 0.40$ are converted to $\lambda_y^+ = 150, 600$ in the local

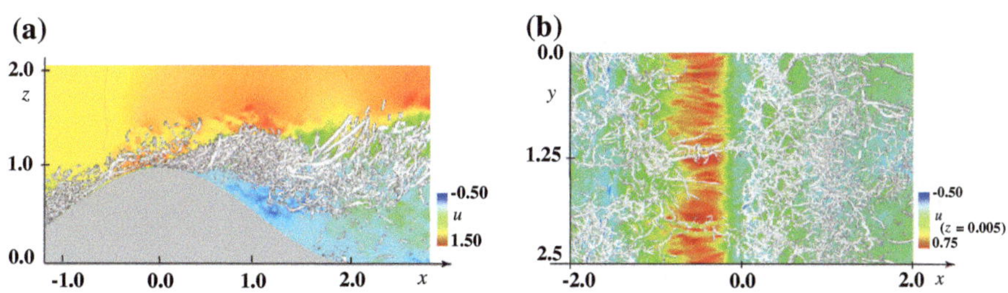

Fig. 3 *Side* and *top view* of the instantaneous flow with iso-surface of the second invariant of deformation tensor, $Q(=Q^*h^{*2}/u_\infty^{*2}) = 0.28$, and contour of u

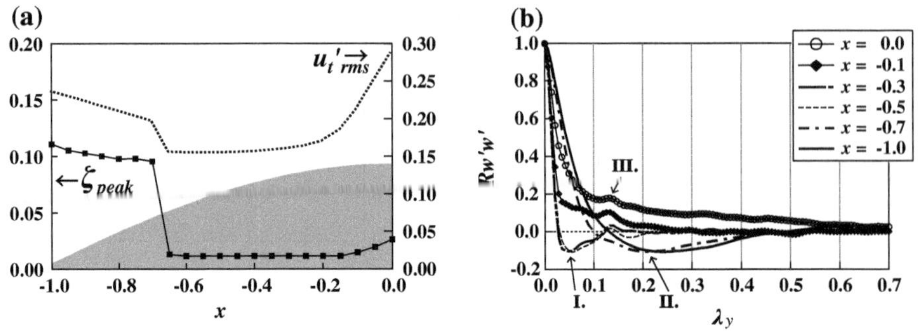

Fig. 4 **a** The maximum value of root mean square of tangential velocity $u'_{t\,rms}$ (*right axis*) and its distance from the wall, ζ_{peak} (*left axis*). **b** Correlations of w' in the spanwise direction at ζ_{peak}

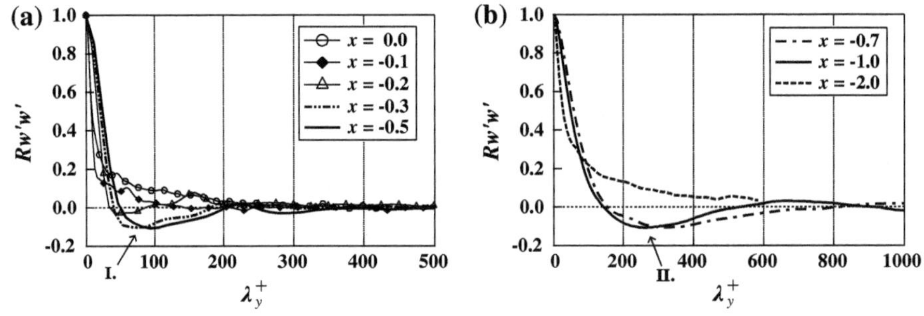

Fig. 5 Correlations of w' in the spanwise direction at ζ_{peak} in the viscous unit; **a** at $x = 0.0, -0.1, -0.2, -0.3$ and -0.5, **b** at $x = -0.7, -1.0$ and -2.0

viscous unit, respectively. The spanwise length scale of $\lambda_y^+ = 150$ is close to that of the streak observed in a fully-developed turbulent channel flow. On the other hand, it is noted that peaks of the third scale $\lambda_y = 0.13$ are not apparent in the rescaling plot, Fig. 5a. From the result, the third structure seems to be independent on the local viscous scale.

Acknowledgments The computations in this study were conducted using the K supercomputer in the Advanced Institute of Computational Science, Riken. This research was mainly supported by Strategic Programs for Innovative Research (SPIRE) of the High Performance Computing Initiative (HPCI) (No. hp120296, hp130001, hp140207 and hp150219).

References

1. Gad-el-Hak, M., Bushnell, D.M.: Separation control: review. J. Fluids Eng. **1**, 5 (1991)
2. Butler, K., Farrell, B.: Three-dimensional optimal perturbations in viscous shear flow. Phys. Fluids **4**(8), 1637–1650 (1992)

3. Visbal, M.R., Gaitonde, D.V.: High-order-accurate methods for complex unsteady subsonic flows. AIAA J. **37**(10), 1231–1239 (1999)
4. Lele, S.K.: Compact finite difference schemes with spectral-like resolution. J. comput. phys. **103**, 16–42 (1992)
5. Aono, H., Nonomura, T., Iizuka, N., Ohsako, T., Inari, T., Hashimoto, Y., Takaki, R., Fujii, K.: Scalar tuning of a fluid solver using compact scheme for a supercomputer with a distributed nemory architecture. CFD Lett. **5**(4), 143–152 (2013)

Turbulent Asymptotic Suction Boundary Layers: Effect of Domain Size and Development Time

Alexandra Bobke, Ramis Örlü and Philipp Schlatter

1 Introduction and Motivation

Boundary layers (BLs) are prevalent in many technical applications and need therefore to be considered in the design of fluid machineries. In order to find a suitable design to e.g. reduce the drag and thereby the fuel consumption, a sufficient knowledge about the physical phenomena has to be at hand. More than a century ago, Ludwig Prandtl suggested a way to stabilise the BL by applying suction through a porous surface or a series of slots. By applying suction, the low momentum fluid is removed and the resulting BL becomes thinner than in the case without suction and thereby more stable, which in turn delays transition and separation.

Hereafter several experiments were carried out to control laminar and turbulent BLs with suction [1, 4]. On the other hand, Mariani et al. [5] performed the first direct numerical simulation (DNS) of a true turbulent asymptotic suction BL (TASBL) and documented the turbulence statistics. They could confirm the results from previous studies [1] of a lower fluctuation level in presence of suction. The dependence on the size of the computational domain in TASBL studies was recently investigated in Ref. [8]. Despite low Reynolds numbers ($Re = U_\infty/V_0 = 333, 400, 500$, with V_0 and U_∞ denoting the suction rate and free-stream velocity, respectively) they could achieve large friction Reynolds numbers with a clear scale separation.

The boundary layer develops in time, indicating a homogenous character in both streamwise and spanwise direction. Contrary to spatially developing, temporally developing BLs are easier to realise and demand less computational resources. Once

A. Bobke (✉) · R. Örlü · P. Schlatter
Linné FLOW Centre, KTH Mechanics, 100 44 Stockholm, Sweden
e-mail: bobke@mech.kth.se

R. Örlü
e-mail: ramis@mech.kth.se

P. Schlatter
e-mail: pschlatt@mech.kth.se

J. Peinke et al. (eds.), *Progress in Turbulence VI*,
Springer Proceedings in Physics 165, DOI 10.1007/978-3-319-29130-7_31

an asymptotic state is reached, the BL thickness e.g. measured in δ_{99} remains statistically constant. In the present work, additionally to the box-size dependency (thereby repeating the study in Ref. [8]), also the temporal development of the TASBL is investigated, and different scaling laws are assessed.

2 Numerical Method and Simulation Cases

Large eddy simulations (LES) were carried out for the flow over a flat plate. Uniform suction was applied in wall normal direction to counteract the momentum loss and the Reynolds number was thereby set through V_0 and U_∞.

The laminar solution can directly be obtained from the Navier-Stokes equation for a constant suction rate and free-stream velocity, i.e. $u = U_\infty(1 - \exp(-yV_0/\nu))$, where ν is the kinematic viscosity. In the asymptotic state, the integral momentum conservation gives a fixed value for the skin friction coefficient through the Reynolds number, viz. $c_f = 2(U_\infty^+)^{-2} = 2/Re$. To catch the large(st) length scales occuring in wall-bounded turbulent flows, big box sizes are required, which makes the simulations computationally expensive. Hence the simulations were performed as LES applying a subgrid-scale (SGS) model, specifically the ADM-RT model based on high-order relaxation of the velocity field, for the unresolved scales utilising the fully spectral method code SIMSON [3] for the computations. The wall-normal directions are discretized with Chebyshev polynomials, while the streamwise and spanwise directions with Fourier compositions. The LES including the SGS model was verified against DNS data (with $\Delta x^+ = 14$ and $\Delta z^+ = 7$) and it was checked that the chosen SGS model describes the flow with good accuracy (not shown here). The grid resolution in inner units for the considered LES with $Re = 333$ is $\Delta x^+ = 57$ and $\Delta z^+ = 28.5$. Local perturbations were applied on the initial, laminar profile with the characteristic displacement thickness δ_0^* used for non-dimensionalisation together with U_∞. Different box sizes (Table 1) were chosen to elucidate the influence on the

Table 1 Performed simulations of the TASBL with constant suction rate ($Re = 333$) for different domain sizes (scaled by the laminar displacement thickness δ_0^*) and resolutions, yielding the given friction Reynolds numbers Re_τ and boundary-layer thicknesses δ_{99} (normalised by δ_0^*)

Case	Domain size	Resolution	Re_τ	δ_{99}
LES333e	1600 × 450 × 800	512 × 301 × 512	2000	109
LES333d	800 × 300 × 400	256 × 201 × 256	2000	111
LES333c	400 × 300 × 200	128 × 201 × 128	1960	107
LES333b	200 × 300 × 100	64 × 201 × 64	1520	83
LES333a	100 × 300 × 50	32 × 201 × 32	1120	61
DNS333b	200 × 300 × 100	256 × 301 × 256	1680	92

large structures and the simulations were run over long time periods to study the time evolution.

3 Results and Discussions

In this study both the dependence on the size of the computational domain and the temporal development towards the asymptotic state are studied, while classical scaling laws [6, 7, 9] are tested on the data from the asymptotic state. The mean streamwise velocity profile in Fig. 1a shows a strong influence of the box sizes on the statistics, while the inner layer exhibits universality irrespective of the box size L_z. The wake region, however, reveals a clear dependence on the box size: the wake strength reduces with increasing L_z and ultimately disappears. Additional simulations were performed to affirm L_z (not L_x) as the influencing factor. This phenomenon cannot be seen in channel flows or turbulent boundary layers, consequently it is essential to choose large boxes and study this case in detail.

A considerably long time is required to reach the asymptotic state: the large structures and the correct (i.e. asymptotic) BL thickness can only be obtained after a long BL development. As apparent from the inner-scaled mean velocity profile in Fig. 1b, the flow quickly adheres to a law of the wall description, while the asymptotic state is only slowly approached in the outer region.

The performance of different scaling laws were tested in order to describe U^+ in the presence of wall transpiration. It is found that the linear law [10],

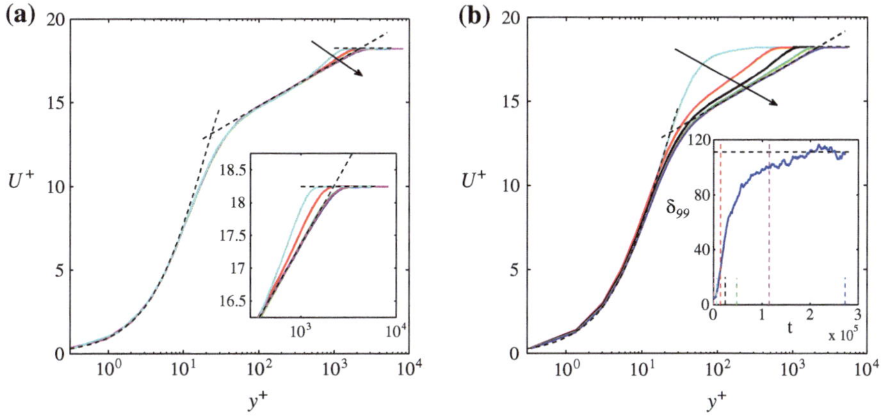

Fig. 1 **a** Viscous-scaled mean streamwise velocity U^+ for increasing box size L_z. **b** Viscous-scaled mean velocity profile for case LES333d at different development times as indicated through vertical *dashed lines* in the time development of the boundary layer thickness, δ_{99}. *Dashed lines* in **a** and **b** represent the modified linear and log law (with modified parameters) [9]

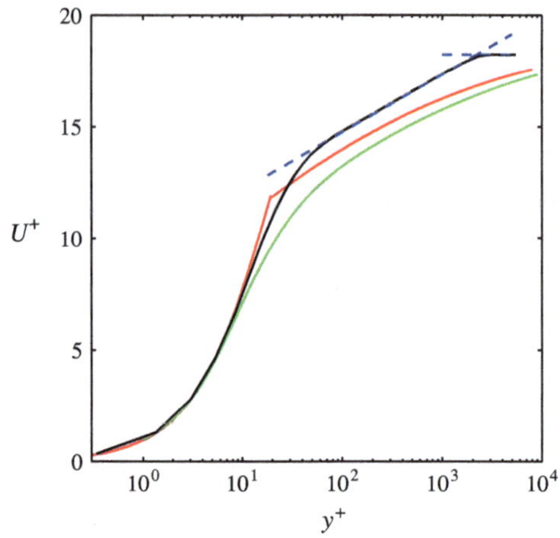

Fig. 2 Scaling laws for U^+. LES333d (*solid black*), empiric log law [9] with modified parameters $\kappa^* = 0.82$, $B^* = 9.2$ (*dashed blue*), modified log law [6] (*solid red*), application of the Prandtl mixing length incl. a damping function [7] (*solid green*)

$$U^+ = (1/V_0^+)(e^{V_0^+ y^+} - 1), \tag{1}$$

represent the mean velocity well in the viscous sublayer. Figure 2 shows that the empirical log law with modified parameters is able to describe the overlap region, while the proposed scaling laws in Refs. [6, 7] fail to describe the overlap and wake region.

4 Conclusions

The investigation of a TASBL at $Re = 333$ emphasised the importance of both a large box size and sufficient development time. With both an increasing box size and development time, the mean profiles exhibit a similar trend, viz. a diminishing wake region. However, profiles in the viscous sublayer and inner layer collapse irrespective of development time and box size, respectively. Once an asymptotic state is reached, large structures can be observed and the separation between small and large structures becomes apparent, yielding high friction Reynolds numbers. Moreover, observations from earlier studies (such as lower fluctuation level when suction is present [5]) could be confirmed. Finally, a more appropriate scaling law needs to be found in order to describe the mean profile of TASBLs. The present conclusions have important ramifications for experiments, viz. a true TASBL requires very large facilities (a rough estimate yields a wind tunnel length of 350 m) in order to reach an asymptotic state.

Disclaimer: Parallel to the present paper, a largely extended and more detailed study based on the present work has been published by Bobke et al. [2].

Acknowledgments The project is financed by the Linné. The simulations were performed on resources provided by the Swedish National Infrastructure for Computing (SNIC) at the High Performance Computing Center North (HPC2N) and National Supercomputer Centre in Sweden (NSC).

References

1. Antonia, R., Fulachier, L., Krishnamoorty, L.V., Benabid, T., Anselmet, F.: Influence of wall suction on the organized motion in a turbulent boundary layer. J. Fluid. Mech. **190**, 217–240 (1988)
2. Bobke, A., Örlü, R., Schlatter, P.: Simulations of turbulent asymptotic suction boundary layers. J. Turbul. **17**, 155–178 (2016)
3. Chevalier, M., Schlatter, P., Lundbladh, A., Henningson, D.S.: SIMSON. Tech. Rep, KTH Mechanics, Sweden (2007)
4. Kay, J.M.: Boundary-layer flow along a flat plate with uniform suction, HM Stationery Office (1958)
5. Mariani, P., Spalart, P., Kollmann, W.: Direct simulation of a turbulent boundary layer with suction, Near-Wall Turbulent Flows. In: So, R.M.C., Speziale, C.G., Launder, B.E (eds.), pp. 347–356 (1993)
6. Piomelli, U., Ferziger, J., Moin, P., Kim, J.: New approximate boundary conditions for large eddy simulations of wall-bounded flows. Phys. Fluids A **1**, 1061–1068 (1989)
7. Rotta, J.C.: Control of turbulent boundary layers by uniform injection and suction of fluid. Tech. Rep, DTIC Document (1970)
8. Schlatter, P., Örlü, R.: Turbulent asymptotic suction boundary layers studied by simulation. J. Phys.: Conf. Ser. **318**, 022020 (2011)
9. R.L. Simpson, Ph.D. thesis, Stanford University (1967)
10. Tennekes, H.: Similarity laws for turbulent boundary layers with suction or injection. J. Fluid. Mech. **21**, 689–703 (1965)

Heat Transfer in a Shallow Cavity

Fatima Madi Arous

Abstract This numerical study is about the heat transfer in a turbulent flow over a shallow cavity. Two different incoming flows are considered: a boundary layer flow and a plane wall jet flow. The latter is characterized by the presence of an outer layer, an additional turbulence source. In the boundary layer case, the Reynolds number and the turbulence intensity were investigated, while in the wall jet case, the ratio of the cavity depth to the jet nozzle height was examined. The results of this study reveal that the cavity flow is very sensitive to the incoming flow characteristics; an increase of the turbulence intensity accelerates the shear layer reattachment and enhances the heat transfer. The Reynolds number increase also improves the heat transfer. In addition, a heat transfer enhancement was observed in the wall jet case.

1 Introduction

Flows characterized by the separation and recirculation phenomena, as flow over cavities and steps or around obstacles, promote mixing fluid and can play an important role in heat transfer. This kind of flows is of a great interest as it is related to various engineering applications such as cooling systems for electronic components, combustion chambers, heat exchangers and solar energy collectors. Turbulators ribs are widely used in several engineering applications such as serpentine cooling air and gas turbine. However, the flow field between two roughness elements is similar to a cavity flow Oka [1]. In the present study, the cavities aspect ratio is AR = 10. The cavities walls temperature was fixed at Tw = 320 K and that of the inlet flow at 300 K. The case studied experimentally by Estve et al. [2] and numerically by Madi Arous et al. [3] has been used for the turbulence model validation.

F.M. Arous (✉)
LMFTA, Faculté de Physique, USTHB, Bab Ezzouar, Algeria
e-mail: madi.fatima@yahoo.fr; fmadi@usthb.dz

J. Peinke et al. (eds.), *Progress in Turbulence VI*,
Springer Proceedings in Physics 165, DOI 10.1007/978-3-319-29130-7_32

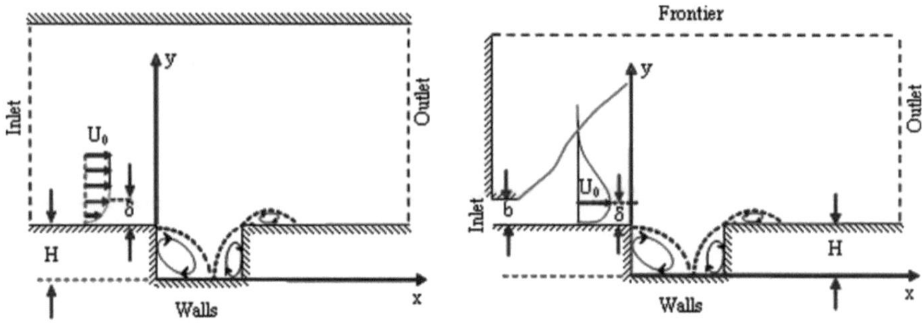

Fig. 1 Incoming flow calculation domaines (*left* boundary layer case—*right* wall jet case)

2 Turbulence modelling and Numerical Procedure

The fluid considered in the present study is incompressible and the flow is assumed two-dimensional and statistically steady. The numerical approach is realized by the standard k-ω model of Wilcox [4]. The Numerical method is based on the finite volume method. The calculation domains are given by Fig. 1. The grids are Cartesian, non-uniform and sufficiently refined near the cavity walls where very high gradients prevail in the viscous sub-layer.

The boundary conditions imposed at the calculation domain frontiers are: uniform profiles at the inlet frontier, constant pressure at the outlet boundary and the upper frontier in the wall jet case, and the no-slip condition at the wall boundaries. The turbulent quantities correspond to the wall function approach. The cavity walls are maintained at a constant temperature $T = 320\,\text{K}$ and the other ones are adiabatic.

3 Results and Discussions

For the validation of the numerical results, we have considered the same incoming flow conditions of [3]. the Reynolds number based on the cavity depth is 67000, the boundary layer thickness is about 20 mm and the turbulence intensity of the external flow is about 2 %. Figure 2a compares computed and measured pressure coefficient at the cavity floor. This figure shows that the numerical results are in good agreement with the experimental ones.

The flow structure is characterized by the presence of an important recirculation zone and two vortices at the cavity corners. We also note the presence of a small vortex above the downstream step (Fig. 2b). The present study shows that the flow structure is less affected by the Reynolds number variation while the turbulence intensity increase causes an important diminution of the principal recirculation zone size leading to the acceleration of the reattachment phenomenon.

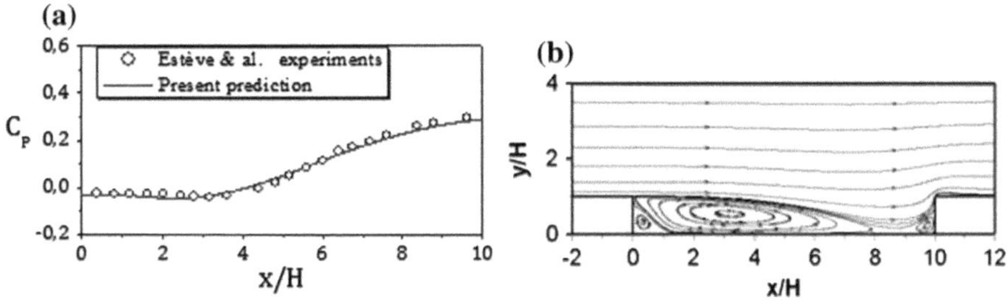

Fig. 2 **a** Pressure coefficient along the cavity floor (Re = 67000, I = 0.02). **b** The mean flow structure (boundary layer incoming flow case—Re = 67000, I = 0.02)

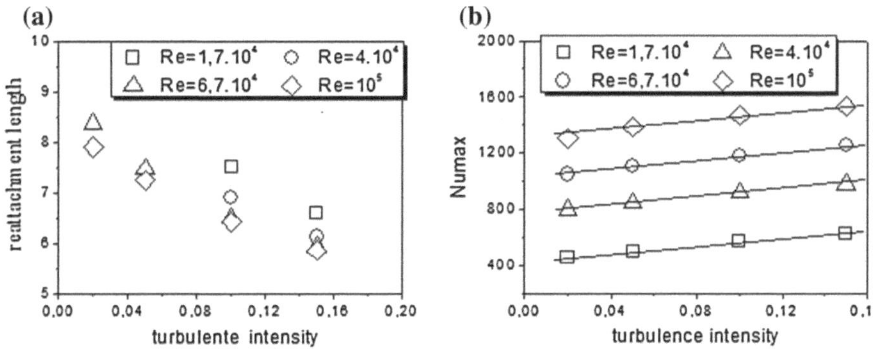

Fig. 3 **a** Evolution of the reattachment length (boundary layer case). **b** Evolution of the maximum Nusselt number (boundary layer case)

Figure 3a shows that the reattachment length decreases with the turbulence intensity increase and it is less effected by the variation of the Reynolds number.

Figure 3b shows that the Numax increases linearly with the turbulence intensity. The increase of the Reynolds number and that of the turbulence intensity improve the heat transfer.

Computations of the wall jet upstream flow cases have been carried out for the same inlet flow conditions of the boundary layer case. Thus, at x = −H, the Re = 67000, the boundary layer thickness is about 20 mm and the turbulence intensity is about 5 %. The height of the outlet jet nozzle is maintained constant b = 50 mm whereas the cavity depth was varied while keeping an aspect ratio AR = L/H = 10. Different cavity depths to nozzle height ratios are considered in this study. Figure 4a regroups the Nusselt number profiles along the cavity floor of the two different incoming flows. This Figure shows a considerable heat transfer enhancement in the wall jet case compared to that of the boundary layer. Figure 4b regroups the Nusselt number profiles along the cavity floor of different H/b ratios. We note that the augmentation of H/b ratio enhances the heat transfer. This same result was found by Abu-Mulaweh et al. [5] where their experimental study has shown that the step height increase enhances the heat transfer in the turbulent flow over a backward facing

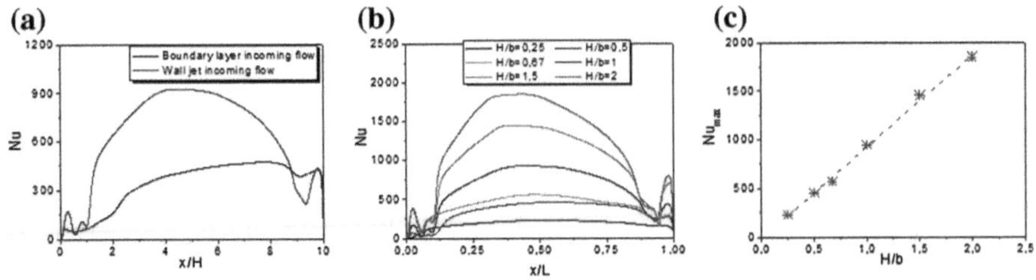

Fig. 4 **a** Evolution of the local Nusselt number along the cavity floor. **b** Evolution of the local Nusselt number along the cavity floor for different H/b ratios. **c** Evolution of the maximum Nusselt number for different H/b ratio

step. Apparently, the earlier reattachment phenomenon leads to an enhancement in a heat transfer. It is very interesting to note a linear increase of the maximum Nusselt number with the cavity depth Fig. 4c.

3.1 Concluding Remarks

The comparison of the numerical results with the experimental ones shows that the k-ω model predicts fairly well this flow pattern. It was found that the flow structure is very sensitive to the upstream flow characteristics. The turbulence intensity increase accelerates the shear layer reattachment to the cavity floor and improves the heat transfer. The reattachment phenomenon seems to be less affected by the Reynolds number. However, the increase of this latter enhances the heat transfer in the cavity. The wall jet incoming flow case is characterized by an earlier reattachment phenomenon and an enhancement of the heat transfer as compared to that of the boundary layer. The augmentation of the cavity depth to nozzle height ratio improves even more the heat transfer.The maximum Nusselt number is reached upstream of the reattachment locations, in the principal recirculation region. This study reveals a linear evolution of the maximum Nusselt number with the turbulence intensity. The interesting result of this study is a linear evolution of the maximum Nusselt number with the cavity depth to the height of the jet exit nozzle.

References

1. Oka, S.: Flow field between two roughness elements in developed turbulent channel flow. Heat and Mass Transfer in Flows with Separated Regions and Measurement Techniques. Pergamon Press, New York (1972)
2. Esteve, M.J., Reulet, P., Millan, P.: Flow field characterisation within a rectangular cavity. In: proceedings of 10th International Symposium on Applications of Laser Techniques to Fluid Mechanics, Portugal (2000)

3. Madi Arous, F., Mataoui, A., Terfous, A., Ghenaim, A.: Jet cavity interaction: effect of the cavity depth. Prog. comput. fluid dyn. **11**, 322–332 (2012)
4. Wilcox, D.C.: Turbulence Modelling for CFD. DCW Industries Inc, La Canada (1998)
5. Abu-Mulaweh, H.I., Chen, T.S., Armaly, B.F.: Turbulent mixed convection flow over a backward facing step- the effect of the step heights. Int. J. Heat Fluid Flow **23**, 758–765 (2002)

Temperature Effects in Hot-Wire Measurements on Higher-Order Moments in Wall Turbulence

Alessandro Talamelli, Fabio Malizia, Ramis Örlü, Andrea Cimarelli and Philipp Schlatter

1 Introduction and Approach

Measuring turbulent fluctuations using hot-wire anemometry is still the method that provides the highest degree of accuracy, in particular when it comes to temporal and spatial resolution. The question of spatial resolution of hot-wires has been treated in the past in a number of publications [1], and corresponding correction schemes, both heuristic [2] and data-driven [3], have been suggested. Nonetheless, there are still unresolved open questions in wall turbulence, such as the proper scaling of the near-wall and secondary peak (or even its existence) in the variance profile of the streamwise velocity component or the scaling behaviour of its higher-order moments [4]. While the effect of temporal and spatial resolution on these quantities has been investigated in a number of studies it was only very recently that the effect of temperature fluctuations—as they are e.g. encountered in non-isothermal flows—on the mean, variance and spectra in wall-bounded turbulent flows when utilising hot-wire anemometry has been documented and corrections were proposed [5].

The aim of the present investigation is to extend the aforementioned efforts on the effect of temperature fluctuations on the skewness and flatness factors as well as to assess the performance of commonly employed temperature compensations.

A. Talamelli (✉) · F. Malizia · A. Cimarelli
Department of Industrial Engineering, Alma Mater Studiorum,
Università di Bologna, Forlì, Italy
e-mail: alessandro.talamelli@unibo.it

F. Malizia
e-mail: fabio.malizia@studio.unibo.it

A. Cimarelli
e-mail: andrea.cimarelli2@unibo.it

A. Talamelli · R. Örlü · P. Schlatter
Linné FLOW Centre, KTH Mechanics, 100 44 Stockholm, Sweden

R. Örlü
e-mail: ramis@mech.kth.se

P. Schlatter
e-mail: pschlatt@mech.kth.se

J. Peinke et al. (eds.), *Progress in Turbulence VI*,
Springer Proceedings in Physics 165, DOI 10.1007/978-3-319-29130-7_33

For this purpose, the response of a hot-wire is simulated through the use of velocity and temperature (i.e. passive scalar) data from a turbulent channel flow generated by means of direct numerical simulations (DNS) at a friction Reynolds number of $Re_\tau = 590$ as done in Örlü et al. [5] to which the reader is referred for details. In particular, the effect of a 2, 4, and 6 K temperature gradient across the channel height will be simulated and the performance of temperature compensations of the instantaneous velocity signals will be demonstrated utilising common techniques, i.e. the mean centreline (or external) temperature $\langle T_{CL} \rangle$, the mean local temperature $\langle T(y^+) \rangle$, and the newly proposed synthetic instantaneous temperature by Örlü et al. [5], henceforth denoted $T_{\text{synt}}(t; y^+)$ given through

$$T_{\text{synt}}(t; y^+) = \langle T \rangle (y^+) + \frac{u(t) T_\tau}{u_\tau} \left(\frac{\langle uu \rangle}{\langle uu \rangle_{max}} \right), \tag{1}$$

where u_τ and T_τ is the friction velocity and temperature, while $\langle \cdot \rangle$ and $^+$ denote the time-average operator and wall units, respectively. In all these exercises u_τ is taken from the simulation and is therefore unaffected by the temperature gradient. This procedure is comparable to an experimental where u_τ is obtained independently from velocity measurements, as e.g. from the pressure drop in internal flows.

2 Results and Discussions

In situations where $\langle T_{CL} \rangle$ is monitored throughout an experiment, it is common to employ this temperature for the compensation of the instantaneous hot-wire (voltage) readings. The effect on the mean velocity profile for temperature gradients of 2, 4, and 6 K is depicted in Fig. 1 together with their relative errors. Although not directly apparent from the mean velocity profile, the relative errors are considerable. Utilising

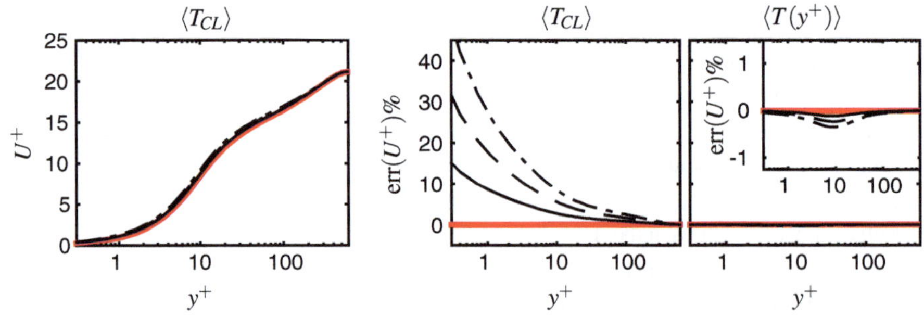

Fig. 1 Inner-scaled mean streamwise velocity U^+ profile from DNS (—) and sensed by virtual hot-wire probe for $\Delta T = 2$ (—), 4 (– –), and 6 K (– · –), which is temperature-compensated by means of $\langle T_{CL} \rangle$, as well as the relative errors when utilising $\langle T_{CL} \rangle$ and $\langle T(y^+) \rangle$

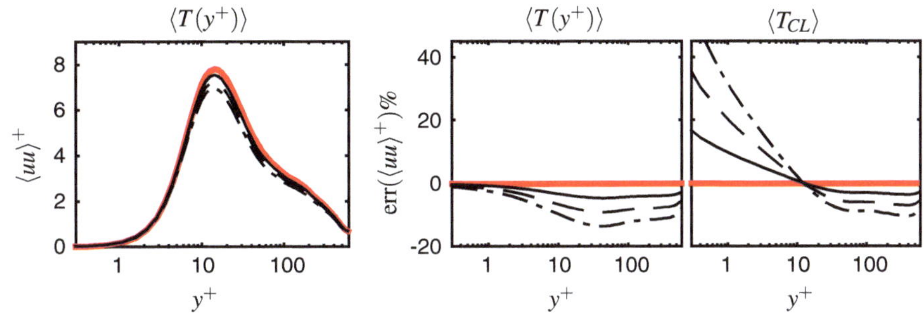

Fig. 2 Inner-scaled variance profile of the streamwise velocity component $\langle uu\rangle^+$ obtained when utilising $\langle T(y^+)\rangle$ for the compensation as well as the relative errors when using $\langle T(y^+)\rangle$ and $\langle T_{CL}\rangle$

instead $\langle T(y^+)\rangle$, as is common practise (*if* one is aware of possible temperature gradients), one is able to recover the correct mean velocity profile.

Based on this result, one would assume that $\langle T(y^+)\rangle$ is superior to $\langle T_{CL}\rangle$ when it comes to the variance profile. Although this is the case as shown in Fig. 2, the obtained variance profile is significantly attenuated around the near-wall peak when utilising $\langle T(y^+)\rangle$, while it is nearly fully recovered when $\langle T_{CL}\rangle$ is used, which makes it interesting for practical purposes, since it is the near-wall peak that is effected the most in absolute terms. Consequently, there is no clear favourite between the two temperatures when it comes to corrections beyond the mean velocity and hence the correction scheme suggested in Ref. [5], i.e. Eq. (1), is compared to the previous ones in the following for the higher-order moments.

Contrary to the previous figures, where relative errors were presented, Fig. 3 documents the absolute errors for higher-order moments, since relative errors become problematic for odd moments due to zero-crossings and values close to zero. As apparent, the utilisation of $\langle T_{CL}\rangle$ is not recommended although it might yield (only) the correct amplitude of the variance near-wall peak; errors (obviously) amplify in the limit of $y^+ \to 0$. This becomes in particular important when assessing the limiting behaviour of the variance, i.e. amplitude of the wall shear stress fluctuations, and skewness or flatness factors. While both $\langle T(y^+)\rangle$ and $T_{\text{synt}}(t; y^+)$ yield the correct mean velocity profile, the advantage of utilising $T_{\text{synt}}(t; y^+)$ becomes apparent when inspecting the absolute errors for the higher-order moments. Consequently, in the absence of simultaneous, spatially and temporally resolved, temperature and (hot-wire) velocity measurements, the correction scheme proposed in Örlü et al. [5] is recommended whenever temperature gradients are unavoidable, be it for the purpose of correcting or assessing errors due to temperature fluctuations.

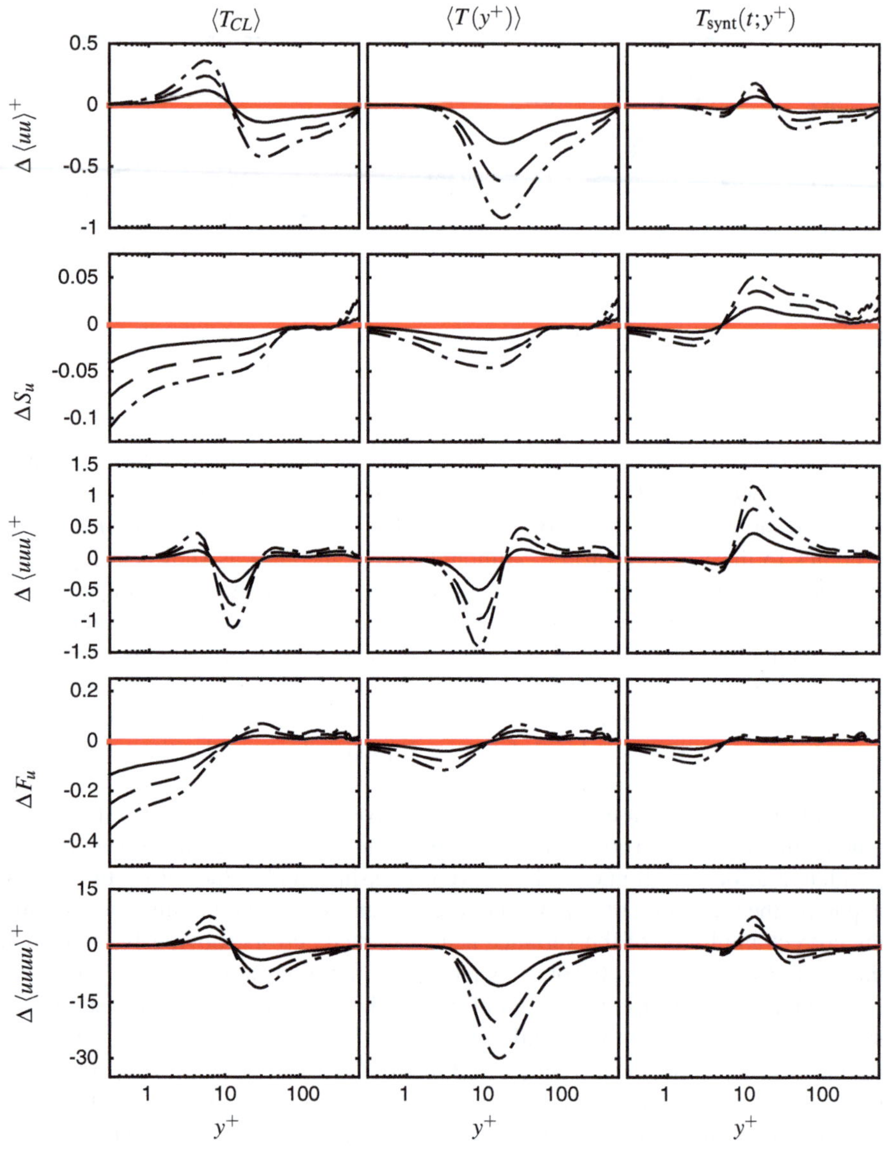

Fig. 3 Absolute errors for the variance, skewness factor, third-order moment, flatness factor and fourth-order moment of the streamwise velocity fluctuations, temperature compensated with $\langle T_{CL} \rangle$, $\langle T(y^+) \rangle$, as well as the method proposed in Örlü et al. [5], i.e. $T_{synt}(t; y^+)$. To appreciate the relative importance of the absolute errors compared to the actual amplitudes see e.g. Ref. [4]

Acknowledgments This research project has been partly supported by the European Community - Research Infrastructure Action under the FP7-INFRASTRUCTURES-2012-1: "European High-performance Infrastructures in Turbulence" INFRA-2012-1.1.20. Computer time was provided by the Swedish National Infrastructure for Computing (SNIC).

References

1. Hutchins, N., Nickels, T.B., Marusic, I., Chong, M.S.: Hot-wire spatial resolution issues in wall-bounded turbulence. J. Fluid Mech. **635**, 103–136 (2009)
2. Smits, A.J., Monty, J.P., Hultmark, M., Bailey, S.C.C., Hutchins, N., Marusic, I.: Spatial resolution correction for wall-bounded turbulence measurements. J. Fluid Mech. **676**, 41–53 (2011)
3. Segalini, A., Örlü, R., Schlatter, P., Alfredsson, P.H., Rüedi, J.-D., Talamelli, A.: A method to estimate turbulence intensity and transverse Taylor microscale in turbulent flows from spatially averaged hot-wire data. Exp. Fluids **51**, 693–700 (2011)
4. Talamelli, A., Segalini, A., Örlü, R., Schlatter, P., Alfredsson, P.H.: Correcting hot-wire spatial resolution effects in third- and fourth-order velocity moments in wall-bounded turbulence. Exp. Fluids **54**, 1496 (2013)
5. Örlü, R., Malizia, F., Cimarelli, A., Schlatter, P., Talamelli, A.: The influence of temperature fluctuations on hot-wire measurements in wall-bounded turbulence. Exp. Fluids **55**, 1781 (2014)

Quantification of Global Intermittency in Stably Stratified Ekman Flow

Cedrick Ansorge and Juan Pedro Mellado

1 Introduction

Cessation of turbulence is a well-recognized feature of the stably stratified boundary layer [1, 2]. In stratified channels, it has been shown that this cessation does not occur as an on–off process but is rather a complex transition from a turbulent to a laminar state [3, 4]. When stratification increases gradually, this transition begins with the localized absence of turbulent eddies in an otherwise turbulent flow. Recently, [5] showed that a very similar transition occurs in stably stratified *Ekman* flow. This localized absence of turbulence bears a striking resemblance to the absence of turbulence on some or all scales, even close to the surface, which is observed in the atmosphere and termed *Global Intermittency* (GI) by [6].

To the authors' knowledge, the recognition of the occurrence of GI is not reflected in a separate study of the turbulent and laminar patches in such a rotating and stratified flow yet. The challenge is twofold: First, sufficiently resolved data in space and time became available just recently ([5] calculated the intermittency factor from DNS data); Second, while the enstrophy-based classical method developed by [7] is well-suited to detect external intermittency in neutrally stratified flows, there are problems with GI because of large gradients in the mean-flow close to the surface [5]. To overcome this issue, we propose here a method combining the intermittency factor with a high-pass filter operation.

We use direct numerical simulation of stably stratified *Ekman* flow. The set-up used here is—along physical aspects of the turbulence collapse in the flow—described in [5]. External parameters of this particular set-up are the free-stream geostrophic wind velocity G, the fluid viscosity ν, the Coriolis parameter f and the buoyancy difference B_0 between the surface and free stream. These parameters combine into two dimensionless groups, a *Reynolds* and *Richardson* number, $Re_\delta = G\delta/\nu$ and

C. Ansorge (✉) · J.P. Mellado
Max-Planck-Institute for Meteorology, 20146 Hamburg, Germany
e-mail: cedrick@posteo.de

J. Peinke et al. (eds.), *Progress in Turbulence VI*,
Springer Proceedings in Physics 165, DOI 10.1007/978-3-319-29130-7_34

$Ri_\delta = B_0\delta/G^2$ respectively. Here, $\delta = u_{\star,\text{neutral}}/f$ is a measure of the boundary layer height in neutral conditions with the corresponding friction velocity $u_{\star,\text{neutral}}$ of the neutrally stratified case.

2 Intermittency Calculation Based on Filtered Fields

When partitioning a flow to turbulent and non-turbulent patches, vorticity has proven useful as it is one defining property of turbulence: Since the seminal works by [7], vorticity is used systematically in the study of external intermittency and the turbulent/non-turbulent interface ([9] and references therein). We also base our partitioning method on vorticity, and use the intermittency function $\gamma(z) = \langle H(\omega - \omega_{\text{threshold}})\rangle$, where H is the Heaviside function, ω is the local vorticity magnitude, and $\langle\cdot\rangle$ denotes a horizontal average. As a threshold we use $6\omega_{\text{rms}}(\delta)$, where $\omega_{\text{rms}}(\delta)$ is the root mean square (rms) of the vorticity at $z = \delta$. Given the self-similarity of vorticity profiles within the boundary layer when the Reynolds number increases, this threshold may also be expressed in terms of a constant fraction of the vorticity inside the buffer layer or at the surface. Ansorge and Mellado [5] showed that the expression of the enstrophy threshold in terms of ω_{rms} leads to intermittency profiles which are independent of the Reynolds number. As demonstrated by a comprehensive body of work following [7], this approach is well-suited to detect external intermittency [8]. The detection of GI based on this method is, however, difficult: The contribution of the mean velocity gradient dominates the turbulent contribution to the enstrophy close to the surface [5].

To overcome the problem of partitioning the flow close to the surface, we consider a horizontally filtered velocity field retaining only small-scale fluctuations. The filter transfer function is

$$F_\delta(k_x) = \frac{1}{2}\left\{\text{erf}\left[\log\left(\frac{|k_x|}{k_\delta}\right)\right] + 1\right\}, \tag{1}$$

where k_x is a wavenumber in the horizontal direction and k_δ is the wavenumber corresponding to a wavelength δ. In this work, we consider two filters: First, $\mathscr{F}_\delta^+$ defined through the filter response $F_\delta(k_x)$. The residual field is equal to the original field filtered with the low-pass filter $\mathscr{F}_\delta^-$ defined through the complementary filter response $1 - F_\delta(k_x)$. That is, we first consider the fields $\mathbf{u}_{\texttt{hi}} = \mathscr{F}_\delta^+\{\mathbf{u}\}$ and $\mathbf{u}_{\texttt{lo}} = \mathscr{F}_\delta^-\{\mathbf{u}\} = \mathbf{u} - \mathbf{u}_{\texttt{hi}}$. The second filter we consider corresponds to the *Reynolds* decomposition.

We analyze here two particular realizations of turbulent Ekman flow obtained from the set-up described in [5] with $Re_\delta = 26{,}450$. First, a neutrally stratified case in a domain approximately the size of the *Rossby* radius, labelled as `ri00`. Second, two stably stratified case at *Richardson* numbers $Ri_\delta = 0.62$, $Ri_\delta = 0.31$ and $Ri_\delta = 0.15$ in the same domain are used and labelled as `ri62`, `ri31` and `ri15` respectively. While the neutral case is in its statistically steady state, the stratified ones are in the phase of recovery after a complete, respectively partial cessation of turbulence. (The cases `ri15` and `ri31` use a neutrally stratified turbulent field with a buoyancy profile as described in [5] as reference. The case `ri62` is spun off the case `ri15` after $tf/2\pi \approx 1/6$ and integrated forward through about 1.5 inertial periods.)

A comparison of the statistics between filtered and unfiltered fields unveils that in the neutrally stratified flow, and also in case `ri31` where the turbulence is sustained in a large are fraction (not shown), the rms of turbulent vorticity ω_{rms} is dominated by contributions from the high-pass-filtered field $\mathbf{u}_{\mathtt{hi}}$ at all heights (thin and thick solid lines in Fig. 1a). Conversely, the vorticity rms residing in the large-scale modes $\mathbf{u}_{\mathtt{lo}}$ (dashed lines in Fig. 1) is less than one third of that contained in the small-scale modes. This result qualifies vorticity of the unfiltered field as a reasonable indicator for the intensity of small-scale processes under neutral stratification.

When the flow is exposed to very strong stable stratification, the above picture changes. Now, in particular close to the surface ($z^- < 0.2$), the vorticity rms is largely explained by the low-pass-filter contribution (orange, solid line in Fig. 1) due to a high level of large-scale background enstrophy. A large portion of this low-pass-filter enstrophy is due to mean shear increasing under stable stratification. Large-scale coherent motions seem to contribute to the high level of vorticity rms at large scales too. Conversely, the small-scale contribution to vorticity rms in the surface layer is

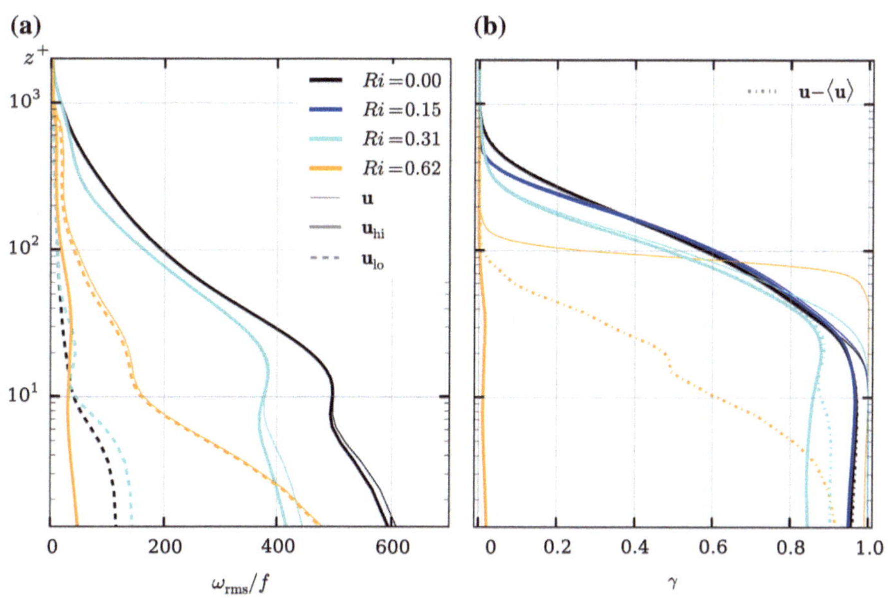

Fig. 1 **a** Vorticity-rms of the unfiltered (*solid*, *thin*), high-pass-filtered (*solid*, *thick*) and low-pass-filtered (*dashed*, *thick*) fields for case `ri00` (*black*), `ri31` (*cyan*) and `ri62` (*orange*). **b** Intermittency factor γ with a threshold vorticity $6\omega_{rms}$ (δ) calculated from a filtered (*solid*, *thick*—$\mathscr{F}_\delta^+$, *dash-dotted*—*Reynolds'* decomposition) and *solid*, *thin*—unfiltered field. Color coding is as in panel **a**; case `ri15` is shown in *blue*

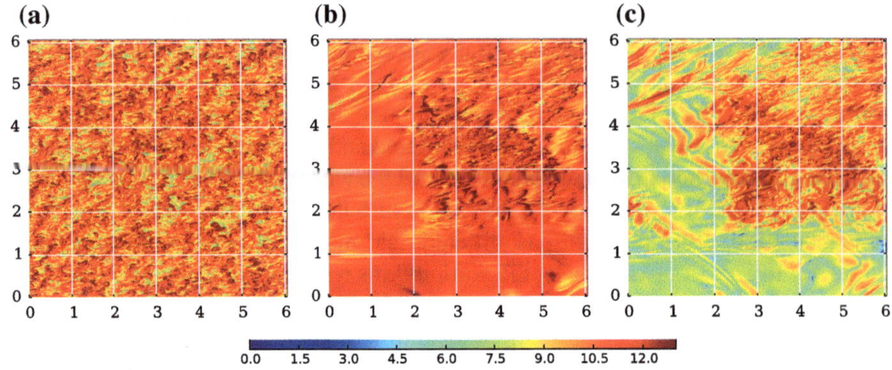

Fig. 2 Logarithm of enstrophy at a level $z^- = 0.04$ normalized by $1/f$ for case `ri00` (**a**) and case `ri62` (**b**, **c**). **a** and **b** show the enstrophy of the field u, **c** plots the enstrophy of the field $u_{\mathtt{hi}}$. Axes are normalized by δ and about $\frac{1}{3} \times \frac{1}{3}$ of the total domain is shown

much smaller and vorticity of the unfiltered field **u** is no longer a good indicator of small-scale activity in the flow.

The significant increase in background large-scale shear under stable stratification is the reason why most of the flow is turbulent with respect to the flow-partitioning method based on unfiltered fields (thin, solid orange line in Fig. 1b). The classical method of measuring external intermittency not only fails to detect the localized absence of turbulence close to the surface that is evident in Fig. 2b, c, but also gives a higher turbulent area fraction close to the surface (up to $z^- = 0.1$ in Fig. 1b) when compared to the neutral reference. When, instead of the full field, high-pass-filtered fields are used to partition the flow, the absence of turbulent motion in the lower surface layer and the buffer layer is well detected (red versus black solid lines in Fig. 1b). This is also seen comparing Fig. 2b, c: While in panel (b) the enstrophy reaches values around $10/f$ throughout the domain, even in regions where there seems to be no mixing activity, the filtering of large-scale motions removes these contributions, and the enstrophy is significantly reduced in regions with less mixing. At the same time, for the neutrally stratified case, the consideration of filtered fields has no impact on $\gamma(z)$.

3 Conclusions

In the inner region of boundary layers, a partitioning of the flow into turbulent and non-turbulent patches based on full instantaneous velocity fields is problematic since the enstrophy may be dominated by the gradient of the mean flow or large-scale structures. Our results indicate that, in neutrally stratified boundary layers, this approach is efficient, while it is not so under strong stable stratification. This problem can be overcome if the large-scale contribution to the enstrophy is removed. We have introduced here a possible method—based on filtered fields of the full flow—to differentiate between turbulent and non-turbulent patches of strongly stably stratified Ekman flow. This approach is sufficient to distinguish turbulent and non-turbulent patches in the inner layer of stably as well as neutrally stratified Ekman flow.

Acknowledgments Support from the Max Planck Society through its Max Planck Research Groups program is gratefully acknowledged. Computational time was provided by the Jülich Supercomputer Centre under that grant `HHH07`.

References

1. Mahrt, L.: Stably stratified atmospheric boundary layers. Annu. Rev. Fluid Mech. **46**, 23–45 (2014)
2. van de Wiel, B.J.H., Moene, A.F., Jonker, H.J.J., Baas, P., Basu, S., Donda, J.M.M., Sun, J., Holtslag, A.A.M.: The minimum wind speed for sustainable turbulence in the nocturnal boundary layer. J. Atmos. Sci. **69**(11), 3116–3127 (2012)
3. Flores, O., Riley, J.J.: Analysis of turbulence collapse in the stably stratified surface layer using direct numerical simulation. Bound.-Layer Meteorol **139**, 231–249 (2011)
4. Brethouwer, G., Duguet, Y., Schlatter, P.: Turbulent-laminar coexistence in wall flows with Coriolis, buoyancy and Lorentz forces. J. Fluid Mech. **704**, 137–172 (2012)
5. Ansorge, C., Mellado, J.P.: Turbulence collapse and global intermittency in the atmospheric boundary layer. Bound.-Layer Meteorol **153**, 89–112 (2014)
6. Mahrt, L.: Stratified atmospheric boundary layers. Bound.-Layer Meteorol **90**(3), 375–396 (1999)
7. Corrsin, S., Kistler, A.L.: Free-stream boundaries of turbulent flows. Technical Report TR1244-3133, John Hopkins University, Washington DC (1955)
8. Pope, S.B.: Turbulent Flows, 771 p. Cambridge University Press, New York (2000)
9. da Silva, C.B., Hunt, J.C.R., Eames, I., Westerweel, J.: Interfacial layers between regions of different turbulence intensity. Annu. Rev. Fluid Mech. **46**, 567–590 (2014)

A Comparison Between DBD and Corona Actuators with Non-Straight Electrodes

Federico Messanelli and Marco Belan

Abstract An experiment about plasma actuators, both corona and DBD (Dielectric Barrier Discharge) types, is performed, in order to maximize their performance as flow control devices on aerodynamic airfoils at high angles of attack by acting on the electrodes shape. In particular, an innovative serrated geometry for the electrodes is investigated. Lift measurements on a NACA0015 are taken in the wind tunnel at 3 different Reynolds numbers ($Re \approx 35{,}000, 50{,}000, 70{,}000$). Results indicate that the new geometry may induce better performance than traditional straight edge configurations, both for corona and DBD. Furthermore, power considerations suggest that serrated geometry corona actuators are more efficient than DBD. Finally, the serrated geometry seems to increase the stability of corona discharge.

1 Introduction

The main objective of our experimental research is to perform a geometrical optimization of plasma actuators [1], both corona and DBD (dielectric barrier discharge) types, in order to maximize their performance as flow control devices on aerodynamic airfoils at high angles of attack by acting on the electrodes shape. For this purpose, an innovative serrated geometry for the electrodes is studied (Fig. 1). A detailed analysis of the effectiveness of this geometrical configuration for corona plasma actuators, and in particular the optimization of the serrated geometry, has never been performed before this work. On the other hand, the serrated geometry has already been investigated for DBD in some previous works, as those of Thomas et al. [2], Berendt et al. [3] and Joussot et al. [4], even though a proper optimization of this geometry has never been accomplished. In particular [3] and [4] deal with the description of the three-dimensional flow field induced by these innovative electrodes.

F. Messanelli (✉) · M. Belan
Politecnico di Milano, Dipartimento di Scienze e Tecnologie Aerospaziali,
Via La Masa 34, Milan, Italy
e-mail: federico.messanelli@polimi.it

J. Peinke et al. (eds.), *Progress in Turbulence VI*,
Springer Proceedings in Physics 165, DOI 10.1007/978-3-319-29130-7_35

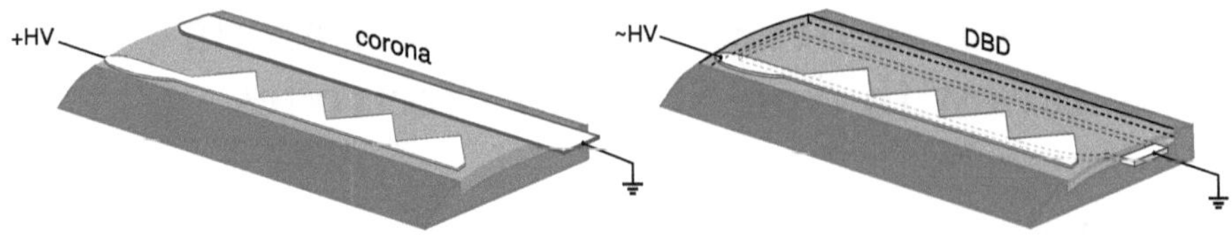

Fig. 1 Schematic view of the actuators with serrated geometry (not to scale for clarity)

2 Experimental Set-Up

The experiments take place in an open circuit wind tunnel with a $300 \times 300\,\text{mm}^2$ test section. Lift measurements are taken on a NACA0015 ($c = 100\,\text{mm}$, $b = 292\,\text{mm}$) by a 3 components aerodynamic balance at 3 Reynolds numbers ($Re \approx 35{,}000, 50{,}000, 70{,}000$), corresponding to freestream velocities of $5, 7.5$ and $10\,\text{m/s}$. End plates are mounted on the airfoil in order to minimize end effects. Turbulators are also located near the leading edge, in order to prevent the formation of laminar separation bubbles. The effect of the turbulators has been checked confronting the $C_L - \alpha$ curves with and without them and comparing these curves with the literature about model airplanes with artificially turbulent boundary layers [5]. Electrical measurements are performed by means of a 60 MHz–25 ks oscilloscope in order to evaluate the power consumption of the actuators. Different parametric tests have been performed, changing the tips length while keeping a constant number of tips and changing the tips number while keeping a constant tips length (Fig. 2). For the sake of comparison, two reference actuators have been used in the tests: a straight wire-to-plate for the corona and a straight plate-to-plate for the DBD. All the actuators are 200 mm wide. The electrodes are made of aluminium tape, 0.12 mm thick. The dielectric is Kapton, 0.075 mm thick. The anode–cathode gap in corona actuators is 15 mm; the gap between the upper and the lower electrodes for the DBD is 0 mm. The lower electrode is insulated by a silicone spray, in order to prevent the discharge on the lower side of the actuator. All the actuators are located at 10 % of the chord. The actuators effectiveness is evaluated by means of the percentage increase

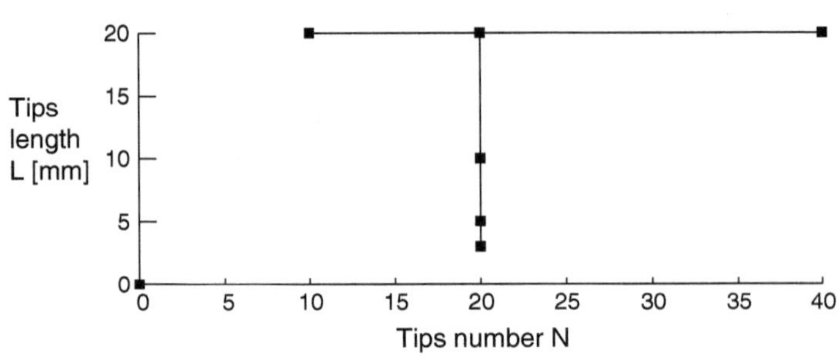

Fig. 2 Space of the investigated parameters

of maximum lift coefficient $\Delta C_{L_{max}}$ and by a non-dimensional efficiency parameter $\varepsilon = \frac{U \cdot \Delta L}{W}$, which takes into account their power consumption W.

3 Experimental Results

The new corona actuators show higher discharge stability: the serrated geometry helps avoiding glow-to-arc transition and favours the ignition of the discharge, even in unfavourable conditions of high relative humidity and non optimal actuators cleanliness. Furthermore, the results indicate a performance improvement increasing the length (Fig. 3) as well as the number (Fig. 5) of tips. The behaviour is asymptotic beyond a certain length at the lowest Reynolds. Probably the same asymptotic behaviour could be registered at higher Reynolds, but beyond the lengths tested. This will be investigated in further works. We believe that beyond a certain tips length the electric field is no longer affected by the boundary conditions imposed by the geometry upstream of the tips. About the tips number, we remark that corona discharge originates locally from the anode tips and expands like a fan towards the cathode. Thus, increasing the tips number gives an improvement, because there are more starting points for the corona and finally a wider discharge area. For DBD, the serrated geometry introduces also an induced transversal velocity, giving rise to a pair of counter-rotating vortices in each zone between adjacent tips. The actuator starts to work as a vortex generator only beyond a certain tips length (Fig. 4). On the other hand, in Fig. 5 an optimal geometry within those tested as function of the tips number has been identified: in this case, it is likely that the tip dimensions create vortices of the best size with respect to the boundary layer thickness.

If we compare the most performing corona and DBD within those tested, we may notice that, while lift improvements are substantially the same (Fig. 6), the efficiency expressed by the non-dimensional parameter ε is much better for the corona than for the DBD (Fig. 7). In fact, power consumption is in the order of 3 W for the corona and 10 W for the DBD.

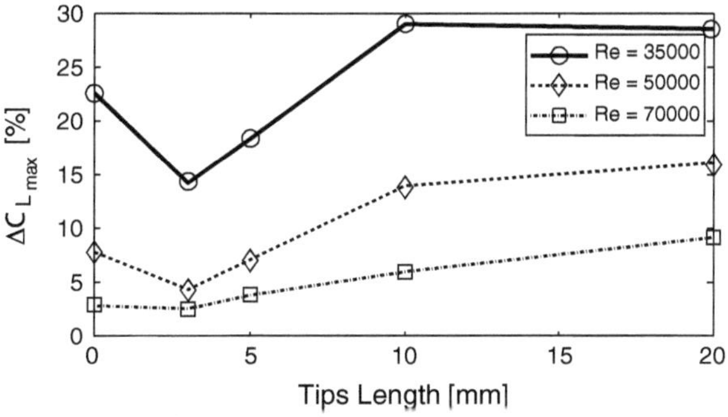

Fig. 3 $\Delta C_{L_{max}}$ versus tips length L at different Reynolds number for corona actuators

Fig. 4 $\Delta C_{L_{max}}$ versus tips length L for corona and DBD actuators at $Re \approx 35{,}000$

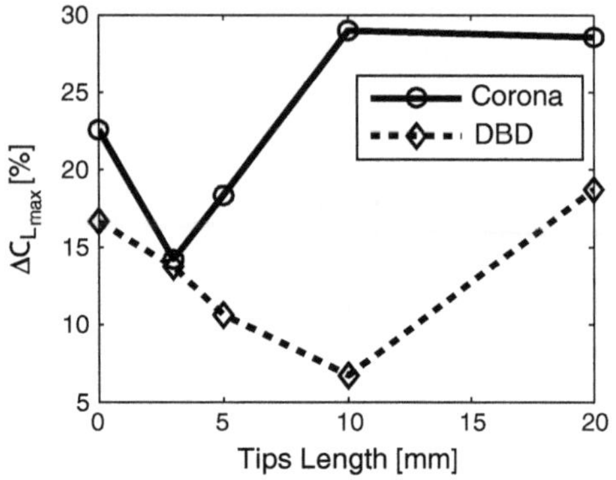

Fig. 5 $\Delta C_{L_{max}}$ versus tips number N for corona and DBD actuators at $Re \approx 35{,}000$

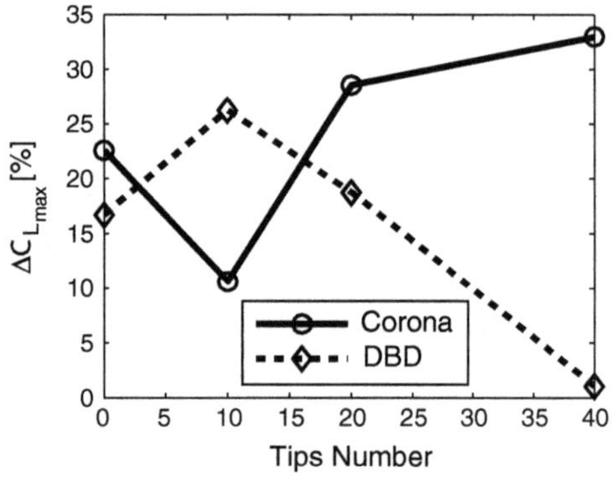

Fig. 6 $\Delta C_{L_{max}}$ versus Reynolds for the most performing actuators having the same span and tips length (tips length = 20 mm, tips number is 10 for DBD and 40 for corona, the tips width decreases as 1/N)

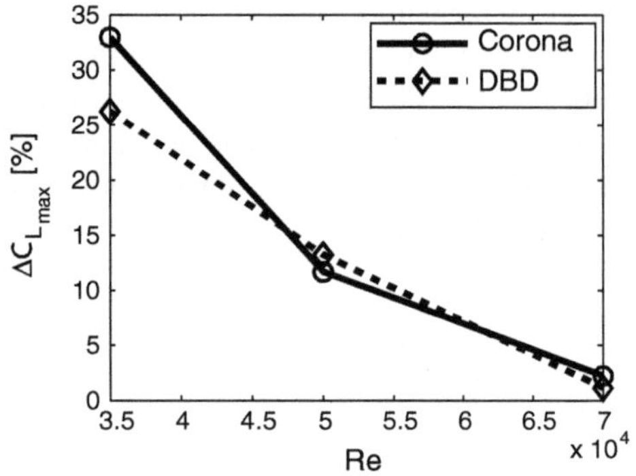

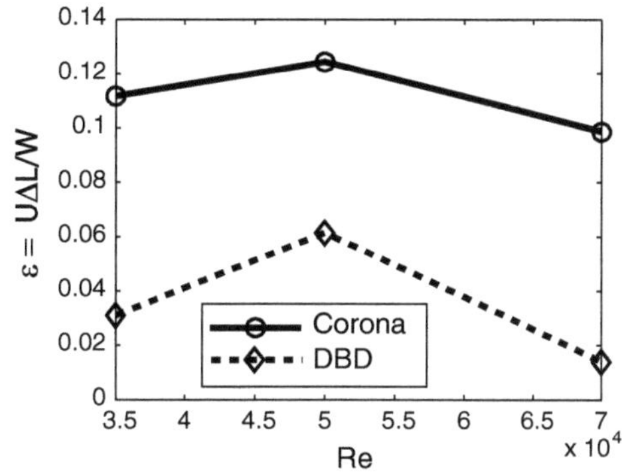

Fig. 7 Parameter ε versus Reynolds for the most performing actuators having the same span and tips length (tips length = 20 mm, tips number is 10 for DBD and 40 for corona, the tips width decreases as 1/N)

4 Conclusions

In conclusion, an optimized serrated geometry may induce better performance than traditional configurations, both for corona and DBD actuators. In particular, the main advantage for corona actuators could be the increased stability and ease of ignition, with respect to the standard straight geometry. Furthermore, corona actuators require less power than DBD to obtain similar lift improvements. Future research will investigate more geometries, aiming to prove the effectiveness of serrated geometry at higher freestream velocities.

References

1. Moreau, E.: Airflow control by non-thermal plasma actuators. J. Phys. D: Appl. Phys. **40**(3), 605–636 (2007)
2. Thomas, F.O., Corke, T.C., Iqbal, M., Kozlov, A., Schatzman, D.: Optimization of dielectric barrier discharge plasma actuators for active aerodynamic flow control. AIAA J. **47**(9), 2169–2178 (2009)
3. Berendt, A., Podliski, J., Mizeraczyk, J.: Comparison of airflow patterns produced by DBD actuators with smooth or saw-like discharge electrode. J. Phys.: Conf. Ser. **301**(1), 012018 (2011)
4. Joussot, R., Leroy, A., Weber, R., Rabat, H., Loyer, S., Hong, D.: Plasma morphology and induced airflow characterization of a DBD actuator with serrated electrode. J. Phys. D: Appl. Phys. **46**(12), 12 (2013)
5. Althaus, D.: Profilpolaren für den Modellflug. Neckar-Verlag, Villingen-Schwenningen (1980)

Part III
Pipe Flows

The Final Design of the Long Pipe in CICLOPE

G. Bellani and A. Talamelli

Abstract One of the fundamental characteristics of turbulent flows is that the ratio of large to small scales increases drastically with increasing Reynolds number. In standard laboratory apparatus, the fine structure of turbulence becomes exceedingly small, as compared to the size of standard probes, leading to large experimental uncertainties. Therefore an accurate description of the mutual interactions becomes complicate and many questions remains unanswered. The Center for International Cooperation for Long Pipe Experiment (CICLOPE) was established to design a pipe-flow facility (Long Pipe) that, thanks to its large size, has the potential to eliminate these uncertainties. Here we present the final design of the Long Pipe, which is now complete and operative.

1 Background

One of the fundamental characteristics of turbulent flows is that the ratio of large to small scales increases drastically with increasing Reynolds number. Furthermore, the interactions between large and small scales are strong and complex [1, 2]. Thus, in order to design efficient predictive models, it is important to provide an accurate qualitative and quantitative description of these complex interactions, especially in the near-wall region, where shear is generated. At present, this description is incomplete because the fine structure of turbulence becomes exceedingly small in standard laboratory flow-loops, as compared to the size of standard probes, leading to large

G. Bellani
CIRI Aeronautica - Univeristà di Bologna, Forlì, Italy

A. Talamelli (✉)
Department of Industrial Engineering, Alma Mater Studiorum,
Università di Bologna, 47100 Forlì, Italy
e-mail: alessandro.talamelli@unibo.it

J. Peinke et al. (eds.), *Progress in Turbulence VI*,
Springer Proceedings in Physics 165, DOI 10.1007/978-3-319-29130-7_36

experimental uncertainties [3]. Consequently, many questions are still unanswered [4, 5].

The Center for International Cooperation for Long Pipe Experiment (CICLOPE) was established to design a pipe-flow facility that eliminates these uncertainties [6]. The unique characteristic of this pipe-flow loop (the Long Pipe) is its large size ($D = 0.9$ m, $L = 110.8$ m). This size is such that the smallest turbulent eddies (ℓ_*) that are generated when the facility operates at high-speed (i.e. high-Reynolds number), are larger than the spatial-resolution capabilities of standard hot-wire probes. Hence statistics of the turbulent velocity field can be measured with unprecedented accuracy in a wide range of (high) Reynolds numbers. The design guidelines and potential impact of this new infrastructure has been discussed in several research papers [6–8]. In this paper we present the final design of the Long Pipe, which is now complete and operative. The Long Pipe is located in Predappio (Italy), and belongs to the University of Bologna. Funding opportunity to access the facility is provided by the European network EuHIT (www.euhit.org).

2 The Facility

The facility is a 110.8 m long pipe embedded in a closed-loop circuit. The design of the aerodynamic components is inspired by some of the best existing research wind-tunnels, such as the MTL and BL at KTH, Stockholm [9, 10] and the NDF at IIT, Chicago [11]. A closed loop design is chosen in order to ensure stability of the flow conditions, and low turbulence level. The loop includes a heat exchanger to control temperature within the range of ± 0.1 °C, an axial fan and a flow-conditioning assembly (settling chamber, honeycombs, screens and contraction, see Fig. 1).

The Pipe. The core of the flow-loop is a long pipe of constant cross-section. The pipe is 110.8 m long (see Fig. 2d), with an inner diameter of 900 mm, resulting in a L/D of about 123. The pipe is made of twenty-two 5 m long carbon fiber modules, plus one 1.5 m long element, produced using the technology of filament-winding. This technology produced a surface roughness of $k_{rms} < 0.1$ µm ($k_{rms}/\ell_* < 0.01$), and a diameter precision of 900 ± 0.2 mm. These values are well below the required

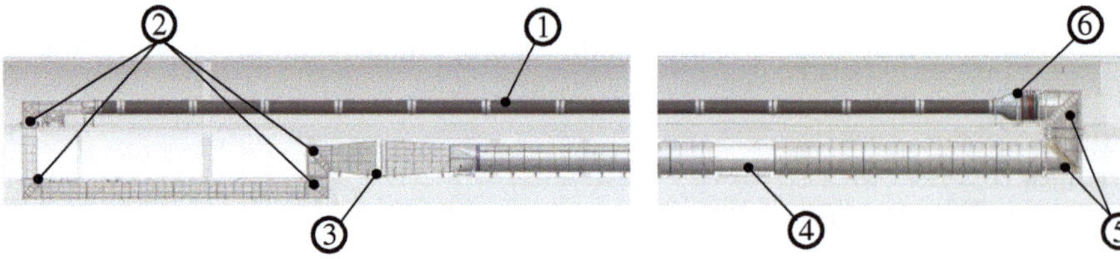

Fig. 1 Overview of the Long Pipe (LP) facility: *1* The carbon-fiber pipe. *2* Rectangular expanding corners. *3* Heat exchanger. *4* Fan assembly. *5* Round corners. *6* Settling chamber/Convergent assembly

Fig. 2 **a** The axial fan. **b** Heat exchanger with fast diffuser and expanding corners. **c** Pipe inlet, convergent and settling chamber. **d** The Long Pipe

specifications [6]. Extremely precise alignment is obtained thanks to 3-degrees of freedom supporting elements, combined with a high-precision optical measurement system (Leica TS50). All pipe elements are equipped with seven pressure ports and four access holes of diameter of 150 mm distributed radially to provide access for measurement instruments. Equalization with ambient pressure is provided at the 1.5 m main test section (see Fig. 2c) at the end of the pipe, but it can be moved upstream thanks to a flexible design of the junctions. A 300 mm wide connection ring is used to connect the first element of the LP with the contraction. This space can be used to insert grids or other devices to manipulate the inlet conditions.

Corners and diffusers. The return duct includes a series of shape converters, diffusers and corners. It first runs under the floor of the main laboratory to allow access to the laboratory, and then rises above the floor of the main tunnel trough an "S"-shape made of two expanding corners (see layout in Fig. 1). The circuit includes four rectangular expanding corners and two non-expanding circular ones positioned after the fan and before the settling chamber. The expanding corners have been designed using the optimal parameters found in the BL wind tunnel at KTH [9], with an expansion ratio of: 1.08, 1.16, 1.34, 1.26 from the first to the fourth corner, respectively.

Part of the expansion occurs in two main diffusers. The first two-dimensional diffuser is located after the first shape converter, it is 3 m long and has an expansion ratio of 1.39. The second main vertical diffuser is located between corner 1 and 2. It achieves an expansion rate of 1.54 in a length of 3.44 m. To minimize the risk of separation, this diffuser includes one horizontal and three vertical guide vanes. A third, fast diffuser is used to further slow-down the flow before the heat exchanger (see Fig. 2b).

Heat exchanger. The heat exchanger, located downstream of the "S"-shape corners, is designed to control the temperature of the flow with a target accuracy of ±0.1 °C. The heat exchanger has a cross section of 6.2 m^2 (2.3 × 2.7 m), which is the largest dimensions that could be fitted along the return duct to reduce the velocity as much as possible, and minimize the pressure drop across it. It is connected to the return duct using a large split diffuser and a contraction. The heat exchanger produces pressure losses of only 230 Pa at 38 m^3/s for a total capacity of 340 kW, which corresponds to the maximum power allowed for the fan. This heat exchanger allows to keep the flow at the exit of the heat exchanger at a temperature of 15 °C at full speed. The electro-valve can be controlled remotely by any type of control system. The shape of the return duct is converted downstream of the heat exchanger from rectangular to circular and remains the same for the rest of the circuit.

The fan. A four-stages axial fan provides a maximum volume flow rate of 38 m^3/s (at 1100 RPM), corresponding to a velocity of 60 m/s in the test section. At this speed, the fan is able to provide a total pressure head of 6500 Pa (2350 Pa of which in the pipe section), which corresponds to the total estimated pressure drop in the flow loop computed assuming a complete loss of dynamic pressure in the expansions. The fan is made of two counter-rotating parts mounted in series. Each part includes two propellers mounted on a common motor powered by a dedicated inverter. The fan has a diameter of 1.8 m, and a total length of 4.2 m. Each propeller has 11 blades made of extruded aluminum (see Fig. 2a). The power system is composed of two 200 kW inverters from Rockwell Automation able to synchronize both motors and to regulate RPM with an accuracy of 0.001 % using encoders mounted on the motor shafts. To minimize electromagnetic noise, connections, shielding and grounding solutions were designed with great care when developing the electricity distributor. The inverters are enclosed in a Faraday cage. Furthermore, the shielded cables connecting them to the motors are enclosed in 3 mm thick steel conduits.

Settling chamber. The settling chamber assembly is composed of the turbulence manipulators, the settling chamber and the contraction. The turbulence manipulators include: 1 honeycomb section and 5 screens with decreasing mesh size. The contraction has a contraction ratio of 4:1, which is the maximum allowed by geometrical constrains of the installation site. The design of these elements are based on the MTL wind tunnel at KTH, Stockholm [10].

Acknowledgments We are grateful to J.D. Rüedi for his fundamental contribution in the design phase, as well as to the many researchers involved in CICLoPE: P.H. Alfredsson, C. Casciola, A.V. Johansson, I. Marusic, P. Monkewitz, H. Nagib, K.R. Sreenivasan. A special thought to T. B.

Nickels (1966–2010), Cambridge University, for his support to CICLoPE since the first moment it was conceived. G. Piraccini, P. Proli and A. Bassi are kindly acknowledged for their help during the design and construction phase of the facility. Financial support is provided by EuHIT.

References

1. Marusic, I., Mathis, R., Hutchins, N.: Predictive model for wall-bounded turbulent flow. Science **329**, 193–196 (2010)
2. Cimarelli, A., De Angelis, E., Casciola, C.M.: Paths of energy in turbulent channel flows. J. Fluid Mech. **715**, 436–451 (2013)
3. Örlu, R., Alfredsson, P.H.: Comment on the scaling of the near-wall streamwise variance peak in turbulent pipe flows. Exp. Fluids **54**, 1431 (2013)
4. Smits, A.J., McKeon, B.J., Marusic, I.: High-Reynolds number wall turbulence. Annu. Rev. Fluid Mech. **43**, 353–375 (2011)
5. Smits, A.J., Marusic, I.: Wall-bounded turbulence. Phys. Today **66**(9), 25 (2013)
6. Talamelli, A., Persiani, F., Fransson, J.H.M., Alfredsson, P.H., Johansson, A.V., Nagib, H.M., Rüedi, J.D., Sreenivasan, K.R., Monkewitz, P.A.: CICLoPE-a response to the need for high Reynolds number experiments. Fluid Dyn. Res. **41**, 021407 (2009)
7. Kim, J.: Progress in pipe and channel flow turbulence. J. Turbul. **13**, 45 (2011)
8. Talamelli, A., Bellani, G., Rossetti, A.: The long pipe in CICLoPE: a design for detailed turbulence measurements. Springer Proc. Phys. **149**, 127–131 (2014)
9. Lindgren, B., Österlund, J., Johansson, A.V.: Measurement and calculation of guide vane performance in expanding bends for wind-tunnels. Exp. Fluids **24**, 265–272 (1998)
10. Lindgren, B., Johansson, A.V.: Evaluation of a new wind tunnel with expanding corners. Exp. Fluids **36**, 197–203 (2004)
11. Nagib, H., Hites, M., Won, J., Gravante, S.: Flow quality documentation of the national diagnostic facility. 18th Aerospace Ground Testing Conference AIAA, pp. 94–2499 (1994)

A New High-Order Method for Simulating Turbulent Pipe Flow

Peter Lenaers, Philipp Schlatter, Geert Brethouwer and Arne V. Johansson

1 Introduction and Problem Formulation

Fully developed incompressible turbulent flow in a smooth pipe is an important, but yet canonical problem in fluid mechanics. During recent years, large experimental [1] and numerical [2] interest has shown a number of open questions, in particular related to the scaling of mean and fluctuating quantities, but also to explain similarities and differences to the other canonical wall flows (channels, boundary layers). Most previous numerical efforts towards simulating pipe flows were based on low-order discretisation schemes, motivated by the potential difficulties in the pipe centre, and the need to resolve the numerical singularity. Nevertheless, also a number of high-order methods, based on either radial Chebyshev discretisation, or spectral elements, have been proposed, and used successfully.

In the present work, we describe a new method that solves the incompressible Navier–Stokes equations on a polar grid, with axial and azimuthal Fourier (periodic) discretisation, and high-order finite-difference derivatives in the radial direction. As opposed to most other methods, the pressure nodes are collocated with the velocity points, thus not requiring any inter- and extrapolation between two meshes. As will be described in the following, the equations are formulated in primitive-variable form, and the arising pressure-poisson equation is solved with consistent boundary conditions derived using an influence-matrix method [3]. The method is an adaptation of the recent scheme by Lenaers et al. [4], and further details about the present method can be found in the corresponding thesis [5].

The formulation is in cylindrical (z, r, θ) coordinates with assumed periodicity in z, leading to the two homogeneous directions z and θ. The governing equations thus contain three velocities (u, v, w) and the pressure p. No-slip conditions at the walls at $r = R$ are imposed, however, the method can also deal with other

P. Lenaers · P. Schlatter (✉) · G. Brethouwer · A.V. Johansson
Linné FLOW Centre, KTH Mechanics, 100 44 Stockholm, Sweden
e-mail: pschlatt@mech.kth.se

J. Peinke et al. (eds.), *Progress in Turbulence VI*,
Springer Proceedings in Physics 165, DOI 10.1007/978-3-319-29130-7_37

conditions [4]. Using a Fourier ansatz and high-order compact finite differences, a semi-discretisation can be obtained, which is further discretised in time using an explicit Runge–Kutta scheme.

Regarding the formulation of the derivatives close to the pipe axis, two specific issues appear: Firstly, the grid lines converge to a single point, which requires special prescription of axis conditions, and secondly, the azimuthal grid is becoming very fine leading to severe time-step limitations of the explicit integration. Two somewhat standard approaches have been chosen; to obtain consistent derivatives at the centre, parity conditions are formulated for each quantity (velocity components, pressure, divergence) and Fourier mode. These allow then the formulation of new derivative matrices. Special care has to be taken of terms multiplied by $1/r$ and $1/r^2$ as described in Ref. [5]. Alleviating the strict time-step limitations was achieved by reducing gradually the number of azimuthal Fourier modes as the centre is approached. This method appears unproblematic as the *physical* spanwise grid spacing is approximately retained, and thus spurious reflections are neither expected nor observed a posteriori.

The treatment of the pressure follows the principle outlined in Refs. [3, 4]. A pressure-Poisson equation is derived, which replaces the continuity equation in the interiour of the domain. The in principle unknown pressure boundary conditions are obtained by observing a linear relation between the wall pressure and the velocity (for which the boundary conditions are known). The system can then be solved as a superposition of particular and homogeneous solutions, together with a few specific adaptations necessary to deal with discretisation and commutation errors as outlined in Ref. [4]. So far, the method is general for all radial derivative matrices. In order to achieve an efficient method, compact finite differences are chosen, which lead to banded instead of full matrices. Whereas for the momentum equations no problems appear, the Poisson equation in cylindrical equations cannot be written exactly in banded form, as opposed to the channel [4]. However, using an iteration by regrouping some of the terms, an efficient solution still retaining time complexity N_r (i.e. scaling with the wall-normal resolution) could be found.

2 Results

The above method has been implemented, and a number of validation runs have been performed to check the accuracy and efficiency; in particular, the growth rates of instability waves have been carefully checked. For the current paper, only results pertaining to standard turbulent pipe flow at $Re_\tau = 180$ are presented. The results are compared to the literature data by El Khoury et al. [2]. In the meantime we have also computed higher Reynolds numbers, for which the agreement is equally good.

Figures 1 and 2 show standard statistics of wall turbulence. No difference to the literature data can be observed, even in higher-order statistics (not shown). In particular, the behaviour at the centre of the pipe clearly shows that no attenuation of

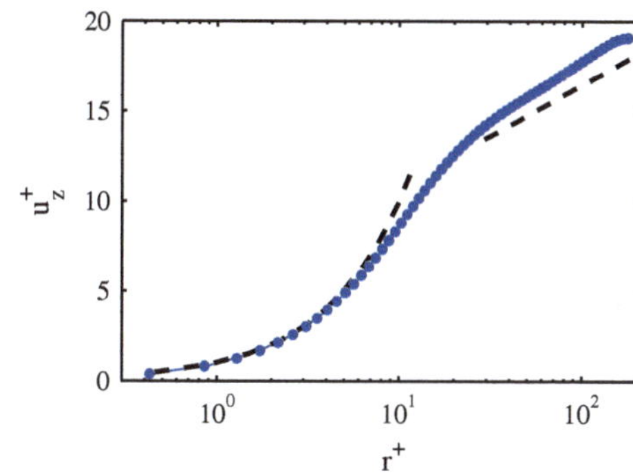

Fig. 1 Mean axial velocity with the *full line* representing results from the present method, and *circles* the results by El Khoury et al. [2]. The log law is indicated by the *black dashed lines*

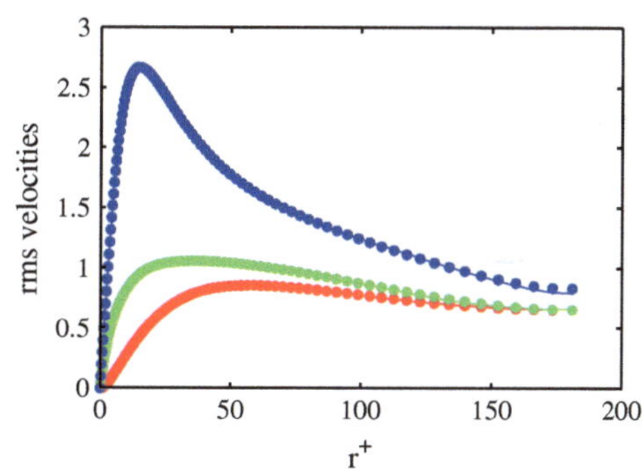

Fig. 2 Turbulent fluctuations throughout the pipe. The present method is indicated by the *full line*, and the *circles* correspond to data by El Khoury et al. [2]

the fluctuations due to the axis conditions is taking place; similarly, the reduction of Fourier modes as the axis is approached cannot be seen in the statistics.

Further insight into the adequacy of the method is given by Fig. 3, and the corresponding enlargements in Figs. 4 and 5. The near-wall region clearly shows the appearance of the well-known velocity streaks, and the corresponding streamwise vortices involved in the turbulence re-generation. In the present method, the wall boundary conditions are fulfilled exactly at $r = R$. On the other hand, in the centre $r = 0$, shown in Fig. 5, shows no special distortions of the velocity field. This indicates that the formulation of the discretisation using the parity conditions over the pipe axis, indeed lead to a smooth field. Note that the in-plane velocity in the axis position is clearly non-zero.

To summarise, the proposed computational method for incompressible flow in a pipe geometry allows the solution of the coupled continuity and momentum equations on a collocated mesh in primitive variables. Due to the influence matrix method, both the boundary conditions as well as the (discrete) divergence-free conditions are fulfilled up to machine precision in the whole velocity field. The operators for all radial derivatives, including the Poisson equation, are compact high-order schemes. To retain time complexity N_r (instead of N_r^2 for full matrices), the Poisson equation is formulated as a sequence of two first-order equations, with a few iterations. The present method is unique in the sense that it allows exact fulfillment of boundary

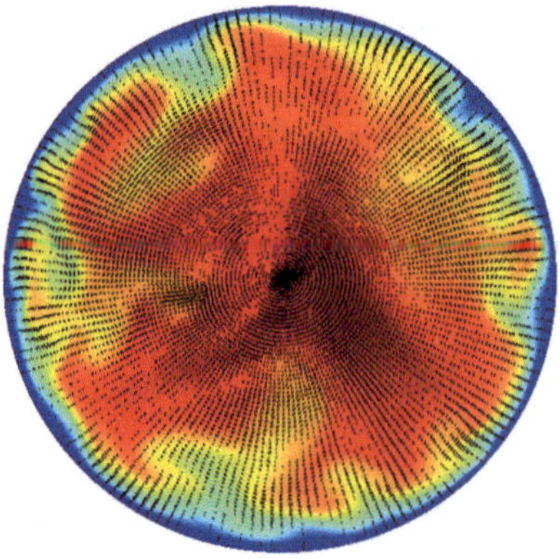

Fig. 3 Colours of the axial velocity with superimposed vector field of the in-plane velocities, at an arbitrary axial position and time, for $Re_\tau = 180$

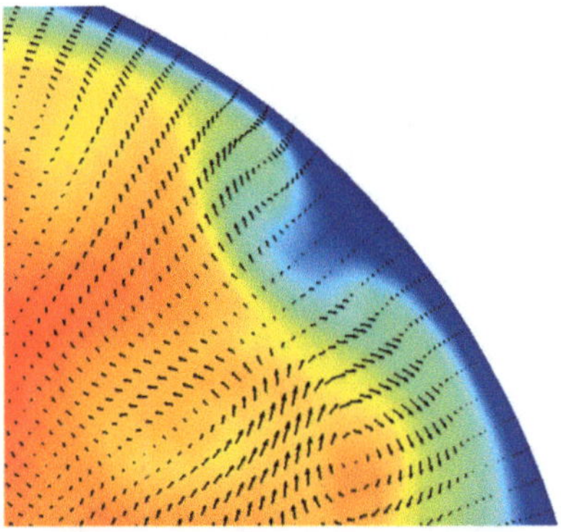

Fig. 4 Enlargement of Fig. 3 in the near-wall region. The typical velocity deformations (streaks) close to the wall can be seen, together with the corresponding streamwise vortices

Fig. 5 Enlargement of Fig. 3 in the centre region. Note that the mesh in physical space does not show the reduction in azimuthal Fourier modes. The in-plane velocity close to the centre is varying smoothly over the axis due to the imposed parity conditions

conditions and continuity on non-staggered meshes, alleviating the need for interpolations and other filtering operations. In particular for high-Reynolds number cases, these properties are believed to be crucial for a faithful representation of both the small and large-scale flow simultaneously.

References

1. Talamelli, A., Persiani, F., Fransson, J.H.M., Alfredsson, P.H., Johansson, A.V., Hagib, H.M., Rüedi, J.-D., Sreenivasan, K.R., Monkewitz, P.: CICLoPE - a response to the need for high Reynolds number experiments. Fluid Dyn. Res. **41**, 021407 (2009)
2. El Khoury, G.K., Schlatter, P., Noorani, A., Fischer, P.F., Brethouwer, G., Johansson, A.V.: Direct numerical simulation of turbulent pipe flow at moderately high Reynolds numbers. Flow Turbul. Combust. **91**, 475–495 (2013)
3. Kleiser, L., Schumann, U.: Treatment of incompressibility and boundary conditions on 3-D numerical spectral simulations of plane channel flow. Proc. 3rd GAMM Conf. Numer. Methods Fluid Mech. **59**, 165–173 (1980)
4. Lenaers, P., Schlatter, P., Brethouwer, G., Johansson, A.V.: A new high-order method for the simulation of incompressible wall-bounded turbulent flows. J. Comput. Phys. **272**, 108–126 (2014)
5. Lenaers, P.: A new high-order method for direct numerical simulations of turbulent wall-bounded flows, PhD Thesis, KTH Mechanics (2014)

Structure Investigation in Pipe Flow at High Reynolds Numbers

Emir Öngüner, El-Sayed Zanoun, Franziska König and Christoph Egbers

Abstract For linear stable shear flows, turbulent pipe flow has been investigated for different Re-numbers. Turbulence can be achieved in natural or artificial methods. For transitions some perturbations are needed to trigger turbulence and some structures as puffs and slugs can be observed. In last decades the so called large-scale motions (LSM), which are composed of detached eddies with wide range of azimuthal scales in the outer layer, are identified. Advanced versions of LSMs, the very large-scale motions (VLSM), have radial scales. The VLSMs are concentrated around a single azimuthal mode and make a smaller angle with the wall compared to the LSM. These above mentioned phenomena will be investigated at high Reynolds numbers in the pipe facility Cottbus-Large Pipe at BTU Cottbus-Senftenberg (CoLa-Pipe) which provides a bulk Reynolds number of $Re_m \leq 1,5 \times 10^6$. Zimmer et al. [1] and König et al. [2] provide an outline for conditions of fully developed turbulent flow state with natural as well as artificial transition. Considering these fully developed flow conditions at CoLa-Pipe, next investigations will be primarily focused on the structures in boundary layer in terms of LSM and VLSM by using hot wire anemometry and PIV. The main aim of this work will be analysing the lengths of structures at high Re-numbers in terms of their wavelengths and comparing with those of low Re-numbers regions.

E. Öngüner (✉) · C. Egbers
Department of Aerodynamics and Fluid Mechanics, Brandenburg University of Technology, 03046 Cottbus, Germany
e-mail: emir.oenguener@b-tu.de

C. Egbers
e-mail: egbers@tu-cottbus.de

E.-S. Zanoun
Department of Mechanical Engineering, Benha University, P.O. Box 12512, New Benha, Egypt
e-mail: elsayed.zanoun@aol.com

F. König
Department of Metallurgy and Materials Engineering, Brandenburg University of Technology, 03046 Cottbus, Germany
e-mail: franziska.koenig@b-tu.de

J. Peinke et al. (eds.), *Progress in Turbulence VI*,
Springer Proceedings in Physics 165, DOI 10.1007/978-3-319-29130-7_38

1 Introduction to CoLa-Pipe Turbulence Facility

The unique pipe facility CoLa-Pipe (Cottbus Large Pipe) at the Department of Aerodynamics and Fluid Mechanics, Brandenburg University of Technology Cottbus-Senftenberg, is designed and built to investigate fully developed pipe flow at high Reynolds numbers ($Re_m \leq 1,5 \times 10^6$). The type of this facility is closed-return with two available test sections providing a length-to-diameter ratio of $L/D \approx 148$ ($D_i = 190 \pm 0.23$) and $L/D \approx 79$ ($D_i = 342 \pm 0.32$ mm) (Fig. 1). The power unit of CoLa-Pipe facility is working with a nominal power of 45 kW contains a powerful radial blower connected to the pipe on its suction side, a three-phase motor, and a frequency converter which provides a flow rate of 0.05–2.5m^3/s and can be controlled by changing the frequency of the radial blower blades utilizing the frequency converter unit. With these properties and in conjunction with the inner diameter of 0.19 m of the lower test section the power assembly provides a maximum velocity of 80 m/s at the contraction exit.

2 Turbulent Structures in Pipe Flow

In order to understand the structures of wall bounded flows and the behavior of these structures the enery distribution according to their wavenumber domain should be observed. These structures in the turbulent wall region and outer flow have been

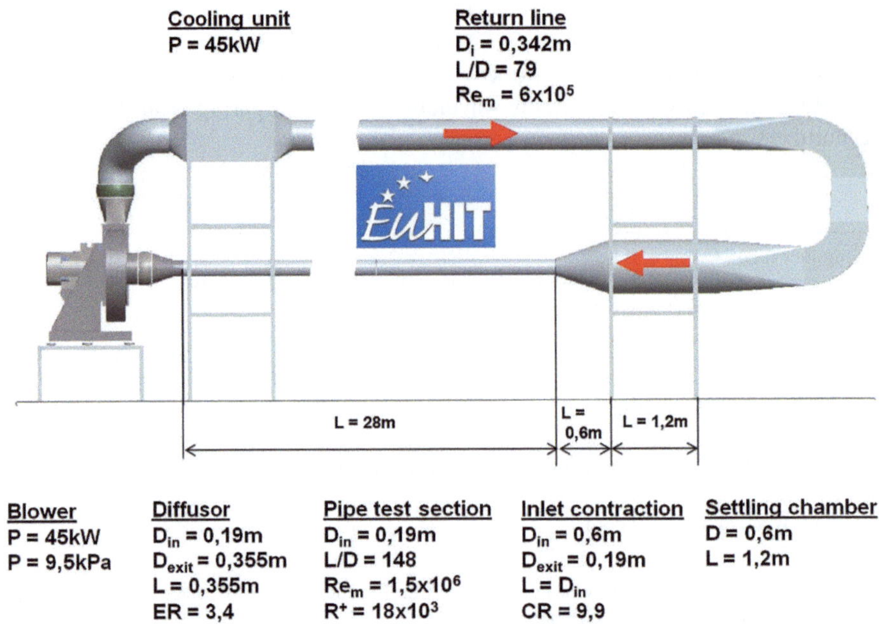

Fig. 1 Cottbus-large pipe at BTU Cottbus-Senftenberg (CoLa-Pipe)

observed in pipes by many authors, Kim and Adrian [3], Guala et al. [4], Monty et al. [5], Bailey and Smits [6] and Rosenberg et al. [7], where the structures are extending up to $20R$ (R considered as pipe radius), in the streamwise direction and as well as in turbulent wall region of boundary layers (Hutchins and Marusic [8], Balakumar and Adrian [9]) extending up to 20δ (δ considered as boundary layer thickness), in length. They are usually characterized as large-scale motions (LSM). Kim and Adrian [3] gives also the first assumption in literature about possible sizes of these structures; if the streamwise length of the structure is $2R$-$3R$ the turbulent structure is called LSM and if its size between $8R$-$16R$ it is mentioned as very large-scale motion VLSM. These interpretations remained until today as an unclear assumption.

3 Methods and Results

The main investigation method of these structures mentioned previously is based on observing energy spectra of fluid on specific radial locations. As explained in Guala et al. [4] the very large-scale motions are energetic, typically containing half of the turbulent kinetic energy of the streamwise component, and they are unexpectedly active, typically containing more than half of the Reynolds shear stress. Previous research show us that Townsend [10] and Grant [11] inferred that large-scale motions existed, even close to the wall and that they contribute to the fraction of the turbulent kinetic energy. Literature survey shows us that very large-scale motions have been almost neglected. There is still no commonly accepted definition for the sizes of these structures. Therefore a quantitative evaluation of the energy and the Reynolds stress associated with each scale has not been clearly established. Vallikivi [12] uses the whole pre-multiplied spectrum and the observable peaks are obtained to determine the wavenumbers for each radial location. For the CoLa-Pipe investigation same method is used and the results can be shown in Fig. 2. It is also clearly observable that the growing character of these structures can be interpreted as almost linear. The shear Reynolds numbers for this experimental study are following: $Re_\tau = 2036, 2749, 3483, 4184, 4912, 5629, 6340, 7059$. At highest Re-Number there is a turbulent motion identified with a non-dimensionalized streamwise length of $\lambda_{max}/R = 47$. Considering the pipe radius at this facility as $R = 9.5\,\text{cm}$, this structure has an streamwise extension of nearly 4.5 m.

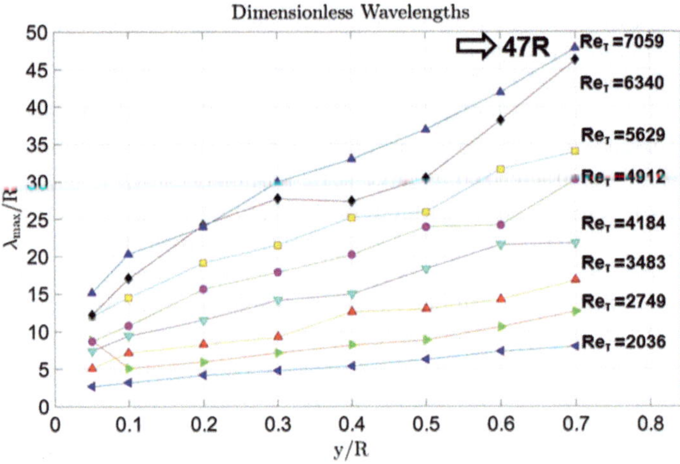

Fig. 2 Non-dimensionalized wavelengths of turbulent structures by R in CoLa-Pipe at different high re-numbers

References

1. Zimmer, F., Zanoun, E.-S., Egbers, C.: A study on the influence of triggering pipe flow regarding mean and higher order statistics. J. Phys. Conf. Ser. **318**, 032039 (2011)
2. Koenig, F., Zanoun, E.-S., Jehring, L., Egbers, C.: The CoLaPipe—a high reynolds number pipe test facility. In: 4th International Conference on Jets, Wakes and Separated Flows, ICJWSF2013, September 17–21. Nagoya, Japan (2013)
3. Kim, K.C., Adrian, R.J.: Very large-scale motion in the outer layer. Phys. Fluids **11**, 417–422 (1999)
4. Guala, M., Hommena, S.E., Adrian, R.J.: Large-scale and very-large-scale motions in turbulent pipe flow. J. Fluid Mech. **554**, 521–542 (2006)
5. Monty, J.P., Stewart, J.A., Williams, R.C., Chong, M.S.: Large-scale features in turbulent pipe and channel flows. J. Fluid Mech. **589**, 147–156 (2007)
6. Bailey, S.C.C., Smits, A.J.: Experimental investigation of the structure of large- and very-large-scale motions in turbulent pipe flow. J. Fluid Mech. **651**, 339–356 (2010)
7. Rosenberg, B.J., Hultmark, M., Vallikivi, M., Bailey, S.C.C., Smits, A.J.: Turbulence spectra in smooth- and rough-wall pipe flow at extreme Reynolds numbers. J. Fluid Mech. **731**, 46–63 (2013)
8. Hutchins, M., Marusic, I.: Large-scale influences in near-wall turbulence. Phil. Trans. R. Soc. A **365**, 647–664 (2007)
9. Balakumar, B.J., Adrian, R.J.: Large- and very-large-scale motions in channel and boundary-layer flows. Phil. Trans. R. Soc. A **365**, 665–681 (2007)
10. Townsend, A. A.: The turbulent boundary layer. In: Boundary Layer Research (ed. H. Goertler), vol. 1, Springer, Heidelberg (1958)
11. Grant, H.L.: The large eddies of turbulent motion. J. Fluid Mech. **4**, 149–190 (1958)
12. Vallikivi, M.: Wall-bounded turbulence at high Reynolds numbers, PhD thesis, Princeton University (2014)

Heat Transfer in Turbulent Boundary Layers of Pipe Flow: A Wavelet Transforms Approach

Makrand A. Khanwale, C.S. Sona and Channamillkarjun S. Mathpati

Abstract Transport of energy in turbulent boundary layers of pipe is characterised by two simulatneous, complimentary processes *vis-a-vis* turbulent advection and conduction. Both the aforementioned processes are manifestations of passive scalar transport in the turbulent boundary layer. Since passive scalar transport is intimately related to flow structure evolution, a wavelet based method is used to characterise flow structure evolution. This work deals with the calculation of turbulent advective and conductive heat fluxes for a high temperature heat transfer molten salt FLiNaK.

Heat transfer of a molten fluoride salt, FLiNaK with a eutectic composition of 46.5 %LiF-11.5 %NaF-42 %KF (mol %) in the turbulent boundary layer of pipe is characterised by the transport of passive scalar (Temperature). Energy coupled large eddy simulation (LES) were carried out to obtain transient velocity and temperature (passive scalar) data in the domain. Two complimentary fluxes namely turbulent advective and conductive are a manifestation of this passive scalar distribution. Calculation of conductive flux from the aforementioned transient velocity and temperature data is relatively straightforward which involves knowledge of thermal conductivity (k), whose temperature dependence is well documented [1]. But, the calculation of turbulent advective flux involves knowledge of local heat transfer coefficient (HTC), which has a extremely complex dependence on evolution of local turbulent flow structures. We deal with this issue using multi-scale wavelet methodology namely Wavelet transform modulus maxima (WTMM) [2, 3], to disseminates the flow struc-

M.A. Khanwale (✉)
Cain department of Chemical Engineering, Louisiana State University,
Baton Rouge, LA 70803, USA
e-mail: makrand_khanwale@hotmail.com

C.S. Sona · C.S. Mathpati
Department of Chemical Engineering, Insitute of Chemical Technology, Matunga,
Mumbai, Maharashtra, India
e-mail: sonuchiyyarath@gmail.com

C.S. Mathpati
e-mail: cs.mathpati@ictmumbai.edu.in

J. Peinke et al. (eds.), *Progress in Turbulence VI*,
Springer Proceedings in Physics 165, DOI 10.1007/978-3-319-29130-7_39

ture time-scale information which can be used along with a more realistic version of the formalism known as small eddy model given in [4, 5] to calculate local HTC.

To compare the obtained results with experiment, experimental measurements of local HTCs are needed. But the literature reports only overall HTCs calculated experimentally [6–9], the reason being the impracticability of measurement at temperatures exceeding 650° C, which is the temperature range in which FLiNaK operates. Therefore, a representative local HTC has to be selected to be compared with the experimental data. The local HTC at a spatial location (y^+), where conductive flux completely decays to a negligible value, is used as a representative HTC to be compared with the experimental HTC. To reaffirm the accuracy of calculation energy balance is verified at every spatial location, at which the fluxes are calculated.

Many studies reported the thermo-physical properties of FLiNaK, which account for their temperature dependency. The thermo-physical properties of FLiNaK used in the simulation are given in Table 1. A more detailed description of the previous work focusing on the Thermo-physical properties of FLiNaK can be found in [10].

We carried out energy coupled large eddy simulations in a round pipe for Reynolds number of 25500. The computational domain was divided into 0.7, 1.1 and 1.2 million grid points to check the grid independence. The results found similar for 1.1 and 1.2 million grid points. Hence, 1.1 grid points were selected for further simulations. In order to resolve viscous sub-layer accurately, 5 grid points were placed below y^+ of 1 and 15 grid points below y^+ of 5. The turbulent kinetic energy goes through a maximum below y^+ of 20 and hence, 27 grid points were placed in this region in order to properly resolve this behaviour. The energy coupled large eddy simulations were validated using DNS data of Fukagata et al. [13]. One equation dynamic sub-grid kinetic energy model was used as sub-grid scale model given by [14]. This model has better accuracy compared to Smagorinsky model (1963). Since the thermo-physical properties of the FLiNaK are strong function of temperature, those variations were incorporated in the simulation. Figure 1 shows the comparison of non-dimensionalized mean axial velocity profile. The variation from DNS data is less than 5 %. High quality transient velocity and temperature data was recorded for further analysis. Further, the continuous wavelet transform of the raw velocity data can be calculated as

$$T^{(a,b)}(t) = \int u(t)\psi(t)\mathrm{d}t \tag{1}$$

where $T^{(a,b)}(t)$ are the continuous wavelet coefficients and $u(t)$ is the fluctuating velocity time-series and ψ is the continuous wavelet kernel function. The wavelet function ψ may be selected as the nth derivative of a Gaussian (DOG) function so that the local polynomial trends up to $(n-1)$th order are eliminated.

The residual was studied for extracting the masked singularities. The maxima of the absolute values of the wavelet coefficients are found out at each scale a. These local maxima of $|T^{(a,b)}|$ are grouped to form maxima lines across the scales a, which represent the singularity locations. The observed maxima lines delineate the structures. In WTMM plot, the maxima lines depict loci of local discontinuities due to the flow structures occurring at various scales. These lines show the singularity

Table 1 Properties of FLiNaK used in the former studies, temperature T is in kelvin (K)

Property	Grele and Gedeon (1954) [6]	Hoffman and Lones (1955) [7]	Vriesema et al. (1979) [8]	Ambrosek et al. (2009) [11]	Ferng et al. (2012) [12]	Present work [3]
Density, kg/m^3	2555–0.6 T	2555–0.6 T	2729.4–0.73 T	2729.4–0.73 T	2579.3–0.624 T	2729.4–0.73 T [11]
Thermal conductivity, W/mK	4.5	4.5	1.3	0.6, 1.0 and 0.36 + 0.00056 T	0.36 + 0.00056 T	0.36 + 0.00056 T [1]
Specific heat, kJ/kgK	1.89	2.09	1.89	1.88	1.88	1.88 [11]
Viscosity, kg/ms	$2.5 \times 10^{-5} e^{(4790/T)}$	$2.5 \times 10^{-5} e^{(4790/T)}$	$1.1 \times 10^{-4} e^{(3379/T)}$	$4 \times 10^{-5} e^{(4170/T)}$	$0.225 – 6.93 x 10^{-4} T + 8.12 x 10^{-7} T^2 – 4.25 x 10^{-10} T^3 + 8.36 x 10^{-14} T^4$	$4 \times 10^{-5} e^{(4170/T)}$ [11]

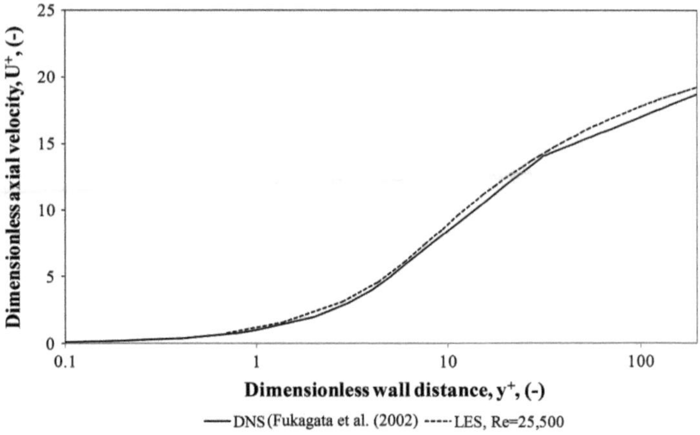

Fig. 1 Comparison of mean axial velocity profile with DNS data

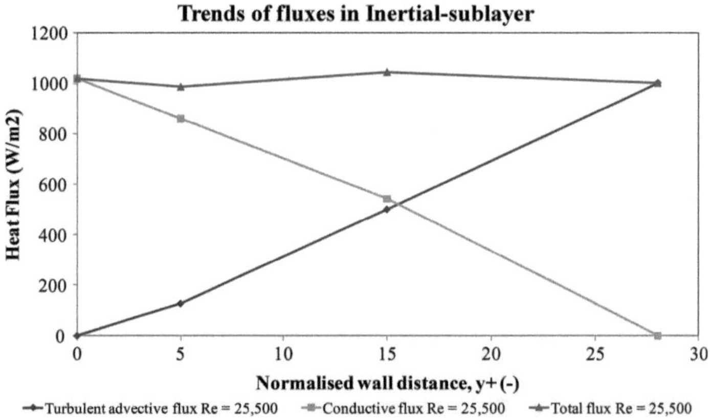

Fig. 2 Trends of turbulent advective and conductive flux for Re = 25,500, and comparison of their summation with input flux

behaviour, scale-wise at each location and bring out the presence of small structures within larger ones. The nature of their evolution and interaction in time domain is also seen. The multiple branching of singularity loci lines towards higher scales show the complex break-up phenomena. Age of these flow structures has been calculated as the difference between the singularity start and end times. Calculated ages are further used with the small eddy formalism to yield advective flux at each y^+. The conductive fluxes at each y^+ can be calculated using Fourier's law. For more details on calculation please see [10].

The turbulent advective flux obtained from the transient velocity data and the conductive fluxes obtained from the temperature data at various spatial locations in the inertial boundary layer are illustrated in Fig. 2. From Fig. 2 it can be noticed that

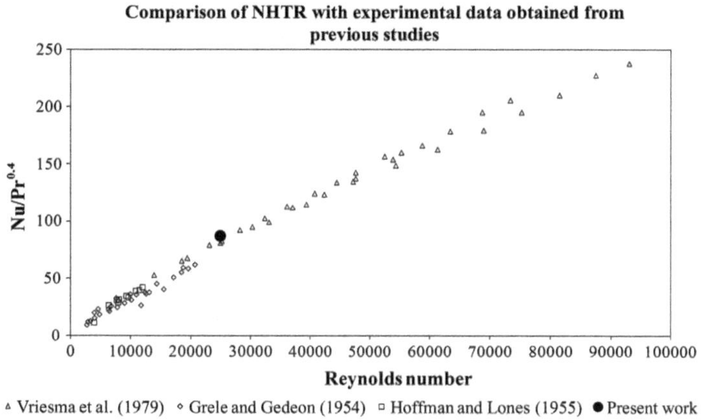

Fig. 3 Normalised Heat transfer ratio for Re = 25,500 (•), in comparison with experimental data of Vriesema et al. [8] (△); Grele and Gedeon [6] (◇); Hoffman and Lones [7] ([])

the sum of the conductive and the turbulent advective flux at every spatial location in the inertial Sub layer have less than 5 % error when compared to the input wall flux of $1000\,\mathrm{W/m^2}$.

The Nusselt number to Prandtl number ratio based on the heat transfer coefficients (HTC), calculated using small eddy formalism coupled with WTMM at a y^+, where conductive flux decays to negligible value is called as Non-dimensional Heat transfer ratio (NHTR). Figure 3 shows the comparison of NHTR with experimental data. A very good match with a error of 3 % is seen and confirms the accuracy of the calculated HTCs.

The authors acknowledge the financial support from the Department of Science and Technology, India and DAE-ICT centre for chemical engineering and research. The Authors are also grateful to Dr. Ashwin W. Patwardhan, for his expert help and suggestions, which contributed significantly to the improvement of the manuscript.

References

1. Smirnov, M.V., Khokhlov, V.A., Filatov, E.S.: Electrochimica Acta **32**(7), 1019 (1987)
2. Farge, M.: Annual review of fluid dynamics **24**, 395 (1992)
3. Sona, C., Khanwale, M.A., Mathpati, C.S., Borgohain, A., Maheshwari, N.K.: Applied Thermal Engineering **70**(1), 451 (2014). doi:10.1016/j.applthermaleng.2014.05.043
4. Lamont, J., Scott, D.: An eddy cell model of mass transfer into the surface of a turbulent liquid. AIChE J. **16**(4), 513–519 (1970)
5. Mathpati, C.S., Joshi, J.B.: Industrial and Engineering Chemistry Research pp. 8525–8557 (2007)
6. Grele, M.D., Gedeon, L.: Research Memorandom Report Number NACA RM E53L18 (April, 1954)
7. Hoffman, H.W., Lones, J.: Oak Ridge National Laboratory Report Number ONRL 1777 (April, 1955)

8. Vriesema, B.: WTHD No. 112. Delft Institute of Technology (1979)
9. Ignat'ev, V.V., Keronovskii, S.V., Surenkov, A.I., Shcherbanyuk, O.P., Manchkha, S.P., Smirnov, Y.B.: Atomnaya Energiya **57**(2), 123 (1984)
10. Khanwale, M.A., Sona, C., Mathpati, C.S., Borgohain, A., Maheshwari, N.K.: Applied Thermal Engineering 0 (2014). doi:10.1016/j.applthermaleng.2014.10.049
11. Ambrosek, J., Anderson, M., Sridharan, K., Allen, T.: Nuclear Technology, vol. 165 (2009)
12. Ferng, Y.M., Lin, K.Y., Chi, C.W.: Applied Thermal Engineering **37**, 235 (2012). doi:10.1016/j.applthermaleng.2011.11.021
13. Fukagata, K., Kasagi, N.: J. Comput. Phys. **181**(2), 478 (2002)
14. Ghosal, S.B., Lund, T.S., Moin, P., Akselvoll, K.: J. Fluid Mech. **286**, 229 (1995)

Part IV
Free Flows

Influence of an Extended Non-equilibrium Region on the Far-Field of Grid Turbulence

R.J. Hearst and P. Lavoie

Abstract The turbulence produced by two regular square mesh grids is compared to that produced by a square-fractal-element grid composed of an array of small square fractals. All three grids have approximately the same blockage. One of the regular grids is designed to have the same mesh length, M, as the fractal element grid, while the other matches the maximum bar thickness of the fractal. The transition of the turbulence from a non-equilibrium to a near equilibrium regime is assessed through the scale-by-scale kinetic energy budget and the velocity derivative skewness. It is found that the turbulence produced by all three grids agrees with many of the predictions for equilibrium phenomenology after approximately 20 M, with the regular grids reaching quasi-equilibrium earlier than the fractal. In the far-field, the fractal grid produces comparable or lower Re_λ than the regular grids in both dimensional and non-dimensional measurements of the streamwise position. This is attributed to an extended rapidly decaying non-equilibrium region in the wake of the fractal grid relative to the regular grids.

1 Introduction

Recent grid turbulence experiments have focussed on the 'non-equilibrium' region that is produced immediately downstream of the grid. Interest in this area was garnered by the studies of Vassilicos and co-workers in the wake of space-filling square fractal grids, e.g., [5, 10, 15, 18]. A space-filling square fractal is a single square fractal pattern that occupies the entire wind tunnel cross-section. Recently, observations of non-equilibrium turbulence have also been reported in the wake of an array of different grids, including: multi-scale cross grids [13], fractal element grids [7, 8, 20], and regular grids [11, 19]. It has also been demonstrated that for all these grid

R.J. Hearst · P. Lavoie (✉)
Institute for Aerospace Studies, University of Toronto, Toronto, ON M3H 5T6, Canada
e-mail: lavoie@utias.utoronto.ca

R.J. Hearst
Engineering and the Environment, University of Southampton, Southampton SO17 1BJ, UK
e-mail: r.j.hearst@southampton.ac.uk

J. Peinke et al. (eds.), *Progress in Turbulence VI*,
Springer Proceedings in Physics 165, DOI 10.1007/978-3-319-29130-7_40

geometries the non-equilibrium turbulence evolves into one that better approximates equilibrium turbulence[1] sufficiently far downstream [7, 8, 11–13, 22].

Observations of quasi-equilibrium turbulence in the wake of grids has been characterized by a homogeneous field with an energy decay rate in the range $-1 \gtrsim n \gtrsim -1.4$, for a power-law decay of the form $\langle u^2 \rangle \sim (x - x_0)^n$, and constant normalized dissipation scaling, $C_\varepsilon \sim \varepsilon L/u'^3$, where ε is the dissipation rate of turbulent kinetic energy and L is the integral length scale. Non-equilibrium turbulence contrasts on all three of these points. The transverse flow fields experience non-zero transverse transport of turbulent kinetic energy and production, the energy decays at an accelerated rate, $n \lesssim -2$, and C_ε grows rapidly [7, 8, 16, 18, 20].

The transition between these regions has been characterized in several ways. Isaza et al. [11] recalled from Batchelor [1] that for equilibrium turbulence, the velocity derivative skewness,

$$S(\partial u/\partial x) = \frac{\langle (\partial u/\partial x)^3 \rangle}{\langle (\partial u/\partial x)^2 \rangle^{3/2}}, \tag{1}$$

should be constant, and hence areas where $S(\partial u/\partial x)$ varied were associated with non-equilibrium turbulence. Hearst and Lavoie [7] discriminated between the regions by identifying where the flow became homogeneous in transverse planes and noting this coincided with a change in the decay rate, n. Later, these same authors used the scale-by-scale kinetic energy budget to determine the difference between the flow regions [8]. The scale-by-scale kinetic energy budget for grid turbulence is given by [4],

$$-\langle (\delta u)(\delta q)^2 \rangle + 2\nu \frac{\mathrm{d}}{\mathrm{d}r} \langle (\delta q)^2 \rangle - \frac{U}{r^2} \int_0^r s \frac{\partial}{\partial x} \langle (\delta q)^2 \rangle \, \mathrm{d}s = \frac{4}{3} \langle \varepsilon \rangle r, \tag{2}$$

where $\delta\alpha = \alpha(x + r) - \alpha(x)$, $\langle (\delta q)^2 \rangle = \langle (\delta u)^2 \rangle + \langle (\delta v)^2 \rangle + \langle (\delta w)^2 \rangle$, $\langle (\delta u)(\delta q)^2 \rangle = \langle (\delta u)^3 \rangle + \langle (\delta u)(\delta v)^2 \rangle + \langle (\delta u)(\delta w)^2 \rangle$, and s is a dummy integration variable. Equation (2) may be represented simply as $G + D + I = C$. G represents energy transfer through advection. D represents the energy transfer through molecular diffusion. I accounts for the influence of longitudinal inhomogeneity resulting from the decay of turbulence behind the grid. Finally, C represents dissipation. Equation (2) is satisfied for equilibrium turbulence but not for non-equilibrium turbulence due to the presence of production and transverse transport [8]. Hence, assessing the validity of (2) at each location in the flow can distinguish regions that approximate the two phenomenologies.

It would appear that one of the primary benefits of the fractal geometry is that it creates a physically large region of non-equilibrium turbulence relative to other grid geometries [7]. As such, the characteristically high $Re_\lambda = \langle u^2 \rangle^{1/2} \lambda/\nu$ region close to the grid extends over a significant downstream range. However, there has not

[1]One of the most cited distinctions between true Richardson-Kolmogorov equilibrium phenomenology and the 'quasi-equilibrium' turbulence measured in the far-field of regular grids has been a lack of a $k^{-5/3}$ velocity spectrum.

been a detailed investigation of the far-field repercussions of extending the near-grid non-equilibrium region. Here, we first investigate the scale-by-scale energy budget and $S(\partial u/\partial x)$ to discriminate the non-equilibrium and quasi-equilibrium regions. We then compare two regular grids with a fractal element grid at the same $Re_M = U_0 M/\nu$, where U_0 is the velocity immediately upstream of the grid, to identify how extending the non-equilibrium region influences the relative magnitude of Re_λ in the far-field.

2 Experimental Details

Three different passive grids were investigated for the present study. The first was the square-fractal-element grid used in [7, 8], which is referred to here as $Fs39$. This grid consists of a 12×8 array of 3 fractal iteration square fractal elements mounted to a background mesh with $M = 100$ mm, and thickness $\tau_0 = 6.7$ mm. $Fs39$ has $\sigma = 0.39$ blockage, and is described in more detail in [7, 8]. A regular square mesh grid, $Sq39$, with $M = 100$ mm, $\tau_0 = 22.0$ mm, and $\sigma = 0.39$, was deliberately designed to match M and σ to $Fs39$. A second regular grid, $Rd38$, with $M = 32$ mm, $\tau_0 = 6.8$ mm, and $\sigma = 0.38$ was designed to approximately match τ_0 and σ of $Fs39$. As such, all three grids represent comparable initial conditions in one or more parameters. The grids are shown in Fig. 1.

Measurements were conducted downstream of the grids in a 1.2 m $\times$ 0.8 m $\times$ 5.0 m wind tunnel using constant temperature hot-wire anemometry. To compare the three grids, measurements were conducted with a nano-scale thermal anemometry probe (NSTAP) [21] at Re_M = 28,500 with a 1.23:1 secondary contraction installed 0.57 m downstream of the grids. A secondary contraction has been used to decrease

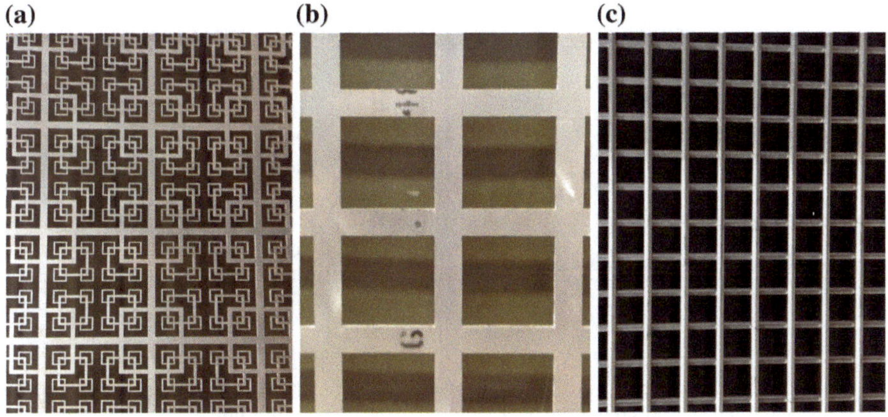

Fig. 1 Photographs of a nominally 260 $\times$ 340 mm area of each of the 1200 $\times$ 800 mm grids; **a** $Fs39$, **b** $Sq39$, **c** $Rd38$

anisotropy in the flow [3, 14], and was used here to collapse the anisotropy levels for all grids in the far-field to $u'/v' = 1.10 \pm 0.05$. Measurements were also conducted in the wake of $Fs39$ at $Re_M = 65{,}000$ with a X-wire and without the secondary contraction in order to investigate the region closer to the grid. Measurements were limited to the area downstream of the secondary contraction when it was installed. The measurements at $Re_M = 65{,}000$ were conducted at a higher velocity and acquired for a longer sampling time to converge the peak of $\langle(\delta u)(\delta q)^2\rangle$ to within $\pm 5\,\%$ using the 95 % confidence interval. The sample time was decreased for the $Re_M = 28{,}500$ measurements for efficiency. Downstream distance was measured relative to the advection time of the flow [3], $t = \int_0^x U(s)^{-1}\,ds$, where $U(x)$ is the local mean velocity at a position x. Hence, non-dimensional downstream position is expressed as $U_0 t/M$.

3 Transition to Approximate Equilibrium

The scale-by-scale kinetic energy budget for grid turbulence, (2), was assessed in normalized form given by $G/C + D/C + I/C = G^* + D^* + I^* = B^*$, where B^* is the 'balance' of the energy budget and all variables are a function of the spatial interval r. When $B^* = 1$ for all r, (2) is satisfied. The various terms of the scale-by-scale energy budget are plotted in Fig. 2 in both the near- and far-field of $Fs39$ at $Re_M = 65{,}000$. It is immediately evident that in the far-field (2) is satisfied, while in the near-field there are significant departures from $B^* = 1$. Contours of B^* shown in Fig. 3 clearly identify that by $U_0 t/M = 20$, $B^* \approx 1$. Therefore, based on this analysis,

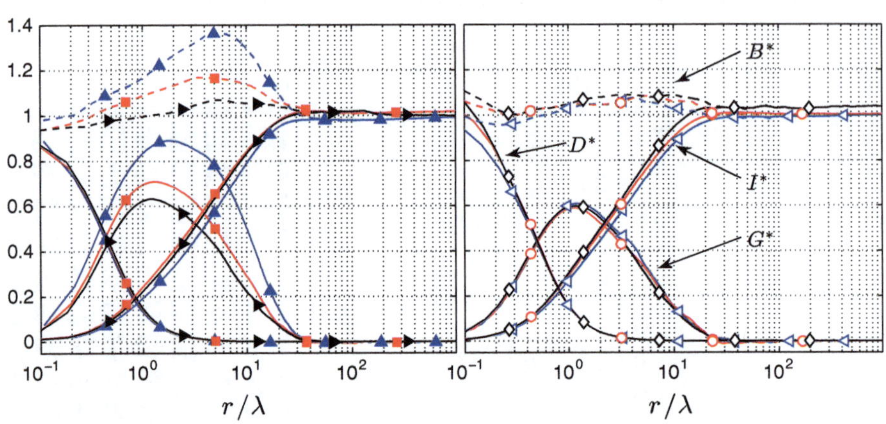

Fig. 2 The normalized scale-by-scale kinetic energy budget for grid turbulence given by (2). The *left* frame and filled symbols represent data acquired in the non-equilibrium region. The *right* frame and empty symbols represent data acquired in the quasi-equilibrium region. (▲) $U_0 t/M = 10.4$, (■) $U_0 t/M = 15.7$, (▶) $U_0 t/M = 18.8$, (◁) $U_0 t/M = 28.6$, (○) $U_0 t/M = 36.3$, (◇) $U_0 t/M = 46.1$

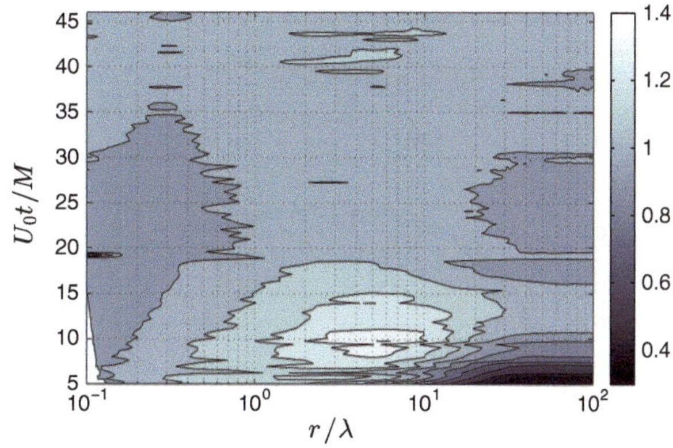

Fig. 3 Contours of the balance, B^*, of the normalized scale-by-scale kinetic energy budget for grid turbulence

the quasi-equilibrium turbulence region is $U_0t/M \geq 20$, and anything ahead of this may be considered non-equilibrium turbulence.

This analysis is covered in more detail for this same grid by Hearst and Lavoie [8], where they also show that there is measurable transverse transport and production in the non-equilibrium region that becomes negligible in the far-field. To confirm that the scale-by-scale energy budget results hold away from the primary axis of measurement, Hearst and Lavoie [8] also showed that the similarity form of (2) held in transverse planes in the far-field, but not in the non-equilibrium near-field.

The NSTAP measurements with a secondary contraction were designed to compare the wakes of the three grids, and were not sampled for a sufficient time to adequately resolve the terms of (2). The NSTAP measurements are more representative of the statistical analysis available for typical grid turbulence measurements. Inasmuch as this is the case, we seek a second means of assessing the transition between non-equilibrium and quasi-equilibrium turbulence that corroborates the scale-by-scale kinetic energy budget results.

To address this issue, the velocity derivative skewness, $S(\partial u/\partial x)$, for all four test cases is plotted in Fig. 4. Recall that for equilibrium turbulence, $S(\partial u/\partial x)$ is constant. From the figure, it is clear that the $Fs39$ results have an evolving $S(\partial u/\partial x)$ that becomes constant by $U_0t/M \approx 20$, in agreement with the scale-by-scale kinetic energy budget results. Interestingly, within the measurement range, both regular grids produce approximately constant $S(\partial u/\partial x)$, suggesting that all regular grid measurements are within the quasi-equilibrium range.

The magnitude of $S(\partial u/\partial x)$ shows a dependence on the initial conditions, as the constant far-field value is different for each test case. Sreenivasan and Antonia [17] showed that $S(\partial u/\partial x)$ has a Re_λ dependence, likely related to the initial conditions, which may account for some of the differences in the far-field values. The large step change between $Fs39$ with and without the secondary contraction is likely related to the significant change in Re_M between the experiments, and the improvement in spatial resolution of the NSTAP for the secondary contraction case; accurate gradient estimation is highly dependent on spatial resolution [2, 6]. When corrections [2] for

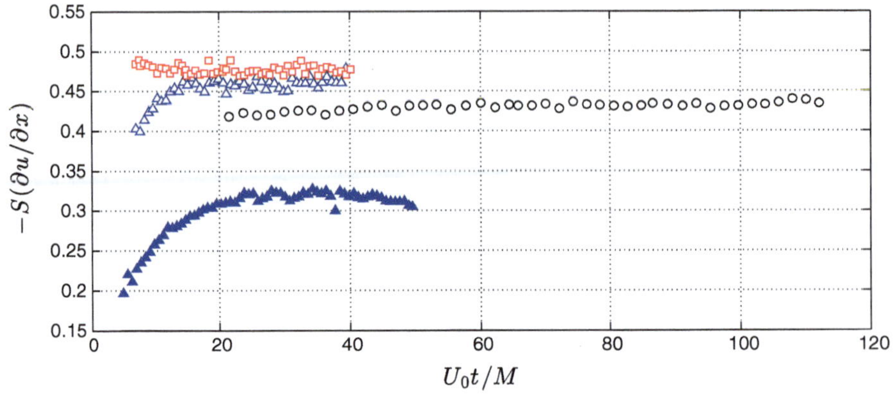

Fig. 4 Evolution of the velocity derivative skewness in the wake of (△) $Fs39$, (□) $Sq39$, and (○) $Rd38$; empty symbols represent measurements with a secondary contraction, and filled symbols represent data acquired without the secondary contraction

resolution are applied to the measurements without the secondary contraction, then the far-field value becomes ~ -0.43, which is in better agreement with contemporary estimates of $S(\partial u/\partial x)$ in grid turbulence [9, 11].

4 Reynolds Number in the Far-Field

The evolution of Re_λ is shown in Fig. 5 for $Fs39$, $Sq39$, and $Rd38$ at $Re_M = 28,500$ with a secondary contraction. Immediately apparent is that $Sq39$ produces the highest Re_λ turbulence for all U_0t/M. It also appears that $Rd38$ produces higher Re_λ than $Fs39$ for the U_0t/M where they overlap. This is perhaps surprising, given that it

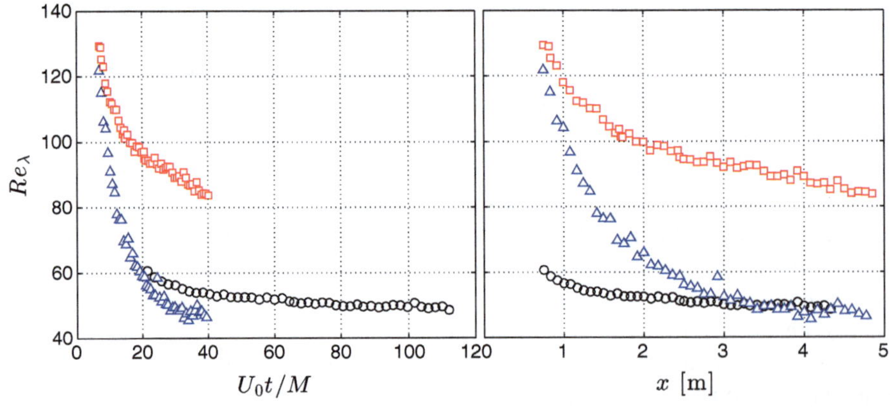

Fig. 5 Evolution of the local Reynolds number in the wake of (△) $Fs39$, (□) $Sq39$, and (○) $Rd38$. All data shown were acquired with the secondary contraction

Table 1 Power-law decay parameters for all grids

Grid	Re_M	Fit range $[U_0t/M]$	U_0t_{min}/M	U_0t_0/M	n	χ [%]
$Rd38$	28,500	[21.5, 113.2]	27.8	+3.5	−1.23	0.88
$Sq39$	28,500	[14.2, 40.5]	15.6	+3.0	−1.32	0.29
$Fs39$	28,500	[6.9, 19.2]	8.6	−8.0	−2.92	0.79
$Fs39$	28,500	[19.6, 39.8]	19.6	+8.0	−1.25	1.26

The χ criterion is a measure of the rms difference between the measured data and the power-law fit; see [7] for details

is typically believed that fractals produce higher Re_λ than regular grids [10, 19]. The evolution of Re_λ is also plotted against dimensional units in Fig. 5, and Re_λ is comparable between $Rd38$ and $Fs39$ beyond 3.5 m. Hence, regardless of the choice of normalization, in dimensional units, the Re_λ produced by $Rd38$ and $Fs39$ is comparable sufficiently far downstream.

It is likely that the far-field flow of $Fs39$ has comparable Re_λ to $Rd38$ because of the extended region of rapid decay experienced by the wake of $Fs39$. This is due to the longer non-equilibrium region in its wake relative to the other grids. Evidence for this hypothesis is provided in Table 1 where power-law decay exponents are estimated based on the methodology proposed in [7]. In the near-field of $Fs39$, there is a rapidly decaying region that is absent from the wake of the other two grids within the investigation range. In this extended rapidly decaying region, most of the turbulent kinetic energy injected at the turbulence generation by the grid is dissipated, leaving less energy in the far-field. This results in lower far-field Re_λ than the regular grids, whose rapidly decaying non-equilibrium region is within the unobserved region closer to the grid.

5 Conclusions

Measurements were performed in the wakes of three turbulence generating grids with approximately the same blockage but different geometries. One grid was a square-fractal-element grid, $Fs39$, which featured an array of fractal elements mounted to a background mesh. The other two grids, $Sq39$ and $Rd38$, were regular grids with M and τ_0, respectively, matched to $Fs39$. The non-equilibrium and quasi-equilibrium regions of the flow were identified by analysis of the scale-by-scale kinetic energy budget for grid turbulence and the evolution of the velocity derivative skewness. Both methodologies suggested that the flow begins to approximate equilibrium turbulence near $U_0t/M \approx 20$ for $Fs39$. In the far-field, it was found that $Fs39$ produced the lowest Re_λ for constant Re_M. This result was verified in both non-dimensional and dimensional units, and was thus not a consequence of a chosen form of normalization. It was determined that the low far-field Re_λ for $Fs39$ was a result of an extended rapid decay region in the wake of the fractal which is associated with the non-equilibrium region. It thus appears that while fractals may offer the benefit of high Re_λ in the near-field compared to regular grids, by the time the flow has evolved into quasi-equilibrium turbulence, it has already depleted the majority of the turbulent kinetic energy injected by the turbulence generating mechanisms at the grid, resulting in a less energetic far-field.

Acknowledgments We acknowledge the financial support of the Natural Sciences and Research Council of Canada (NSERC).

References

1. Batchelor, G.K.: The theory of homogeneous turbulence. Cambridge University Press, Cambridge (1953)
2. Burattini, P., Lavoie, P., Antonia, R.A.: Velocity derivative skewness in isotropic turbulence and its measurement with hot wires. Exp. Fluids **45**, 523–535 (2008)
3. Comte-Bellot, G., Corrsin, S.: The use of a contraction to improve the isotropy of grid-generated turbulence. J. Fluid Mech. **25**(4), 657–682 (1966)
4. Danaila, L., Anselmet, F., Antonia, R.A.: An overview of the effect of large-scale nonhomogeneities on small-scale turbulence. Phys. Fluids **14**, 2475–2484 (2002)
5. Gomes-Fernandes, R., Ganapathisubramani, B., Vassilicos, J.C.: Evolution of the velocity gradient tensor in a spatially developing turbulent flow. J. Fluid Mech. **756**, 252–292 (2014)
6. Hearst, R.J., Buxton, O.R.H., Ganapathisubramani, B., Lavoie, P.: Experimental estimation of fluctuating velocity and scalar gradients in turbulence. Exp. Fluids **53**, 925–942 (2012)
7. Hearst, R.J., Lavoie, P.: Decay of turbulence generated by a square-fractal-element grid. J. Fluid Mech. **741**, 567–584 (2014)
8. Hearst, R.J., Lavoie, P.: Scale-by-scale energy budget in fractal element grid-generated turbulence. J. Turb. **15**(8), 540–554 (2014)
9. Hearst, R.J., Lavoie, P.: Velocity derivative skewness in fractal-generated, non-equilibrium grid turbulence. Phys. Fluids MS #15-0279-L(accepted) (2015)
10. Hurst, D., Vassilicos, J.C.: Scalings and decay of fractal-generated turbulence. Phys. Fluids **19**(035103) (2007)
11. Isaza, J.C., Salazar, R., Warhaft, Z.: On grid-generated turbulence in the near- and far field regions. J. Fluid Mech. **753**, 402–426 (2014)
12. Krogstad, P.A., Davidson, P.A.: Freely decaying, homogeneous turbulence generated by multi-scale grids. J. Fluid Mech. **680**, 417–434 (2011)
13. Krogstad, P.A., Davidson, P.A.: Near-field investigation of turbulence produced by multi-scale grids. Phys. Fluids **24**(035103) (2012)
14. Lavoie, P., Djenidi, L., Antonia, R.A.: Effects of initial conditions in decaying turbulence generated by passive grids. J. Fluid Mech. **585**, 395–420 (2007)
15. Mazellier, N., Vassilicos, J.C.: Turbulence without Richardson-Kolmogorov cascade. Phys. Fluids **22**(075101) (2010)
16. Nagata, K., Sakai, Y., Inaba, T., Suzuki, H., Terashima, O., Suzuki, H.: Turbulence structure and turbulence kinetic energy transport in multiscale/fractal-generated turbulence. Phys. Fluids **25**(065102) (2013)
17. Sreenivasan, K.R., Antonia, R.A.: The phenomenology of small-scale turbulence. Annu. Rev. Fluid Mech. **29**, 435–472 (1997)
18. Valente, P.C., Vassilicos, J.C.: The decay of turbulence generated by a class of multiscale grids. J. Fluid Mech. **687**, 300–340 (2011)

19. Valente, P.C., Vassilicos, J.C.: Universal dissipation scaling for nonequilibrium turbulence. Phys. Rev. Lett. **108**(214503) (2012)
20. Valente, P.C., Vassilicos, J.C.: The non-equilibrium region of grid-generated decaying turbulence. J. Fluid Mech. **744**, 5–37 (2014)
21. Vallikivi, M., Hultmark, M., Bailey, S.C.C., Smits, A.J.: Turbulence measurements in pipe flow using a nano-scale thermal anemometry probe. Exp. Fluids **51**, 1521–1527 (2011)
22. Zhou, Y., Nagata, K., Sakai, Y., Suzuki, H., Ito, Y., Terashima, O., Hayase, T.: Relevance of turbulence behind the single square grid to turbulence generated by regular- and multiscale-grids. Phys. Fluids **26**(075105) (2014)

Anisotropy of Multiscale Grid Turbulence

Paolo D'Addio and Paolo Orlandi

Abstract The effect of the shape of the grid on the decay of turbulence is studied through DNS. We performed simulations of regular, two-scale and fractal grids. The multi-scale grids are characterized by a strong anisotropy in the near field, and these achieve isotropy at the large scales much farther with respect to the single-scale grid. The decay of turbulent kinetic energy (TKE) is affected by the shape of the grid. In the anisotropic region of the fractal grid the decay exponent $m = -2.89$ agrees with the values recently reported in literature. Once the flow achieves isotropy the decay is no longer influenced by the geometry of the grid, and the exponent recovers a value in the classical range $-1.4 \leq m \leq -1$.

1 Introduction

The unusual high turbulence intensities and R_λ generated by fractal grids have been first measured by *Hurst and Vassilicos* [3]. They asserted that by varying the geometry of the grid, it is possible to control the evolution of turbulence. They found that the TKE decays with a power law, with an exponent $m \approx -2.5$, different from that measured behind regular grids ($-1.4 \leq m \leq -1$). Successively Hearst and Lavoie [2] argued that the original set-up of [3] does not fully respect the geometrical condition between the wind tunnel length H and the mesh size M, which only for $H/M \gg 1$ should produce reasonable homogeneity. Therefore they designed a grid where the three iteration square fractal elements used by [3], were repeated several times over a regular single-scale grid. They found a decay rate with $m \approx -2.79$ in the region where the flow is inhomogeneous, in good agreement with results of [3]. Thanks to their set-up, they investigated the homogeneous region, measuring a decay exponent of $m \approx -1.39$, in accordance with the values usually measured behind regular grids.

P. D'Addio (✉) · P. Orlandi
Dip. Mecc. Aer., Sapienza University of Rome, Rome, Italy
e-mail: paolo.daddio@uniroma1.it

P. Orlandi
e-mail: orlandi@dma.ing.uniroma1.it

J. Peinke et al. (eds.), *Progress in Turbulence VI*,
Springer Proceedings in Physics 165, DOI 10.1007/978-3-319-29130-7_41

They thus reconciled fractal-generated turbulence measurements with classical grid-turbulence measurements, pointing out that the non-classical behaviour of turbulence is confined in the anisotropic region. The differences between the decay of fractal and traditional grids have been attributed to the influence of the multi-scale initial conditions. To understand better how the shape of the grids affects the flow, Direct Numerical Simulations of the flow generated by regular, two-scale and fractal grids, are here presented and discussed.

2 Numerical Experiments

The details of the numeric and of the immersed boundary method used to reproduce the interaction between the flow and the solid grids can be found in Orlandi and Leonardi [5]. The grid are shown in Fig. 1. G_{4-4-S1} and $G_{32-8-S1}$ have a solidity of $S1 = 0.57$ and the fractal G_{F4-S2} has $S2 = 0.257$. $G_{32-8-S1}$ is a two scale grid with a mesh ratio equal to 4, close to that used by Veeravalli and Warhaft [6]. The fractal grid is composed by four iterations of the basic square shape. Further details about the geometrical parameters of this grid can be found in Hurst and Vassilicos [3]. The simulations were performed at Reynolds number $Re_{L_0} = U_1 L_0/\nu = 18840$ in a computational domain $L_1 \times L_2 \times L_3 = 36\pi \times 2\pi \times 2\pi$ discretized with $n_1 \times n_2 \times n_3 = 2304 \times 192 \times 192$ points ($L_2 = L_3 = L_0$). Periodicity is assumed in y and z. To reproduce the thickness of the smallest element of the fractal grid the number of points is doubled in y and z. In the plots the coordinate is normalized with L_0 starting from the position of the grid x_G: $x^* = (x - x_G)/L_0$.

3 Isotropy

To assess the achievement of isotropy at large scales it is worth looking at the ratio of the turbulent stresses in the stream-wise and in an orthogonal direction to the grid. The theoretical value for $T_u(x)/T_v(x) = \langle u^2 \rangle^{1/2} / \langle v^2 \rangle^{1/2}$ for isotropic turbulence should

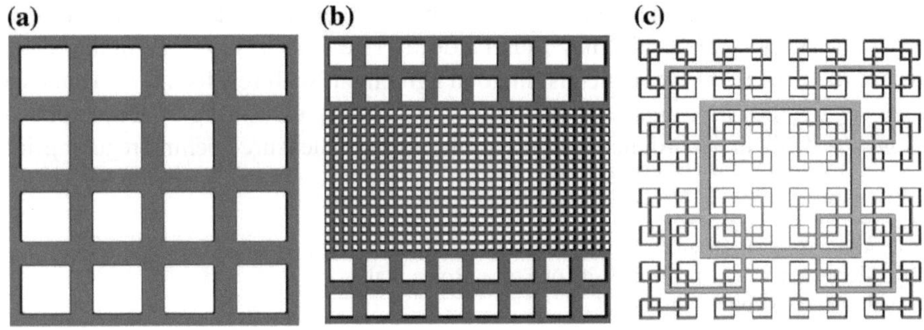

Fig. 1 Geometry of the grids: **a** G_{4-4-S1}; **b** $G_{32-8-S1}$ and **c** G_{F4-S2}

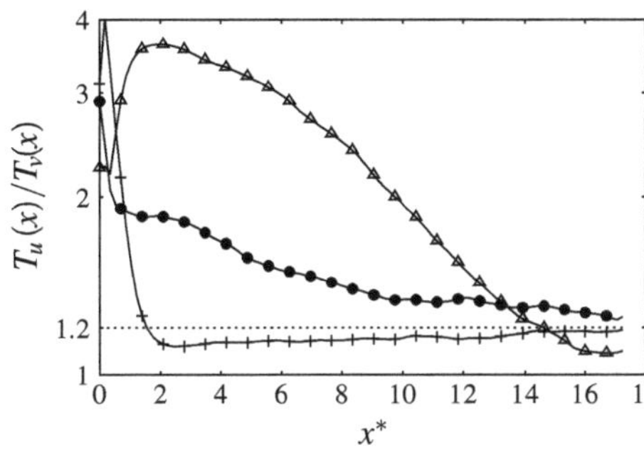

Fig. 2 Turbulence intensity $T_u(x)/T_v(x) = \langle u^2 \rangle^{1/2} / \langle v^2 \rangle^{1/2}$ as function of the stream wise non-dimensional coordinate x^*. Symbols: (+) G_{4-4-S1}, ($\triangle$) $G_{32-8-S1}$, ($\bullet$) G_{F4-S2}

be equal to 1. $T_u(x)$ and $T_v(x)$ are obtained by averaging in the plane of the grid $(y - z)$ and in time. Figure 2a shows that the shape of the grid strongly influences the distance where isotropy is achieved. In the near-field the flow is anisotropic and the value of $T_u(x)/T_v(x)$ strictly depends on the geometry. Grid turbulence measurements by Comte-Bellot and Corrsin [1] show that the flow can be considered homogeneous if $T_u(x)/T_v(x) \approx 1.2$. The present results in Fig. 2a show that for the single-scale grid G_{4-4-S1} this condition is achieved at $x^* \approx 2$, and that $G_{32-8-S1}$ generates a flow with the highest anisotropy level, with the peak of $T_u(x)/T_v(x)$ at $x^* \approx 2$. The location of the peak far from the grid, and the high anisotropy, are due to the formation of two different scales that persist until $x^* \approx 14$. The peak of $T_u(x)/T_v(x)$ for the fractal grid is located at the grid and the anisotropy persists much longer than that for the single-scale grid. The ratio $T_u(x)/T_v(x)$ decreases monotonically, but more gradually than $G_{32-8-S1}$ and it approaches the value of 1.2 only at the end of the domain.

4 Power-Law

The decay of turbulent kinetic energy decays in x as: $\langle q \rangle = A \left(\frac{x}{L_0} - \frac{x_0}{L_0} \right)^m$, where x_0 is the virtual origin and m the decay exponent. m and x_0 are obtained by the least-squares algorithm and are given in Table 1. The fitting ranges were chosen following the procedure of Hearst and Lavoie [2], according to which a first power law is sought in the inhomogeneous region, and a second in the homogeneous region, that is where the variations of $\langle U(y) \rangle$ around the mean value $\langle \overline{U} \rangle$ are $\pm 1\,\%$ (The over-hat accounts for the averages in y, z and time). The value of the decay exponent m in the homogeneous regions of G_{4-4-S1} is within the range $-1.4 \leq m \leq -1$ found in the past in many experiments (Mohamed and LaRue [4]). The same values are found for the multi-scale grids in their range of homogeneity, that are respectively $[15:18]$ for $G_{32-8-S1}$ and $[12:18]$ for G_{F4-S2}. The inhomogeneous field of $G_{32-8-S1}$ is formed by two different flows, one is dominated by the large scales and the other by the small ones. This implies that near the grid these two flows decay independently

Table 1 Power law fitting parameters

	Fit range $[x/L_0]$	m	x_0
G_{4-4-S1}	[4 : 18]	−1.41	1.48
$G_{32-8-S1}$	[15 : 18]	−1.17	6.92
G_{F4-S2}	[4 : 12]	−2.89	−12.39
G_{F4-S2}	[12 : 18]	−1.45	9.45

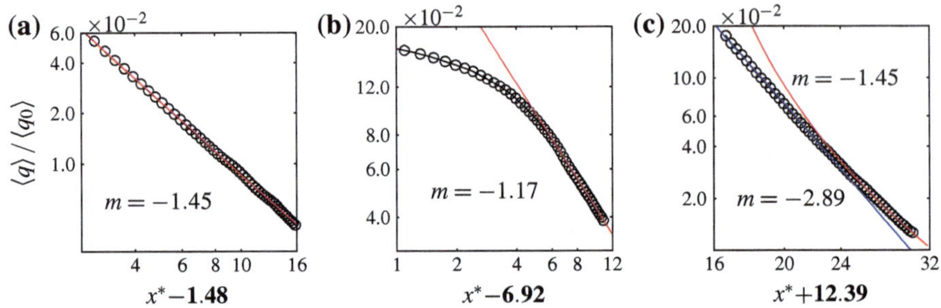

Fig. 3 Decay of the turbulent kinetic energy $\langle q \rangle / \langle q_0 \rangle$ plotted versus $x^* - x_0/L_0$ in log–log scale. Only every 2 points is plotted to reduce clutter. **a** G_{4-4-S1}. **b** $G_{32-4-S1}$. **c** G_{F4-S2}: only the plot with x_0/L_0 in the range [4 : 14] is shown

with different values of m and x_0. Therefore it has not been possible to find a power-law, with a single value of m and x_0. For the fractal grid, we found a power law in the inhomogeneous range [4 : 12] with $m = -2.89$, in agreement with the values previously reported by [2] and [3]. In the homogeneous range [12 : 18], the decay exponent recovers a traditional value $m = -1.50$, close to that of G_{4-4-S1}. The decay of $\langle q \rangle$ for all the grids is shown in Fig. 3a–c.

References

1. Comte-Bellot, G., Corrsin, S.: The use of a contraction to improve the isotropy of grid-generated turbulence. J. Fluid Mech. **25**, 657–682 (1966)
2. Hearst, R.J., Lavoie, P.: Decay of turbulence generated by a square-fractal-element grid. J. Fluid Mech. **741**, 567–584 (2014)
3. Hurst, D., Vassilicos, J.: Scaling and decay of fractal-generated turbulence. Phys. Fluids **19** (2007)
4. Mohamed, M.S., LaRue, J.C.: The decay power law in grid-generated turbulence. J. Fluid Mech. **219**, 195–214 (1990)
5. Orlandi, P., Leonardi, S.: DNS of turbulent channel flows with two- and three-dimensional roughness. J. Turbul. **7–53**, 1468–5248 (2006)
6. Veeravalli, S., Warhaft, Z.: The shearless turbulence mixing layer. J. Fluid Mech. **207**, 191–229 (1989)

Flow Field Topology of Impinging Jets with Fractal Inserts

Gioacchino Cafiero, Stefano Discetti and Tommaso Astarita

Abstract An experimental investigation of the flow field features of a round air jet equipped with a fractal (FG) and a regular (RG) grid insert impinging on a flat surface is carried out by means of 2D-2C Particle Image Velocimetry (PIV). The results are compared to those for a round jet without any grid (JWT). The test Reynolds number is set to 10,000. The average flow fields and the turbulent kinetic energy distributions are presented. In particular, the effect of the presence of the fractal grid on the turbulence intensity distribution and on the planar component of the Reynolds stress is analyzed. Some differences between the location of the maximum of the turbulence intensity profile and the data reported in the literature are found. A possible interaction process between the wakes of the grids and the growing shear layer of the jet might be responsible of this discrepancy. A comparison between the flow field and the heat transfer results obtained by the authors in a previous work is also carried out. What is underlined is that both an higher turbulence level and a much stronger axial velocity cause an increment in the heat transfer rate.

1 Introduction

Even though many turbulence theories relate the vortex cascade process with an underlying fractal structure (see [1–3]), the interest on the turbulence generated by fractal elements has grown only in recent years. Queiros-Conde and Vassilicos [4] and Staicu et al. [5] measured the turbulence statistics in the wake of fractal tree-like

G. Cafiero (✉) · T. Astarita
Dipartimento di Ingegneria Industriale, Università Degli Studi di Napoli Federico II,
21 Via Claudio, 80125 Napoli, Italy
e-mail: gioacchino.cafiero@unina.it

T. Astarita
e-mail: tommaso.astarita@unina.it

S. Discetti
Aerospace Engineering Group, Universidad Carlos III de Madrid,
28911 Leganes, Spain
e-mail: sdiscett@ing.uc3m.es

J. Peinke et al. (eds.), *Progress in Turbulence VI*,
Springer Proceedings in Physics 165, DOI 10.1007/978-3-319-29130-7_42

generators. Though their investigations were not conclusive in distinguishing the effects of the finite size of the fractal tree and their self-similar structure, they pointed out how turbulence with a much more elongated production region can be generated by using a fractal stirrer with respect to the case of traditional "regular grid" turbulators. This stimulated on one side the development of fractal forcing techniques for numerical simulations [6, 7], while on the other side it led to the first systematic investigation of the wind tunnel turbulence generated by fractal grids. Hurst and Vassilicos [8] tested a total of 21 planar fractal grids from 3 different families: fractal cross grids, fractal I grids and fractal square grids. The interest in the first two families of fractal grids in the following years has been quite limited. Geipel et al. [9] used fractal cross grids to improve the turbulent mixing in opposed jet flows; Kinzel et al. [10] applied the same type of grid to increase the turbulence intensity and the local *Re* in shear-free turbulence under the influence of system rotation.

Mazellier and Vassilicos [11] proposed a scaling based on the interaction of the wakes of the different bars. Differently from the case of classical regular grids, the bars of the grid have different size and spacing. As a consequence, their wakes interact at different streamwise locations (see [11] for a more detailed discussion). They introduced the wake-interaction lenghtscale as $x^* = L_0^2/t_0$. It can be defined also for regular grids (RGs), where L_0 is replaced by the effective meshlength. The data collected by Jayesh and Warhaft [12] in the region $1 \leq x/M \leq 30$ (where M is the grid meshlength) highlight the existence of a highly inhomogeneous turbulence in the near-field ($x/M < 3$, corresponding to approximately $x_{peak}/x^* \approx 0.55$, only slightly larger than $x_{peak}/x^* \approx 0.45$, typical of square fractal grids SFGs), where production is dominant, with a peak of turbulence intensity and the subsequent well-known power law decay. What actually makes the difference between regular and fractal grids is the value of x^*; for regular grids this value is typically small, thus leading to a turbulence intensity peak higher than the one which characterizes fractal grids, but very close to the grid, and consequently more difficult to be exploited. Recently, Cafiero et al. [13] exploited the feature of the turbulence peak located relatively far from the grid to enhance the heat transfer of impinging jets injecting turbulent energy into the jet potential core. The comparison between the heat transfer achieved by the jet with a fractal turbulator and that obtained by a free flowing jet, or a jet equipped with an equivalent regular grid, highlights the striking improvement obtained using fractal configuration under the same power input. However, whether the enhancement was related to the wake-interactions, to the local distribution of velocities (due to the non-uniform blockage ratio) or to the shear layer wakes interaction it was still open for further investigation.

In this work the comparison of the flow field generated placing both a fractal and a regular grid in correspondence of the nozzle exit section is investigated and compared to the one obtained in absence of any grid (JWT). In the next section the experimental apparatus used for this experiments is briefly introduced; the mean flow field characteristics are analyzed and discussed in the Data analysis section. In the final section, some conclusions are drawn and the flow field data are compared to those regarding the heat transfer features of impinging fractal jets, obtained in [13].

2 Experimental Setup and Data Processing Technique

The experimental apparatus is schematically described in Fig. 1. The air provided by a fan is conditioned in temperature with a heat exchanger and then the mass flow rate is measured with a Venturi tube. The air passes through a plenum chamber (in which the total pressure is measured), a straight pipe, which length is $6d$ (where d is the exit diameter, equal to 20 mm) and, if present, through the insert and a terminating cap. The air impinges normally on a glass plate (230 mm long, 270 mm wide, 4 mm thick). The separation (indicated with the symbol h) of the nozzle exit section from the target plate is adjusted by using a precision translation stage to move the jet plenum chamber along its own axis perpendicular to the impinged surface. The flow is seeded with olive oil particles of $1\mu m$ diameter thus ensuring that the Stokes number is much lower than one. A double cavity Quantel Evergreen EVG00200 laser (wavelength 532 nm, pulse intensity 200 mJ, maximum frequency 15 Hz) is used to illuminate the flow field. The laser beam is shaped into a planar sheet using an optical lenses system made of two spherical lenses $f = -25.4$ mm and $f = 50$ mm respectively) and a cylindrical lens ($f = 50$ mm). Sequences of tracer particles are collected using a sCMOS camera, Andor Zyla 5.5 Mpixel equipped with a Vivitar 100 mm Macro objective, $f_{\#} = 22$. The distance of the camera from the investigated plane is set so that the recorded area dimensions are equal about to $1.8d$ along the nozzle axis and $2d$ in the orthogonal direction (i.e. parallel to the impinged plate). The test Reynolds number based on the nozzle exit section diameter is set to about 10,000 for the three tested cases.

The acquired images are processed using a final window size equal to 48×48 pixel with 75 % overlap. The raw images are interpolated using a third order spline method as described in [14]; the velocity field is then interpolated using a spline fitting [15]. A Blackmann filtering both of the velocity and of the correlation map field are then applied [16].

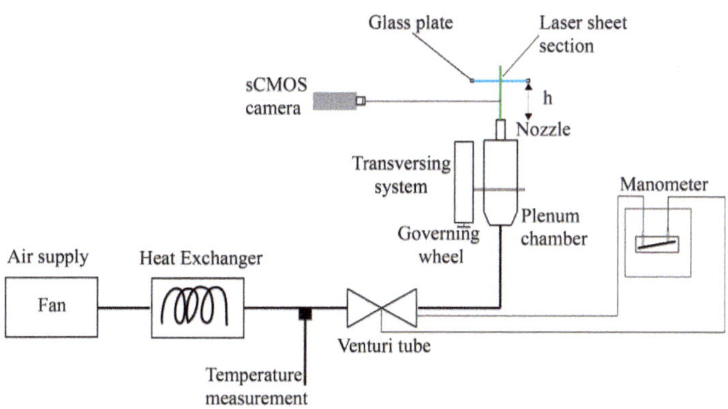

Fig. 1 Experimental apparatus schematic representation

3 Data Analysis

The mean axial component of the velocity vector over 1,000 realization for the case of FG, RG and JWT is reported in Fig. 2. The streamwise evolution of the flowfield in the three cases is quite different. In fact for the FG, due to the uneven blockage ratio, the jet is issued through the central iteration of the grid and through the gap that is formed between the grid bar and its edge. On the other side, the presence of the grid iteration causes a strong recirculation region. The effect of the grid bar is still evident up to the impingement region, where three separated jets impinge onto the glass surface. The behaviour in the RG case is quite different. In the vicinity of the grid, the jet behaves in a multichannel fashion; however, beyond $Y/d = 0.6$ the axial velocity is uniform as the effect of the grid is smeared out by turbulent diffusion. In this region, the flow field results to be practically equivalent to the one of a JWT. In this last case, the potential core is perturbed by the presence of the impinged plate, which causes the jet deceleration. Figure 3 reports the axial distribution of the *tke* (turbulent kinetic energy) normalized with respect to the maximum velocity of the jet. As already discussed in [11], the presence of the regular grid generates a very

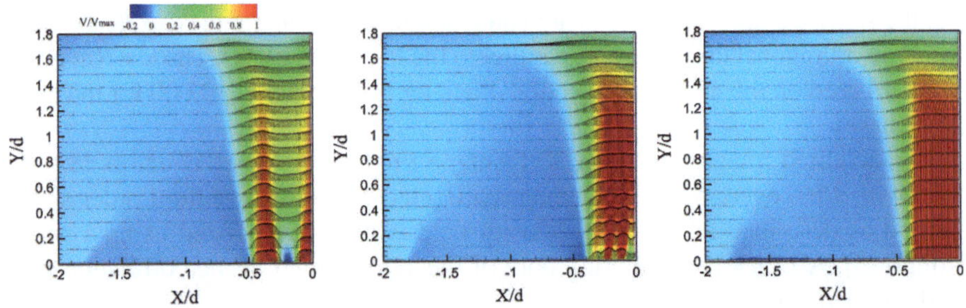

Fig. 2 Mean axial velocity normalized with respect to the maximum velocity in the FG, RG and JWT cases

Fig. 3 Tubulent kinetic energy profile along the jet axis

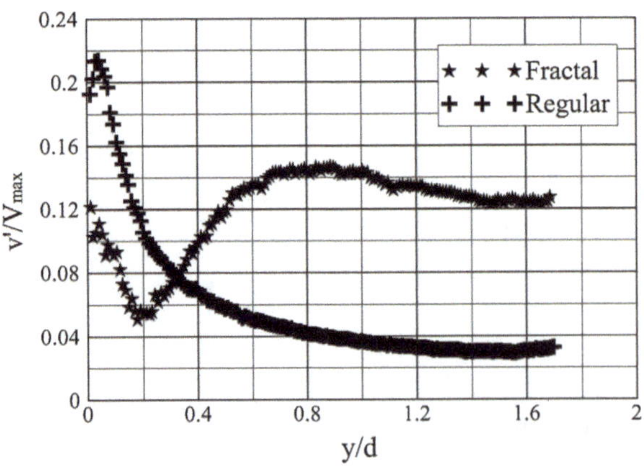

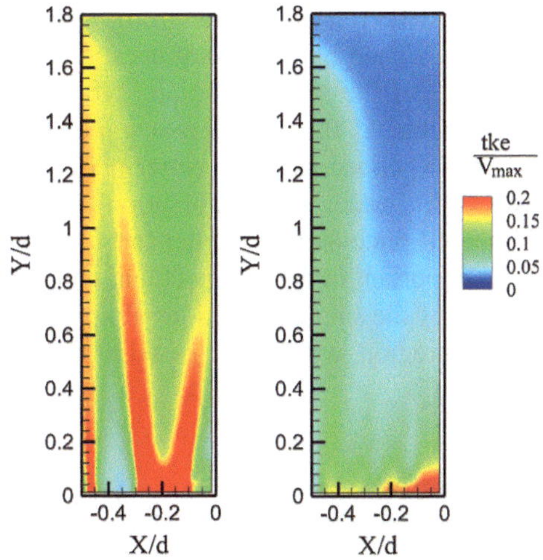

Fig. 4 Tubulent kinetic energy distribution in the FG and RG cases

strong production but also a very fast decay. The FG, on the other side, presents an elongated production region, a local maximum (lower than the RG case) and finally a decay which is much smoother than the former case. The higher turbulence level, along with the higher axial velocity play a key role in the strong heat transfer enhancement which can be detected in the FG case (as extensively described in [13]).

The spatial distribution of the turbulent kinetic energy in the FG and RG cases is reported in Fig. 4; it further underlines the *multiscale* interaction of the wakes of the fractal grid in opposition to the single scale of the RG case. In the former case, the wakes shed from the largest bar iteration meet at $\approx Y/d = 0.8$, where the maximum of the *tke* can be detected along the nozzle axis. For the RG case, there is a region of local maximum within $0.2d$ beyond the grid and then an uniform dissipation as Y/d increases. It is worth noting that, according to [11], the maximum in the *tke* distribution should be located at about $Y/d = 2.25$. This is not true in the present case, where, the presence of the impinged plate from one side, and the interaction with the jet shear layer from the other both act to shift the maximum towards the nozzle exit section (thus exasperating the production rate).

4 Conclusions

An experimental analysis of the flow field features of impinging jets equipped with fractal inserts is carried out. As already pointed out by the authors in an another work [13], this flow field is of particular interest for several engineering applications, such as the heat transfer and mixing enhancement. Differently from what reported in the literature in the case of wind tunnel tests, the location of the maximum cannot be predicted using the correlation proposed by [11]. It is the authors' opinion that this must be related to the strong interaction of the jet growing shear layer with the wakes shed by the fractal iterations bars. This effect is further enhanced by the presence of the impinged plate, which exasperates the production rate. Although not conclusive, this work gives some answers to the questions arisen in a previous work [13] concerning the better performances of the FG with respect to both RG and JWT. In fact, the combination of the strong axial velocity in proximity of the impinged plate and the higher turbulence level which characterize the FG case are both important in the enhancement which is achieved in this case. However, the role played by the smaller iterations along with a characterization of what happens in the wall jet region, will be addressed to future studies.

References

1. Mandelbrot, B.B.: Intermittent turbulence in self-similar cascades: divergence of high moments and dimension of the carrier. J. Fluid Mech. **62**, 331–358 (1974)
2. Mandelbrot, B.B.: The Fractal Geometry of Nature. Freeman, San Francisco (1982)
3. Sreenivasan, K.R., Maneveau, C.: The fractal facets of turbulence. J. Fluid Mech. **173**, 357–386 (1986)
4. Queiros-Conde, D., Vassilicos, J.C.: Turbulent wakes of 3-D fractal grids in Turbulent Flows and Other Dynamical Systems. Cambridge University Press, Cambridge, UK (2001)
5. Staicu, A., Mazzi, B., Vassilicos, J.C., van de Water, W.: Turbulent wakes of fractal objects. Phys. Rev. Lett. E **66**, 066306 (2003)
6. Mazzi, B., Okkels, F., Vassilicos, J.C.: A shell-model approach to fractal-induced turbulence. Eur. Phys. J. B/Fluids **28**, 243–251 (2002)
7. Mazzi, B., Vassilicos, J.C.: Fractal-generated turbulence. J. Fluid Mech. **502**, 65–87 (2004)
8. Hurst, D., Vassilicos, J.C.: Scaling and decay of fractal-generated turbulence. Phys. Fluids **19** 035103 (2007). doi:10.1063/1.2676448
9. Geipel, P., Goh, H.K.H., Lindstedt, P.R.: Fractal generated turbulence in opposed jet flows. Flow Turbul. Comb. **85**, 397–419 (2010)
10. Kinzel, M., Wolf, M., Holzner, M., Lthi, B., Tropea, C., Kinzelbach, W.: Simultaneous two-scale 3D-PTV measurements in turbulence under the influence of system rotation. Exper. Fluids **51**, 75–82 (2011)
11. Mazellier, N., Vassilicos, J.C.: Turbulence without Richardson-Kolmogorov cascade Physics of Fluids **22** 075101 (2010). doi:10.1063/1.3453708
12. Jayesh, Warhaft Z.: Probability distribution, conditional dissipation, and transport of passive temperature fluctuations in grid? generated turbulence. Phys. Fluids **4**, 2292–2307 (1992)
13. Cafiero, G., Discetti, S., Astarita, T.: Heat transfer enhancement of impinging jets with fractal-generated turbulence. Int. J. Heat. Mass Trans. **75**, 173–183 (2014)
14. Astarita, T.: Analysis of interpolation schemes for image deformation methods in PIV: effect of noise on the accuracy and spatial resolution. Exper. Fluids **40**, 977–987 (2006)
15. Astarita, T.: Analysis of velocity interpolation schemes for image deformation methods in PIV. Exper. Fluids **45**, 257–266 (2008)
16. Astarita, T.: Analysis of weighting windows for image deformation methods in PIV. Exper. Fluids **43**, 859–872 (2007)

Experimental Study on Hot-Wire Spatial Resolution in Turbulent Round Jet

Tommaso Fiorini, Gabriele Bellani, Andrea Cimarelli and Alessandro Talamelli

Abstract Hot-wire spatial-resolution effects are investigated in a round jet, using custom-made probes with Platinum wires of 5 and 2.5 μm of diameter. Characteristic turbulent scales are varied by both moving the probes along the jet centerline ($10 < x/D < 30$), and using two different nozzle diameters at a constant Reynolds number of $Re_D = 7 \times 10^4$. The variance of spatial derivatives shows an attenuation of one order of magnitude greater than the velocity variance, i.e. 10–30 % compared to 1–2 %. Power spectra show this attenuation to be located at high frequencies, in agreement with the hypothesis of spatial-filtering effect.

1 Introduction and Experimental Set-Up

Turbulent jets are recurrent in industrial applications and natural environments. For this reason they have been the object of a multitude of theoretical, experimental and numerical studies over the years (see [1] for a complete review). Many of the experimental studies have been carried out using hot-wires anemometers, because of their relatively high spatio-temporal resolution. However, when turbulent structures become smaller than the wire length, an error is introduced in the measurements due to spatial filtering. This is an extremely common occurrence in high Reynolds number wall bounded flows, see for instance [3–5]. In free shear flows, such as jets, there is no influence from the wall, and resolution requirements are not as severe. Despite this, turbulence statistics might still be affected by the phenomena, especially if velocity gradients are of concern.

We performed the experiments in the Coaxial Aerodynamic Tunnel (CAT) of the University of Bologna, located in Forlì. A complete description of the facility can be found in [6, 7]. A centrifugal fan coupled with screens and honeycombs provides a top-hat velocity profile with 0.3 % turbulence intensity from a 50 mm round nozzle. To obtain a different flow case we also use a PVC adapter that provides a 30 mm outlet with the same top-hat velocity profile of the standard nozzle (Fig. 1). We

T. Fiorini (✉) · G. Bellani · A. Cimarelli · A. Talamelli
Department of Industrial Engineering, Alma Mater Studiorum,
Università di Bologna, 47100 Forlì, Italy
e-mail: tommaso.fiorini3@unibo.it

J. Peinke et al. (eds.), *Progress in Turbulence VI*,
Springer Proceedings in Physics 165, DOI 10.1007/978-3-319-29130-7_43

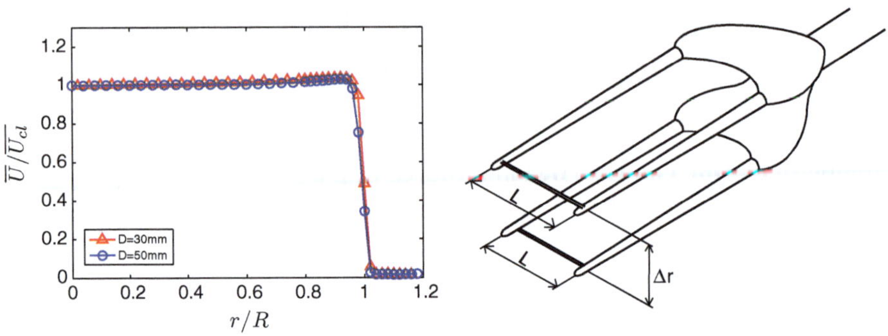

Fig. 1 On the *left*, the normalized mean velocity exit profile of the two nozzles. On the *right*, schematic of the two-wire probe; L is the length of the wires and Δr is the distance between them

Table 1 Experimental parameters

Parameter	Case 1	Case 2
Nozzle diameter, D	50 mm	30 mm
Exit velocity, U_j	22.1 m/s	36.9 m/s
Reynolds number, Re_D	70×10^4	70×10^4
Characteristic frequency, U_j/D	442 Hz	1230 Hz
Acquisition frequency, f_a	20 kHz	60 kHz
low-pass filter, LPF	10 kHz	30 kHz
Acquisition time, T	120 s	40 s
Measurements' points, x/D	10, 15, 20, 25, 30	10, 15, 20, 25, 30
L_1/λ	0.31, 0.29, 0.25, 0.23, 0.22	0.60, 0.57, 0.54, 0.57, 0.64
L_2/λ	0.15, 0.14, 0.13, 0.11, 0.11	0.30, 0.28, 0.27, 0.28, 0.34
L_1/η	16.6, 14.3, 11.9, 9.9, 8.5	29.7, 25.3, 20.9, 17.6, 15.6
L_2/η	8.3, 7.2, 6.0, 4.9, 4.2	14.8, 12.6, 10.4, 8.8, 7.8

manufacture hot-wire probes with Wollaston Platinum wires of 5 and 2.5 μm of diameter, aspect ratio is 200 for both wires, resulting in 1 mm (L_1) and 0.5 mm (L_2) long wires, respectively. A special two-wire probe, with parallel wires placed normal to the flow, (see Fig. 1) allows us to measure spatial derivative quantities. The distance Δr between the wires is ≈750 μm.

We measure on the axis of the jet, where uncertainties due to strong turbulence fluctuations are low (see [2]). We perform calibration against a Prandtl tube at the jet exit, both before and after each set of measurements, to check for signal drift. Mean flow temperature is acquired via a thermocouple placed in the flow and used to correct the hot-wire signal. We define two flow cases: one corresponding to the 50 mm (case 1) and one corresponding to the 30 mm outlet (case 2). Reynolds number Re_D is kept constant by adjusting the jet exit velocity, while turbulent scales (i.e. Taylor microscale λ or Kolmogorov scale η) vary according to the geometry, see Table 1.

2 Results and Discussion

The attenuation of the measured velocity variance for the present cases is of the order of 1–2 %. This effect is small and no clear trend can be extrapolated, we therefore focus on the radial and streamwise derivatives of the streamwise velocity: $\partial u/\partial x$ and $\partial u/\partial r$. The axial derivative is estimated using Taylor hypothesis and the mean velocity as convective velocity, while the radial derivative is obtained by taking the difference in instantaneous velocity measured by the two wires and dividing by Δr. The derivatives' variance and the ratio between values measured by different wires are shown in Fig. 2.

Radial derivative variance shows an attenuation of ≈10 %, when measured with the 1 mm wire. Even greater is the attenuation of the axial derivative, up to ≈40 %. For both terms, the attenuation is higher for case 2, characterized by higher values of L/λ. Another observation is that for the axial derivative the attenuation effect decreases with increasing axial distance, while for the radial derivative it increases. The estimated uncertainty (due to the unsteady nature of the flow and the finite acquisition time) of the derivative variance (with 95 % confidence interval) amounts

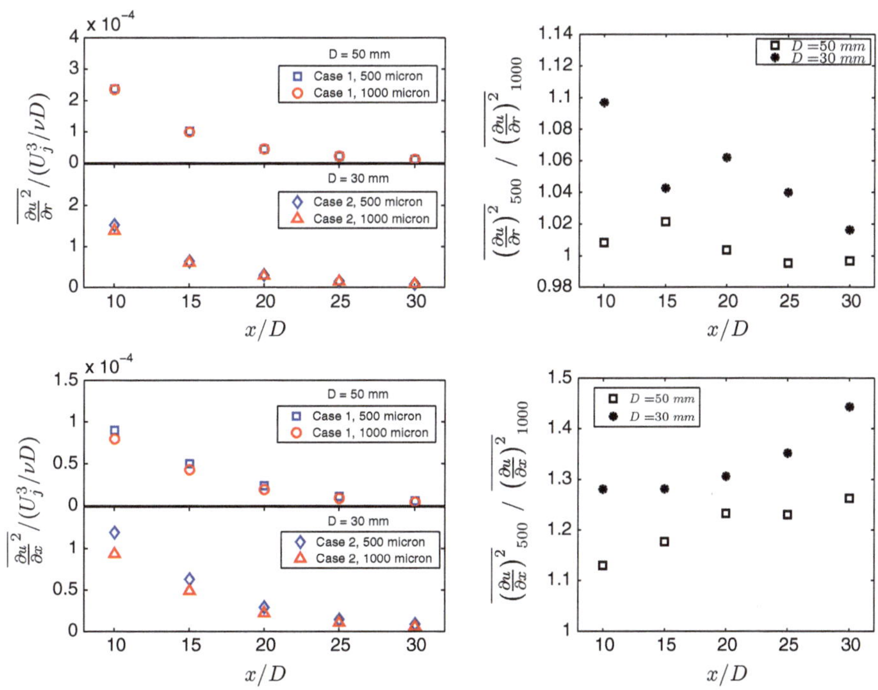

Fig. 2 On the *upper-left* the normalized radial derivative term; On the *upper-right* the ratio between the term measured by different wires. On the *bottom* half of the image the same results are reported for the axial derivative term. Marker size is used as an indicator of uncertainty due to finite acquisition time

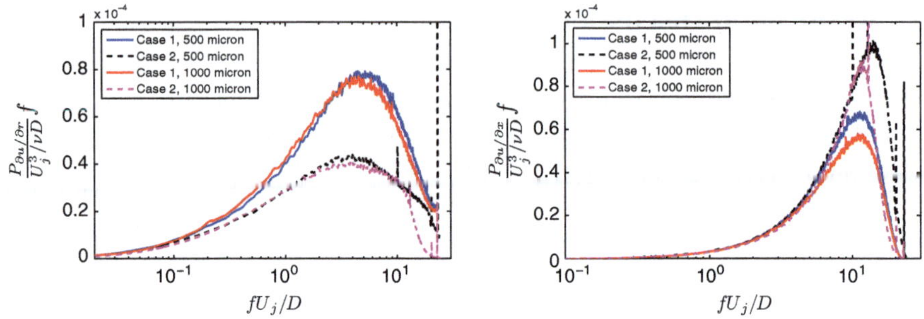

Fig. 3 On the *left* is shown the pre-multiplied PSD of the radial derivative signal and on the *right* the pre-multiplied PSD of the axial derivative signal. Both are for x/D = 10

to 1.5 % close to the nozzle, and up to 5.9 % for the farthest axial position, characterized by longer integral timescales. When measuring the radial derivative term an additional ±30 μm uncertainty on Δr has to be considered, resulting in a 5.3 % uncertainty on the variance of the radial derivative.

To understand how attenuation is distributed among frequencies, and therefore scales, we compute the power spectral density (PSD) of $\partial u/\partial x$ and $\partial u/\partial r$; pre-multiplied PSDs for $x/D = 10$ are shown in Fig. 3. The spectra obtained show good agreement at low frequencies, while at higher frequencies the 1 mm wire shows an attenuation with respect to the 500 μm one. Furthermore, a strong cut-off is observed in the spectrum of the 1 mm wire past 15 kHz. This effect is likely not caused by spatial resolution, but rather by frequency resolution limitation of the anemometer and hot-wire system.

Acknowledgments This research project has been partly supported by the European Community - Research Infrastructure Action under the FP7-INFRASTRUCTURES-2012-1: "European High-performance Infrastructures in Turbulence" INFRA-2012-1.1.20.

References

1. Ball, C.G., Fellouah, H., Pollard, A.: The flow field in turbulent round free jets. Prog. Aerosp. Sci. **50**, 1–26 (2012)
2. Hussein, J.H., Capp, S.P., George, W.K.: Velocity measurements in a high-reynolds-number, momentum-conserving, axisymmetric, turbulent jet. J. Fluid Mech. **258** (1994)
3. Hutchins, N., Nickels, T.B., Marusic, I., Chong, M.S.: Hot-wire spatial resolution issues in wall-bounded turbulence. J. Fluid Mech. **635**, 103–136 (2009)
4. Ligrani, P.M., Bradshaw, P.: Spatial resolution and measurement of turbulence in the viscous sublayer using subminiature hot-wire probes. Exp. Fluids **5**, 407–417 (1987)
5. Örlü, R., Alfredsson, H.: On spatial resolution issues related to time averaged quantities. Exp. Fluids **49**, 101–110 (2010)
6. Burattini, P., Talamelli, A.: Acoustic Control of a Coaxial Jet. J. Turbul. **8** (2007)
7. Segalini, A., Talamelli, A.: Experimental analysis of dominant instabilities in coaxial jets. Phys. Fluids **23** (2011)

Three-Dimensional Instabilities in the Wake of a Wall-Mounted Low-Aspect-Ratio Pyramid

Zahra Hosseini and Robert J. Martinuzzi

Abstract The turbulent wake dynamics of a wall-mounted low-aspect ratio square-based pyramid protruding a nominally thin turbulent boundary layer are investigated for a moderate Reynolds number of 28000. The three-component velocity data are measured using planar stereoscopic Particle Image Velocimetry (PIV), from which the coherent global velocity field dynamics are reconstructed using a surface-pressure sensor-based estimation. An estimation technique is implemented which improves on classical EPOD by accounting for phase delay information between the sensor and velocity data. It is shown that the energy exchange between a slow-varying global and harmonic shedding modes characterise the coherent wake dynamics.

1 Introduction

The flow around wall-mounted tapered bluff bodies protruding a boundary layer generally gives rise to highly three-dimensional quasi-periodic wakes. For low-aspect-ratio *AR* tapered plates [2] and square-based pyramids [6], with apex angles $15° \leq \zeta < 75°$ $(3.8 \geq AR > 0.65)$, the periodicity has been associated with alternate shedding of counter-rotating vortices from the opposing lateral sides of the obstacle. Unlike the case of mildly-tapered obstacles $(\zeta < 15°, AR > 3.8)$ [2, 6, 7], the shedding frequency for low-aspect-ratio bodies is constant over the entire height and scales with the base width rather than local Reynolds number. The inherent three-dimensionality induced by the taper significantly distorts the shed structures resulting in strong modulation of the shedding amplitude and increasing phase jitter with increasing height above the plate. The amplitude modulation, however, is coherent over the entire obstacle height [6] suggesting a strong coupling between a low-frequency global instability and the higher frequency shedding process.

Z. Hosseini · R.J. Martinuzzi (✉)
Department of Mechanical and Manufacturing Engineering,
University of Calgary, Calgary, AB, Canada
e-mail: rmartinu@ucalgary.ca

Z. Hosseini
e-mail: zahra.hosseini@gmail.com

J. Peinke et al. (eds.), *Progress in Turbulence VI*,
Springer Proceedings in Physics 165, DOI 10.1007/978-3-319-29130-7_44

Due to the inherently high phase jitter, a traditional phase-averaging approach based solely on a reference shedding phase cannot capture the amplitude and phase modulations needed to analyse the dynamic coupling between global low-frequency and shedding instabilities. Thus, in this work, a modification to the Extended Proper Orthogonal Decomposition (EPOD) [1] is implemented as a conditional averaging technique to capture aperiodic modulations and account for local cycle-to-cycle variations. This approach is then applied to investigate the dynamic coupling between the most energetic velocity modes.

2 Experimental Set-Up

Time-resolved planar 3C-PIV (LaVision Flow Master) measurements were performed in the wake of a wall-mounted square-based pyramid with aspect ratio $h/d = 0.87$ (apex angle $\zeta = 60°$). Images were processed using the LaVision DaVis 8.2 software. A schematic of the experimental set-up is shown in Fig. 1. The free stream velocity was $U_\infty = 10$ m/s, corresponding to $Re_d = U_\infty d/\nu = 28{,}000$, with turbulent intensity of 0.8 %. The on-coming turbulent boundary layer had a thickness of $\delta/h \approx 0.25$ and shape factor of 1.46. The velocity vectors (u, v, w) were measured in several horizontal $(x-y)$ planes with a sampling frequency of 500 Hz (capturing 10 data points per shedding cycle based on the previously measured mean shedding frequency f). At least 600 shedding cycles were captured for each realization. The fluctuating surface pressure on opposing pyramid side faces and on the flat plate in the wake was recorded, using AllSensors Corp pressure sensors (5INCH-D1-4V-MINI: range ±1.25 kPa, sensitivity: ±0.5 Pa), simultaneously with the velocity data. The pressure-system response was flat to 300 Hz. A typical velocity signal and its Power Spectral Density (PSD) function at a spatial point are shown in Fig. 2 (top). The signal displays a dominant periodic behaviour at

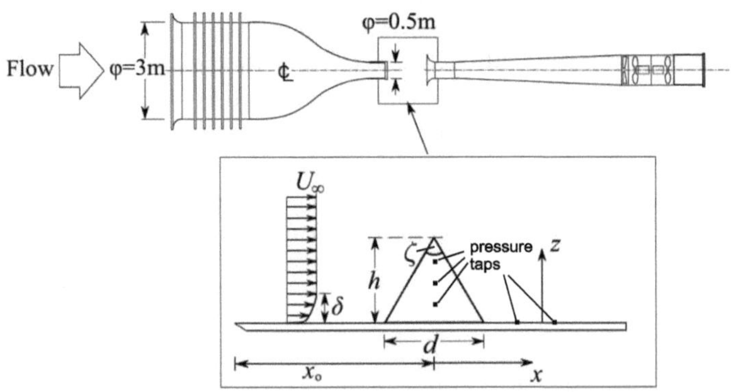

Fig. 1 Schematic of the open-test-section suction type wind tunnel and experimental set-up inside the working section, where $d = 45$ mm, $h = 39$ mm and $\zeta = 60°$

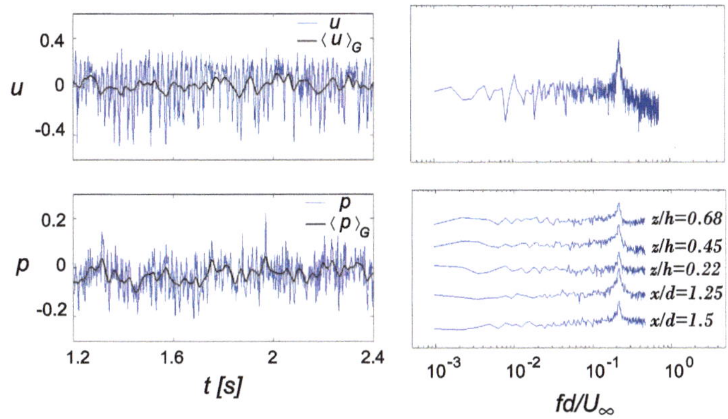

Fig. 2 The velocity (*top*) and pressure (*bottom*) fluctuations, and the PSD functions (log–log scale). *Black line* indicates slow-drift component

non-dimensional frequency of $St = fd/U_\infty = 0.231 \pm 0.005$, and high modulations in both the short-time-average of the signal (black line) and shedding amplitude. The fluctuating surface pressure data (Fig. 2, bottom), show a similar behaviour.

3 Sensor-Based Flow Estimation

The starting point is classical EPOD. Briefly, the spatio-temporal pressure data acquired at locations $\mathbf{x}$ from N_p sensors are expanded onto the POD basis: $p(\mathbf{x}, t) = \sum_{n=1}^{N_p} a_p^{(n)}(t)\phi_p^{(n)}(\mathbf{x})$, where $a_p^{(n)}(t)$ and $\phi_p^{(n)}(\mathbf{x})$ denote the nth temporal coefficient and spatial eigenvectors, respectively. The extended velocity modes are then defined: $\psi_u^{(n)}(\mathbf{X}) = \langle a_p^{(n)}(t)\mathbf{u}(\mathbf{X}, t)\rangle/\lambda_p^{(n)}$, where $\langle\cdot\rangle$ denotes the time-averaging operator; $\lambda_p^{(n)} = \langle a_p^{(n)}(t)a_p^{(n)}(t)\rangle$ and $\mathbf{X}$ denote the velocity domain. The velocity, then, is estimated using:

$$\hat{\mathbf{u}}(\mathbf{X}, t_{est}) = \sum_{n=1}^{N_{mode}} a_p^{(n)}(t_{est})\, \psi_u^{(n)}(\mathbf{X}) \tag{1}$$

where t_{est} denotes the estimation time and $N_{mode} \leq N_p$.

Due to inherent limitations in experiments, the EPOD approach does not guarantee optimality of the estimations. Typically, the data are collected from sensors at a few discrete locations and the ideal placement is unknown a priori. Consequently, a mode-specific lag will exist between the sensor and velocity field data. Generally, $\psi_u^{(n)}(\mathbf{X})$ are non-orthogonal such that the mode-specific velocity-sensor lag cannot be isolated and, hence, cannot be accounted for explicitly.

Herein, to optimise the estimation the velocity data are expanded onto the optimum orthogonal subspace $\phi(\mathbf{X})$, obtained from a POD of the velocity field in each uncorrelated PIV plane. The spatio-temporal correlations of (1), then, reduce to temporal correlations according to:

$$\widetilde{\mathbf{u}}(\mathbf{X}, t_{est}) = \sum_{k=1}^{N_u} \widetilde{a}_u^{(k)}(t_{est})\, \phi_u^{(k)}(\mathbf{X}) \tag{2a}$$

$$\widetilde{a}_u^{(k)}(t_{est}) = \sum_{n=1}^{N_{mode}} a_p^{(n)}(t_{est})\, \frac{\langle a_p^{(n)}(t)\, a_u^{(k)}(t)\rangle}{\lambda_p^{(n)}} \tag{2b}$$

where $\phi_u^{(k)}(\mathbf{X})$ and $a_u^{(k)}(t)$ are, respectively, POD eigenmodes and corresponding temporal coefficients. This simplification enables developing different treatments to recover the phase information of modes with different characteristics. Optimisation is then achieved by accounting for the sensor delay to maximise the pressure–velocity correlations.

For harmonic modes, a modified multi-time-delay approach [3] is used to optimise the velocity-pressure correlations. Briefly, data from each physical sensor is augmented by M virtual sensors containing the same data but delayed by $m\Delta\tau_s$ (for the mth virtual sensor). This approach ensures the synchronisation of the first and higher harmonics. For the slow-drift modes (a_u^{Δ}, a_p^{Δ} for velocity and pressure, respectively), the optimal time delay for synchronisation is the estimated lag of the first maximum of the a_u^{Δ}-a_p^{Δ} cross-correlation function. Omitting this step introduces an artificial and random phase lag between the non-harmonic and harmonic fluctuations; resulting in a misrepresentation of the coherent strain field and the dynamics of the coherent motion. Further details of the methodology can be found in [4].

4 Results and Concluding Remarks

The proposed method offers a systematic way to reconstruct the global field from planar measurements. The improvement gained with this methodology in the rendering of the flow topology and coherent motion is illustrated in Fig. 3a. This figure com-

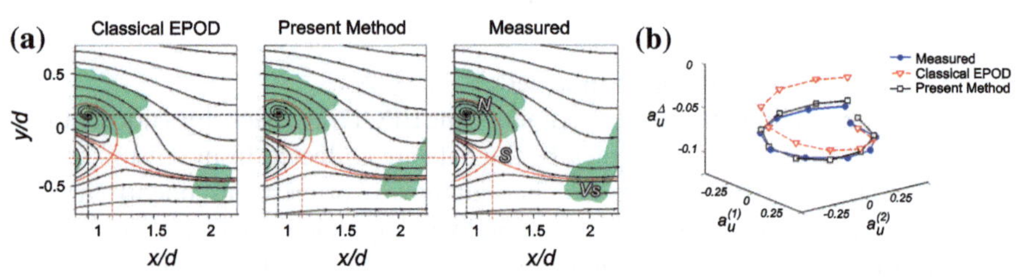

Fig. 3 **a** Sectional streamlines and vortex core regions ($\lambda_2 \leq -0.05$) for measured and estimated velocity vector fields at $z/h = 0.23$. **b** Behaviour of a typical cycle in ($a_u^{(1)}$, $a_u^{(2)}$, a_u^{Δ}) space indicating that modified technique captures better the trajectory compared to classical EPOD

pares the sectional streamlines and the vortex core regions (using the λ_2-criterion of [5]) obtained from the estimated and measured velocity field for a randomly selected PIV-snapshot for the plane $z/h = 0.23$. The shed vortex core (marked ***Vs***) is well estimated using the present methodology, whereas it is incompletely resolved with previous EPOD method. The estimation of the location of the saddle point ***S*** and forming vortex focus ***N*** is also improved with the proposed method. Thus, accounting for the sensor delay better renders the coherent strain field; improving the estimation of the vortex core location and the impact of the non-harmonic contributions to the shedding modulation.

More significantly, the trajectory in the $(a_u^{(1)}, a_u^{(2)}, a_u^{\Delta})$ space is better captured in the modified technique (Fig. 3b). This trajectory is critical to the proper representation of the flow dynamics. Here, the slow-drift coefficient, a_u^{Δ}, is directly correlated with amplitude of the first-harmonic $(\sqrt{(a_u^{(1)})^2 + (a_u^{(2)})^2})$ and essentially describes the envelope (i.e. the cycle-to-cycle modulation) associated with the strength of the shed vortices. Hence, $a_u^{\Delta}(t)$ must be correctly synchronised with the shedding phase.

The global dynamic behaviour of the wake is illustrated in the modal space shown in Fig. 4. Typically, the harmonic oscillation amplitude, $a_u^{(1)}, a_u^{(2)}$, is largest when the slow drift mode a_u^{Δ} approaches a maximum and vice-versa (as locally observed in Fig. 2) indicating an energy exchange between modes. Most notable are events during which a_u^{Δ} abruptly moves between extrema and the harmonic amplitude remains low, indicating an interruption or break-down of the shedding process.

These events indicate a global rearrangement of the shedding structure, which is consistent with earlier observations of a low-frequency global instability [2, 7]. The present behaviour is consistent with a dual-structure shedding model [6], in which the tip and base structures are shed at similar frequencies, but have different phase relations. It is not unreasonable to associate the global behaviour to the interaction of these structures in the base flow region.

The quality of the estimation is sensitive to the choice of the acquisition parameters. For the multi-time-delay implementation, optimum results are obtained by selecting the number of virtual sensors as $M = 4n$, where n is the number of har-

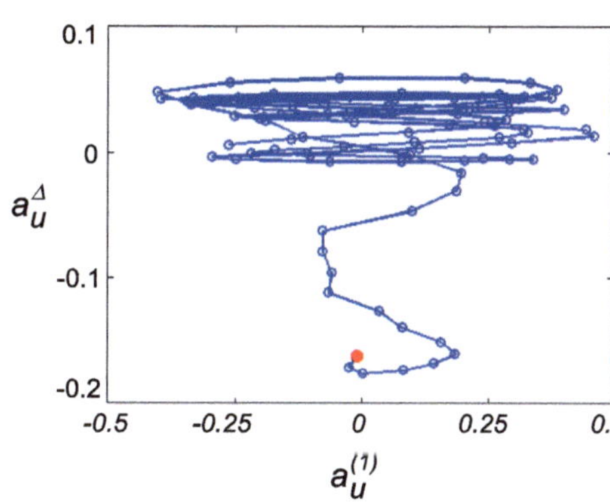

Fig. 4 Dynamic wake behaviour in modal space

monics to be resolved, and a time delay of $\Delta\tau_s = {}^1/_{Mf}$. Increasing the maximum delay beyond the period of the first harmonic (i.e. $M\Delta\tau_s > 1/f$) is generally deleterious to the estimation quality. For the synchronisation of the non-harmonic modes, it is required that the acquisition window is sufficiently long to resolved the first zero-crossing as determined from the autocorrelation function of the relevant modal temporal coefficient. Note that, while convenient, the mathematical method does not require similarity between POD and temporal Fourier modes. Further improvements to this technique can be achieved by considering the quadratic nature of the relationship between the first pressure (sensor) harmonic and higher velocity harmonics as detailed in [4].

References

1. Borée, J.: Extended proper orthogonal decomposition: a tool to analyse correlated events in turbulent flows. Exp. Fluids **35**, 188–192 (2003)
2. Castro, I.P., Watson, L.: Vortex shedding from tapered, triangular plates: Taper and aspect ratio effects. Exp. Fluids **37**, 159–167 (2004)
3. Durgesh, V., Naughton, J.W.: Multi-time-delay LSE-POD complementary approach to unsteady high-Reynolds-number near wake flow. Exp. Fluids **49**, 571–583 (2010)
4. Hosseini Z, Martinuzzi R J, Noack B R Sensor-based estimation of the velocity field in the wake of a low-aspect-ratio pyramid. Exp. Fluids: **56**,13 (2014),
5. Jeong, J., Hussain, F.: On the identification of a vortex. J. Fluid Mech. **285**, 69–94 (1995)
6. Martinuzzi, R.J.: Dual vortex structure shedding from low aspect ratio, surface-mounted pyramids. J. Turbul. **28**, 1–16 (2008)
7. Papangelou, A.: Vortex shedding from slender cones at low Reynolds numbers. J. Fluid Mech. **242**, 299–321 (1992)

The Fine Structure of a Slender Scalar Plume in Sheared Turbulence

Christina Vanderwel and Stavros Tavoularis

Abstract We present measurements of the fine structure of a passive scalar plume in uniformly sheared turbulence generated in a water tunnel. We report on the mixed velocity-scalar statistics of the plume, including the probability density functions of the velocity, scalar, and scalar derivatives, as well as conditional expectations of the velocity and the scalar derivatives, conditioned upon the scalar fluctuations. Such results are particularly relevant to models that are intended to be used for solving the balance equation of the scalar pdf. Specifically, we address the effect of having a highly intermittent scalar field, in which case the scalar pdf is highly skewed and non-Gaussian and the conditional expectations of the velocity components are distinctly non-linear.

Details on the fine structure of turbulent mixing are particularly relevant for the modelling of chemical reactions and combustion. A popular approach for analyzing reactive flows is to consider the balance equation of the scalar probability density function (pdf) [1–3]. Although these equations have certain advantages over conventional Reynolds-averaged balance equations of velocity and scalar moments, they are complicated by the appearance of conditional expectations of the velocity and the scalar dissipation values, conditioned upon the scalar fluctuations. Most previous studies have focused on flows in which the velocity and scalar are nearly homogeneous and jointly-Gaussian, in which case these conditional expectations would be linear functions of the scalar value [2–4]; however, this would not necessarily be the case for inhomogeneous scalar fields as they would appear in industrial and

Article submitted to the iTi Conference on Turbulence, Bertinoro, Italy, September 22–24, 2014.

C. Vanderwel
University of Southampton, Southampton, SO17 1BJ, UK
e-mail: c.m.vanderwel@soton.ac.uk

S. Tavoularis (✉)
University of Ottawa, Ottawa, ON K1N 6N5, Canada
e-mail: stavros.tavoularis@uottawa.ca

J. Peinke et al. (eds.), *Progress in Turbulence VI*,
Springer Proceedings in Physics 165, DOI 10.1007/978-3-319-29130-7_45

environmental flows and even in many canonical flows, such as turbulent plumes, jets, wakes, and mixing layers.

The objective of the present study is to investigate experimentally the fine structure of a passive scalar plume of a solution of Rhodamine 6G dye ($Sc = 2500$) injected isokinetically from a fine tube in uniformly sheared turbulence generated in a water tunnel (Fig. 1). Measurements of the turbulent velocity and concentration fields in cross-sections of the plume were obtained simultaneously with the use of stereoscopic particle image velocimetry (SPIV) and planar laser induced fluorescence (PLIF). The present results complement previous reports of absolute and relative turbulent diffusion of the same plume [5–7], which have also documented the measurement procedure in detail.

The scalar field In the present flow, the turbulence was nearly homogeneous but strongly anisotropic [8]. The present plume was very slender by comparison to the dominant eddies of the turbulence; consequently, the scalar field was highly intermittent. A representative instantaneous concentration map, shown in Fig. 2a, illustrates the sinuous nature of the plume and its extensive meandering. At all points in the cross-section, the scalar intermittency factor γ_c, defined as the portion of time during which the concentration was non-zero, was lower than 0.4. Maps of the mean concentration had the shape of a 2D Gaussian function (Fig. 2b, c) and the standard deviation of the concentration fluctuations exceeded the mean concentration.

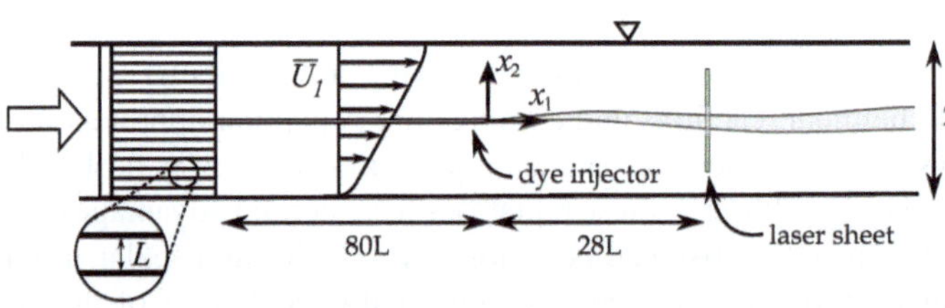

Fig. 1 Schematic of the apparatus ($L = 25.4$ mm). Uniformly sheared flow is generated by a shear generator with linearly-varying porosity inserted at the entrance to the test section

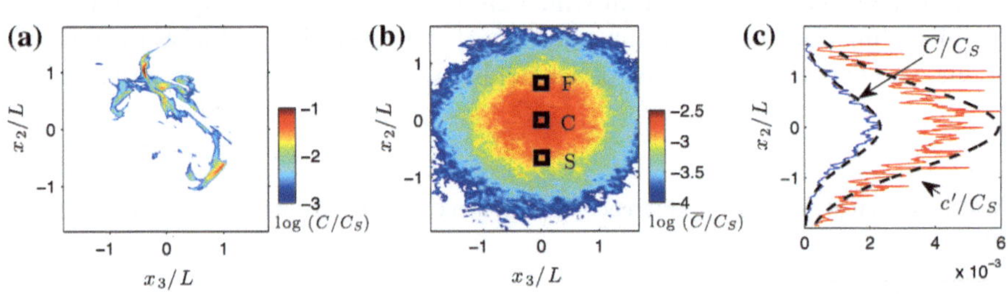

Fig. 2 **a** Representative instantaneous concentration map at $x_1/L = 28$, normalized by the injected dye concentration C_S. **b** Mean concentration map at $x_1/L = 28$, with the three regions of interest indicated by squares. **c** Profiles of the mean concentration and the standard deviation of the concentration fluctuations; dashed lines represent Gaussian profiles

Probability density functions In the following, we will present representative measurements at three locations: on the axis of the plume and at two locations on either side of the axis and along the mean velocity gradient direction, near the two inflection points of the mean concentration profile (Points C, F and S, respectively). Unlike the pdfs of the velocity fluctuations (Fig. 3a), which were Gaussian, the pdfs of the scalar fluctuations (Fig. 3b) were highly skewed and were bounded on the low side by the zero concentration and on the high side by a value that was significantly lower than the injected dye concentration C_S; these observations are consistent with previous measurements in thin plumes [9, 10]. At all three locations, the scalar pdfs nearly coincided, although at the off-centre locations (F and S) the pdfs had slightly higher values in the tails, which can be attributed to the peak generation of scalar fluctuations at the inflection points of the mean scalar field. For small concentration fluctuations ($c/c' < 4$), the scalar pdf could be described by a gamma distribution, whereas, the tails ($c/c' > 5$) were best described by an exponential distribution.

Conditional expectations The conditional expectations of the velocity components conditioned upon the scalar value in the present plume were distinctly non-linear (Fig. 4), unlike those in flows with homogeneous scalar fields in which the joint velocity-scalar pdfs were nearly Gaussian [2–4]. The variation of $\overline{u_2/u_2'|c/c'}$ illustrates that negative scalar fluctuations (i.e., mostly undyed fluid) are associated with

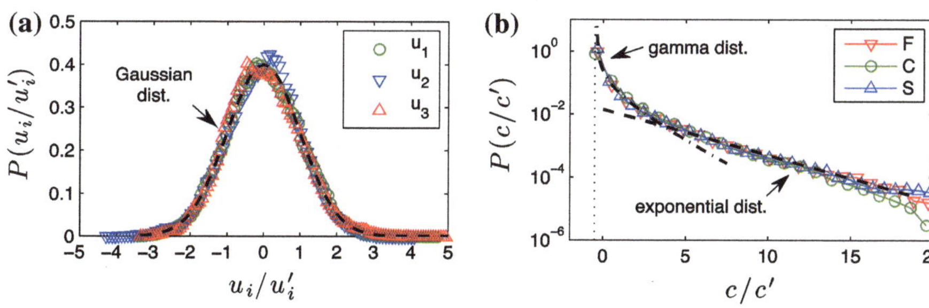

Fig. 3 **a** Pdf of the velocity fluctuations; the *dashed line* represents a Gaussian distribution. **b** Pdf of the concentration fluctuations; the *dotted line* represents undyed fluid and the *dashed* and *dash-dot lines* represent fitted exponential and gamma distributions, respectively

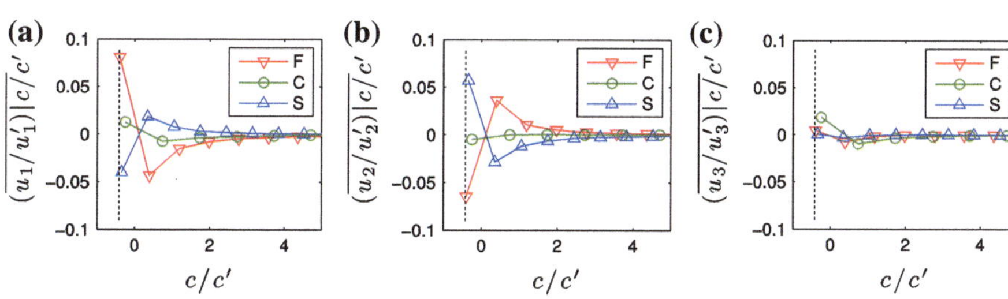

Fig. 4 Conditional expectations of the velocity components, conditioned on the concentration fluctuation value; *dotted lines* represent undyed fluid

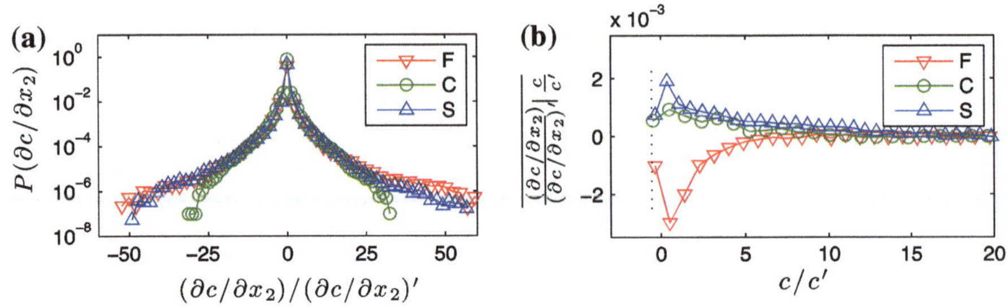

Fig. 5 **a** Pdf of the transverse concentration derivative. **b** Conditional expectations of the transverse concentration derivative, conditioned on the concentration fluctuation value; the *dotted line* represents undyed fluid

motions that originated outside the plume, whereas positive scalar fluctuations are associated with motions that originated largely in the core of the plume. Furthermore, the curves have steep vertical asymptotes on the negative side which are attributed to the fact that the scalar fluctuations are bounded on the negative side by the zero concentration. The variation of $\overline{u_1/u_1'|c/c'}$ had the opposite trends, which is consistent with the sign and strong magnitude of the Reynolds stress $\overline{-u_1u_2}$ in this shear flow [8]. As expected by considering the symmetry of the turbulence and the mean plume, $\overline{u_3/u_3'|c/c'}$ was zero at $x_3 = 0$.

Scalar derivatives Measurements of the scalar derivatives are particularly relevant as surrogates for the scalar dissipation rate [3]. The pdfs of the transverse scalar derivative (Fig. 5) were highly peaked with long nearly-exponential tails, comparable to those measured in flows with less inhomogeneous scalar fields [3, 4, 11]. This is attributed to the small-scale intermittency of the scalar even in flows with Gaussian scalar pdfs. The conditional expectations of the transverse scalar derivative at all measurement locations (Fig. 5b) were also distinctly non-linear functions of the scalar value, although their signs were consistent with the signs of the corresponding local mean scalar derivatives [12].

In summary, these results demonstrate that in a highly intermittent scalar field, the scalar pdf is strongly non-Gaussian and the conditional expectations of the velocity components and the scalar dissipation, conditioned upon the scalar value, are distinctly non-linear.

Acknowledgments Financial support by the Natural Sciences and Engineering Research Council of Canada (NSERC) is gratefully acknowledged.

References

1. Pope, S.: Turbulent Flows, Cambridge University Press, Cambridge (2000)
2. Mydlarski, L.: J. Fluid Mech. **475**, 173–203 (2003)

3. Behnamian, A., Tavoularis, S.: In Eighth International Symposium on Turbulence and Shear Flow Phenomena (TSFP8). Poitiers, France (2013)
4. Ferchichi, M., Tavoularis, S.: J. Fluid Mech. **461**, 155 (2002)
5. Vanderwel, C., Tavoularis, S.: Phys. Rev. E **89**(4), 041005 (2014)
6. Vanderwel, C., Tavoularis, S.: J. Fluid Mech. **754**, 488 (2014)
7. Vanderwel, C., Tavoularis, S.: Exp. Fluids **55**, 1801 (2014)
8. Vanderwel, C., Tavoularis, S.: J. Fluid Mech. **689**, 434 (2011)
9. Rahman, S., Webster, D.: Exp. Fluids **38**(3), 372 (2005)
10. Lavertu, R., Mydlarski, L.: J. Fluid Mech. **528**, 135 (2005)
11. Jayesh, Z.: Warhaft. Phys. Fluids A **4**(10), 2292 (1992)
12. Tavoularis, S., Corrsin, S.: J. Fluid Mech. **104**, 349 (1981)

Wake Dynamics Behind a Normal Thin Flat Plate at Moderate Reynolds Numbers

Arman Hemmati, David H. Wood and Robert J. Martinuzzi

Abstract The wake behind an infinite span (2D) thin flat plate normal to uniform flow was examined using Direct Numerical Simulation (DNS) at $Re = U_0 h/\nu = 1200$ and 2400. Three distinct flow regimes were identified in the wake due to the interruption of regular anti-symmetric Karman shedding. Disruption of the regular shedding (Regime M) was followed by a period of delayed roll-up (Regime L) and a longer duration of high intensity shedding (Regime H) before regular shedding resumes again. Size of the wake mean recirculation region changed with progression of the wake patters from a long-term mean length of 2.90–3.60h (L) and then 2.16h (H). Moreover, the wake turbulence characteristics were effected with the wake evolution. These effects were quantified by variations in the turbulence kinetic energy (TKE) magnitude, production, dissipation and diffusion. Spanwise instabilities were responsible for appearance of the three flow regimes. A projection of these instabilities was observed in the pressure field behind the plate.

1 Introduction

The wake of a two-dimensional (2D) flat plate, infinite span, has been studied only for certain controlled conditions in the past. Experimental studies on the topic [1, 3, 9], however, did not thoroughly examine the wake topology and vortex structure variations. Numerical evaluation [4, 5] of the wake identified a unique flow behavior and changes on flow patterns at $Re = 250 - 750$. Two main flow regimes, H and L, were introduced from these results along with a number of "short-term mean" flow characteristics.

A. Hemmati (✉) · D.H. Wood · R.J. Martinuzzi
Department of Mechanical and Manufacturing Engineering, University of Calgary, Calgary, Canada
e-mail: ahemmati@ucalgary.ca

D.H. Wood
e-mail: dhwood@ucalgary.ca

R.J. Martinuzzi
e-mail: rmartinu@ucalgary.ca

J. Peinke et al. (eds.), *Progress in Turbulence VI*,
Springer Proceedings in Physics 165, DOI 10.1007/978-3-319-29130-7_46

These distinct flow regimes were identified using fluctuations in instantaneous lift and drag [4]. Periods of low amplitude lift variations coupled with small instantaneous drag, compared to the mean drag, was identified as a unique regime (L). This regime is subsequently replaced with a period of high amplitude lift variations and large instantaneous drag (H). Categorization of these regimes, however, remains quantitatively ambiguous. Moreover, long-term lift and drag time series reported [4, 5] indicate that these regimes, as described above, only occupied a small portion of the total recorded time. Moreover, no evidence was provided on the physical flow behavior leading to such drastic changes in the wake topology.

The current study used DNS [2] to reconstruct the wake behind a 2D thin flat plate normal to the uniform incoming flow at $Re = U_0 h/\nu = 1200 - 2400$, based on the plate height, h, and uniform inlet velocity, U_0. Second-order central difference and backward Euler methods were used for spatial and temporal discritization of the flow field, respectively. A non-uniform grid distribution was used with the higher mesh density in close vicinity of the plate and coarser mesh at the outer borders of the computational domain [2], based on recommendations from similar studies [4, 6, 8]. The grid quality ratio, Kolmogorov micro-scale to element length, was below 2 in the critical flow region.

Results obtained from the DNS [2] are verified in comparison to known experimental [1, 3, 9] and numerical [4, 5, 7] studies. The shedding Strouhal number was 0.158, which is comparable to the reported frequencies of 0.16 [1] and 0.161 [4]. Moreover, the mean pressure distribution on the plate surfaces, shown in Fig. 1, matches with the only reliable experimentally obtained results [1]. The mean drag calculated using DNS [2] results was 2.13 ± 0.003, which is comparable to the only reliable reported value of 2.13 [1].

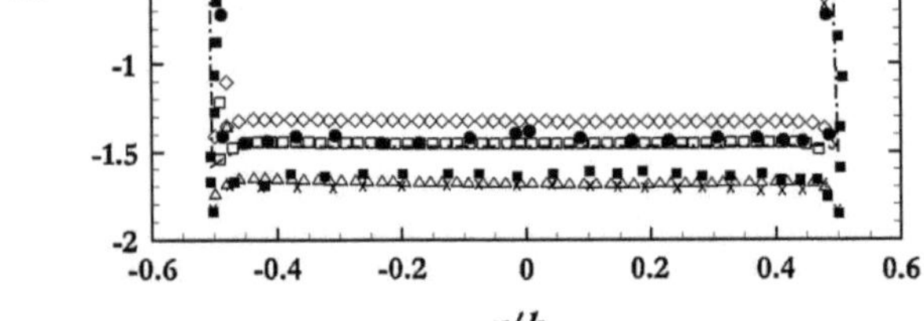

Fig. 1 Span-averaged mean pressure coefficients, $\overline{C_p}$, along the plate surfaces at $Re = 1.2 \times 10^3$ < □ > (Current DNS, Total Mean), < △ > (Current DNS, Regime H), < ◇ > (Current DNS, Regime L); $Re = 7.5 \times 10^2$ [7] < × >; $Re = 2.5 \times 10^2$ [4] < ■ >; $Re = 1.0 \times 10^3$ [5] < − · − > and $Re = 1.5 \times 10^5$ [1] < • >

2 Results and Discussion

Three distinct flow regimes were identified using the instantaneous force results, shown in Fig. 2, and their respective phase angles [2]. Regular shedding cycle of Regime M was interrupted with a period of low intensity shedding, Regime L, identified with very low-amplitude lift variations and smaller drag with respect to the mean field. Subsequently, the shedding cycle was intensified during Regime H, where the high-amplitude lift fluctuations were coupled with larger drag.

Flow topology evolved with interruption of the regular shedding cycles of Regime M. Organization of streamwise ribs was disrupted with breakdown of the first spanwise roller immediately behind the plate. This led to a period of intense stretching and tilting of the ribs, which resulted in a larger recirculation region (by 25 %) and weaker entrainment. This was followed by re-organization of the ribs, aligning them parallel to each other and to the convective stream. The ribs connected top of the forming roller to bottom of the first detached roller, which has reappeared following a short term absence during Regime L. This intensified the vortex shedding process and increased amplitude of lift fluctuations, which had the same frequency as the regular shedding cycle. Drag variations, however, had a lower frequency corresponding to streamwise movement of the minimum pressure point in the base region. This phenomenon had a critical influence on the turbulent kinetic energy production in the wake and led to negative production spikes in close proximity of the plate. Turbulent kinetic energy intensified with higher intensity shedding of Regime H and weakened with lack of organization in the wake during Regime L.

Evidence of a spanwise instability is observed on the pressure field as it is projected on distribution of C_p on the plate back surface. There exists a low frequency spatially-periodic distribution of pressure on the plate back surface during all three regimes. Intensity of this instability is magnified during Regime H (see Fig. 3) and weakened during Regime L. Footprint of this periodicity is evident in both the short-term mean pressure field as well as the total mean field. Changes on intensity of the secondary instability is suspect to be due to its interaction with the regular wake instability. A

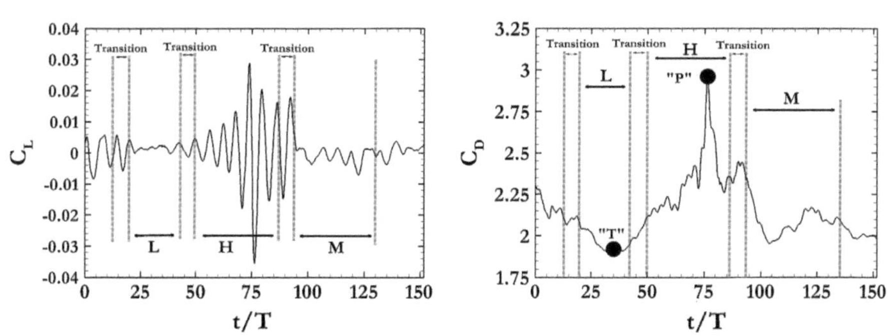

Fig. 2 Plots of instantaneous coefficients of lift, C_L, and drag, C_D at $Re = 1200$

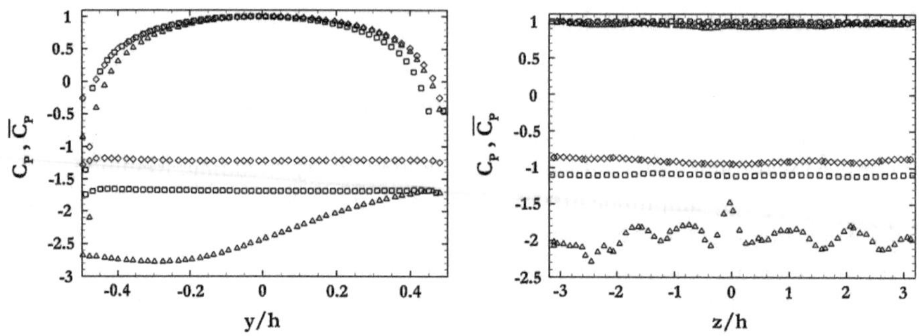

Fig. 3 Normal (*Left*) and Span (*Right*) distribution of instantaneous coefficients of pressure, C_P, on the plate surfaces at $Re = 1200$: $< \square >$ Total Mean, $< \triangle >$ Instance "P" (Max C_D), $< \diamond >$ Instance "T" (Min C_D)

similar behavior is also recorded in the vertical direction on the plate back surface. There exists an anti-symmetric distribution of instantaneous pressure field on the back surface in prefect correspondence to the high intensity shedding cycles. No evidence of such behavior is observed on the mean (short and long term) results, Fig. 1. Thus, vertical instability in the pressure field concludes full cycles with no interruptions.

3 Conclusion

Wake topology behind a normal thin flat plate positioned normal to incoming flow evolves in three distinct stages: regular (M), low intensity (L) and high intensity (H) shedding. Presence of a secondary spanwise instability in the wake with varying intensity is suspected to be responsible for evolution of the wake. High amplitude fluctuations of lift is in correspondence to the regular shedding frequency, while the drag varies at a significantly lower frequency.

References

1. Fage, A., Johansen, F.C.: On the flow of air behind an inclined flat plate of infinite span. Proc. R. Soc. Lond. Ser A, Contain. Pap. Math. Phys. Character **116**(773), 170–197 (1927)
2. Hemmati, A., Wood, D.H., Martinuzzi, R.J.: Characteristics of distinct flow regimes in the wake of an infinitely span normal thin flat plate. Int. J. Heat Fluid Flow (2014). (Submitted)
3. Kiya, M., Matsumura, M.: Incoherent turbulence structure in the near wake of a normal plate. J. Fluid Mech. **190**, 343–356 (1988)
4. Najjar, F.M., Balachandar, S.: Low-frequency unsteadiness in the wake of a normal flat plate. J. Fluid Mech. **370**, 101–147 (1998)

5. Najjar, F., Vanka, S.: Effects of intrinsic three-dimensioanlity on the drag characteristics of a normal flat plate. Phys. Fluids **7**(10), 2516 (1995)
6. Najjar, F.M., Vanka, S.P.: Simulations of the unsteady separated flow past a normal flat plate. Int. J. Numer. Method Fluids **21**(7), 525547 (1995)
7. Narasimhamurthy, V.D., Andersson, H.I.: Numerical simulation of the turbulent wake behind a normal flat plate. Int. J. Heat Fluid Flow **30**(6), 1037–1043 (2009)
8. Rodi, W., Ferziger, J.H., Breuer, M., Pourquie, M.: Status of large eddy simulation: results of a workshop. J. Fluids Eng. **119**(2), 248–262 (1997)
9. Wu, S.J., Miau, J.J., Hu, C.C., Chou, J.H.: On low-frequency modulations and three-dimensionality in vortex shedding behind a normal plate. J. Fluid Mech. **526**, 117–146 (2005)

Part V
Complex Flows

Storm in a Soap Bubble

P. Fischer, C.-H. Bruneau and Y.-L. Xiong

Abstract An original method to implement cartesian computations to solve a spherical problem is introduced in this paper. The problem consists in performing the numerical simulations of a half soap bubble located on a heated plate. The gradient of temperature between the base and the top of the bubble generates plumes at the base that move up. These plumes give rise to eddies that survive for several minutes eventually creating a two-dimensional turbulent thermal convective flow. Our method consists in defining an appropriate stereographic projection in order to use classical numerical scheme defined for two-dimensional Navier–Stokes equations. The results are then analyzed through spherical harmonics decompositions and compared to data obtained with soap bubble experiments.

1 Introduction

Two-dimensional equations [4] are often used as a model for describing the turbulence in the atmosphere. Indeed, atmospheric phenomena are confined in a layer of fluid whose dimensions are only few kilometers in the vertical direction and about thousands of kilometers in the horizontal directions. This implies that most of the kinetic energy of large-scale structures is actually embedded in the horizontal velocities, the vertical component of the velocity being particularly small. Even if this

P. Fischer (✉) · C.-H. Bruneau
University of Bordeaux, IMB. CNRS UMR 5251. INRIA Bordeaux Team MC2,
33400 Talence, France
e-mail: patrick.fischer@math.u-bordeaux1.fr

C.-H. Bruneau
e-mail: Charles-Henri.Bruneau@math.u-bordeaux1.fr

Y.-L. Xiong
Department of Mechanics, Huazhong University of Science and Technology,
Wuhan 430074, China
e-mail: xylcfd@hust.edu.cn

J. Peinke et al. (eds.), *Progress in Turbulence VI*,
Springer Proceedings in Physics 165, DOI 10.1007/978-3-319-29130-7_47

modelization cannot completely describe the complexity of the atmosphere, two dimensional simulations can still be used to mimic large-scale structures motion in the atmosphere. Quite recently H. Kellay et al. [7, 8] designed a new physical experiment: a thermal convection cell composed by a half soap bubble heated at the equator. This device allowed them to study thermal convection and the movement of vortices on the surface of the bubble. They observed the emergence and persistence of isolated vortices that emerge randomly from the equator, grow in size rapidly and persist for relatively long times. They compared the motion of these vortices to hurricane trajectories on Earth and they could show that their isolated vortices and the hurricanes displayed very similar power law scaling [5].

2 Mathematical Model and Equations in Stereographic Coordinates

The present experiment does not correspond to any classical numerical two dimensional turbulence. So our first task was to understand the underneath physics and to write the mathematical equations allowing us to describe our observations. Buoyancy driven turbulence is observed on a very thin hemispherical fluid, the soap bubble. The fluid behavior can be described using the equations for the two-dimensional thermal convection under the Boussinesq approximation as:

$$\begin{aligned} \frac{D\mathbf{U}}{Dt} &= -\frac{1}{\rho}\nabla p + \nu \nabla^2 \mathbf{U} - \beta T \mathbf{g} \\ \nabla \cdot \mathbf{U} &= 0 \\ \frac{DT}{Dt} &= \alpha \Delta T \end{aligned} \tag{1}$$

where $\frac{D.}{Dt} \equiv \frac{\partial .}{\partial t} + (\mathbf{U} \cdot \nabla).$ is the total time derivative, $\mathbf{U}$ denotes the velocity, p the pressure, ρ the mass density, ν the kinematic viscosity, β the coefficient of thermal expansion, $\mathbf{g}$ the gravity field, T the temperature of the fluid and α the coefficient of thermal diffusivity. The laws of physics do not depend on the coordinates system, we consider in this paper 2D orthogonal Cartesian coordinates (x, y) obtained by a stereographic projection onto the equatorial plane. The stereographic projection is a particular mapping that allows us to project a sphere onto a plane. The usual projection is defined on the entire sphere, but we use in this study a version restricted to the northern hemisphere. The Navier–Stokes equations (2) for a soap bubble of radius $r = 1$ can thus be written using stereographic coordinates:

$$\frac{\partial}{\partial t}\mathbf{u} + (\mathbf{u}\cdot\nabla)\mathbf{u} + \frac{2}{1+x^2+y^2}\begin{pmatrix}-xu^2 - 2yuv + xv^2\\ yu^2 - 2xuv - yv^2\end{pmatrix}$$

$$= \nu\left[\frac{(1+x^2+y^2)^2}{4}\nabla^2\mathbf{u} - \mathbf{u} - \left(1+x^2+y^2\right)\left(x\frac{\partial \mathbf{u}}{\partial x} + y\frac{\partial \mathbf{u}}{\partial y} + \begin{pmatrix} y\frac{\partial v}{\partial x} - x\frac{\partial v}{\partial y}\\ x\frac{\partial u}{\partial y} - y\frac{\partial u}{\partial x}\end{pmatrix}\right)\right]$$

$$-\frac{1}{\rho}\left(\frac{1+x^2+y^2}{2}\right)^2\nabla p - \begin{pmatrix}x\\ y\end{pmatrix}\beta g T \qquad (2)$$

$$\nabla\cdot\mathbf{u} = \frac{4\,(xu+yv)}{1+x^2+y^2} \qquad (3)$$

$$\frac{\partial T}{\partial t} + (\mathbf{u}\cdot\nabla)T = \alpha\frac{(1+x^2+y^2)^2}{4}\Delta T. \qquad (4)$$

where $\mathbf{u} = (u, v)$ is the velocity field in the 2D (x, y) coordinates system.

3 Numerical Results

Finite differences schemes are used for solving the equations described in the previous part. The spatial domain is discretized with a staggered uniform Cartesian grid and the time interval is partitioned with an uniform time steps. A second-order Gear scheme is used to solve the unsteady term in the governing equations. The linear terms are solved using second order centered implicit scheme, whereas the convection terms are solved using a third order up flow explicit one. The discrete values of the pressure p and the temperature T are located at the center of each cell and the discrete values of the velocity field are located at the middle of the cell sides. The results obtained with our numerical simulations can be analyzed using spherical harmonics decompositions. The results described in this paper have been obtained with a Rayleigh number $Ra = \frac{g\beta\delta T}{\nu\alpha}$ equal to 10^8 and a Prandtl number $Pr = \frac{\nu}{\alpha}$ equal to 7. The time step for the computations has been chosen equal to 0.001 with outputs every 200 time steps. In the beginning of our simulation, the experiment is similar to a long two-dimensional toric Bénard cell: the cold fluid is heated from below, and the plumes are pushed upward. There is no interaction between the plumes and the center of the domain remains empty. These small structures emerging from the heated equator can be easily spotted on the vorticity field, and are well localized with a peak around wavenumber 100 in the enstrophy and energy spectra. Then the plumes continue to grow and start to interact with each other over time. Power law scaling can then be observed in the entropy ($k^{-4/5}$), energy (k^{-3}) and enstrophy (k^{-1}) spectra. After a while (Fig. 1), we can observe large structures spreading in the vorticity field. These structures are mainly filaments, even if few isolated vortices can

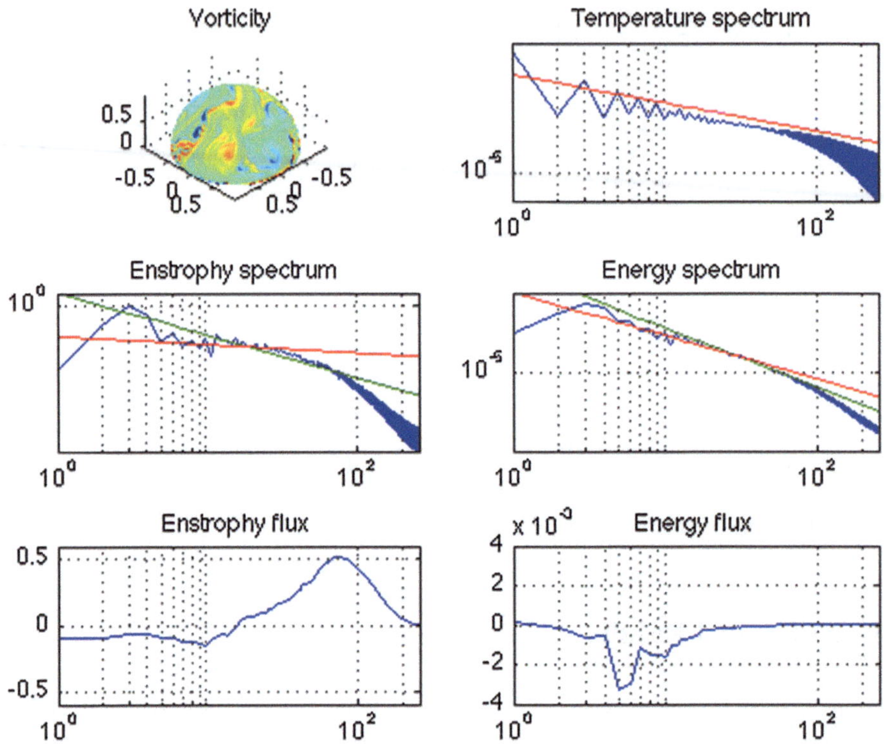

Fig. 1 Numerical results: *curves* are obtained by averaging ten time steps data taken after 80,000 time steps. Indicative slopes: $-4/5$ for the temperature spectrum, $-1/5$ and -1 for the enstrophy spectrum and $-11/5$ and -3 for the energy spectrum

emerge and grow on the surface of the bubble. A new power law scaling seem to take place at large scales: a $-11/5$ slope in the energy spectrum. This power law scaling is confirmed in the enstrophy spectrum by a $-1/5$ power law scaling. A highly oscillating behavior is still observed in the entropy spectrum. According to Mishra et al. [6] the lower branch (composed by the minima of the spectrum) should correspond to a Kolmogorov [3] or a Bolgiano scaling [1, 2]. However, in our numerical results, the observed $-4/5$ slope does not correspond to any scaling described in the literature.

Our numerical results are in good agreement with the experimental results and lead us to conclude that our model and the numerical algorithms are relevant for studying this problem. However, the geometry of the experiments does not allow us to compare our analysis to any well know classical theory.

Acknowledgments This work was funded by an ANR grant *Cyclobulle*. The simulations were run on PLAFRIM platform supported by University of Bordeaux and INRIA Bordeaux Sud-Ouest.

References

1. Bolgiano, R.: Turbulent spectra in a stably stratified atmosphere. J. Geophys. Res. **71**, 2226–2229 (1959)
2. Bolgiano, R.: Structure of turbulence in stratified media. J. Geophys. Res. **71**, 3015–3023 (1962)
3. Kolmogorov, A. N.: The local structure of turbulence in incompressible viscous fluid for very large Reynolds numbers. Dokl. Akad. Nauk. SSSR **30**, 299–303 (1941) (reprinted in Proc. R. Soc. Lond. A **434**, 9–13 (1991))
4. Kraichnan, R.: Inertial ranges in two-dimensional turbulence. Phys. Fluids **10**, 1417–1423 (1967)
5. Meuel, T., Xiong, Y.-L., Fischer, P., Bruneau, C.-H., Bessafi, M., Kellay, H.: Intensity of vortices: from soap bubbles to hurricanes. Sci. Rep. **3**, 3455 (2013)
6. Mishra, P.K., Verma, M.K.: Energy spectra and fluxes for Rayleigh–Bnard convection. Phys. Rev. E **81**, 056316 (2010)
7. Seychelles, F., Ingremeau, F., Pradere, C., Kellay, H.: Thermal convection and emergence of isolated vortices in soap bubbles. Phys. Rev. Lett. **100**, 144501 (2008)
8. Seychelles, F., Amarouchene, Y., Bessafi, M., Kellay, H.: From intermittent to nonintermittent behavior in two dimensional thermal convection in a soap bubble. Phys. Rev. Lett. **105**, 264502 (2010)

Turbulent Boundary Layer Upstream, Over and Downstream a Cylindrical 2D Bump

Julie A. Vernet, Ramis Örlü and P. Henrik Alfredsson

1 Introduction

The quest for more fuel-efficient ground vehicles such as trucks, is directly coupled to reducing the form and/or friction drag without compromising the other. A prototype of a canonical flow on which our understanding of friction drag has been developed is the zero-pressure gradient boundary layer for which drag can be reduced if the laminar boundary layer region can be extended to higher Reynolds numbers (Re). Once the boundary layer has become turbulent there is little hope to reduce drag, except keeping the surface smooth or to employ miniature streamwise oriented riblets on the surface. Most flows of relevance in technical applications are, however, exposed to surface curvature and related pressure gradients, that lead to other phenomena and new challenges when it comes to the design of low-drag bodies. For such cases form drag may be the main contributor to the overall drag and it is especially significant for bodies which experience large regions of separated flows.

The applicability of knowledge from canonical wall-bounded flows is hence limited when it comes to these complex flows and geometries [2]. While the effect of pressure gradient and surface curvature has been the focus of much attention, their combined effect is not a simple superposition and therefore deserves special attention [1, 4].

In an attempt to make a general contribution to this field, the turbulent boundary layer (TBL) developing upstream, over, and downstream a wall-mounted cylinder section is here studied experimentally within the FRANCE (Flow Research on Advanced and Novel Control Efficiency) project. The specific aim of the FRANCE project is to investigate possible methods to control the flow around the front of a truck in order to reduce form drag due to separation at the A-pillar [3]. A parallel numerical project aims to model the flow with and without control.

J.A. Vernet · R. Örlü (✉) · P.H. Alfredsson
KTH Mechanics, Royal Institute of Technology, Linné FLOW Centre,
100 44 Stockholm, Sweden
email: ramis@mech.kth.se

J. Peinke et al. (eds.), *Progress in Turbulence VI*,
Springer Proceedings in Physics 165, DOI 10.1007/978-3-319-29130-7_48

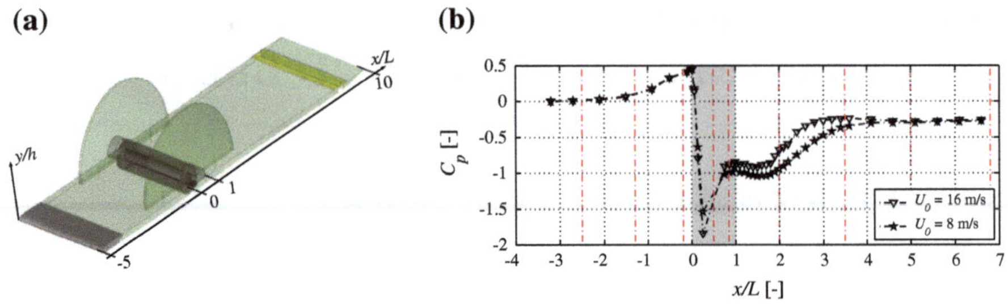

Fig. 1 **a** Schematic of used setup, where the flow direction is aligned with the x-axis. **b** Streamwise evolution of the pressure coefficient $C_p = 2(p_i - p_1)/(\rho U_0^2)$ (p_1 denoting the pressure measured at $x/L = -3.2$ and U_0 the velocity at the inlet of the test section) profile for $U_0 = 8$ and 16 m/s. *Grey area* represents the position of the bump and the *vertical dashed lines* the position where streamwise velocity profiles were measured via hot-wire anemometry

2 Experimental Setup

The experimental setup was designed in such a way that a TBL developing on a flat plate approaches an obstacle mounted on the wall, accelerates over it and separates at the downstream side upon which it reattaches again on the flat plate. The setup consists of a 1.50 m long flat plate with an elliptical leading edge and flap at the trailing edge that was mounted horizontally and placed at a height of 0.20 m from the floor of the test-section (0.5 m (height) × 0.4 m (width) cross-sectional area) in the NT2011, open-loop wind tunnel in the Fluid Physics Laboratory, KTH Mechanics. A cylinder with a diameter (L) of 0.1 m is mounted 0.5 m downstream the leading edge of the flat plate such that a 2D bump (with height $h = L/2$) protrudes out of the plate as schematically shown in Fig. 1a. In order to establish a two-dimensional flow over the cylinder it is equipped with end-plates and the initially laminar boundary layer developing on the test plate is tripped to fix the transition location. The plate, including the cylinder, is equipped with pressure taps along the streamwise direction from which the surface static pressure coefficient distribution shown in Fig. 1b was obtained. Streamwise velocity measurements were performed by means of hot-wire anemometry using a single hot-wire probe.

3 Results and Discussion

Since a single hot-wire probe is insensitive to the direction of the flow it will give erroneous velocity measurements where reverse flow occurs. While several workarounds are known for laminar separated flows to infer the separated region from hot-wire data, these are less applicable in turbulent flows. In order to discern the region in which the hot-wire measurements can be trusted several quantities, compared in

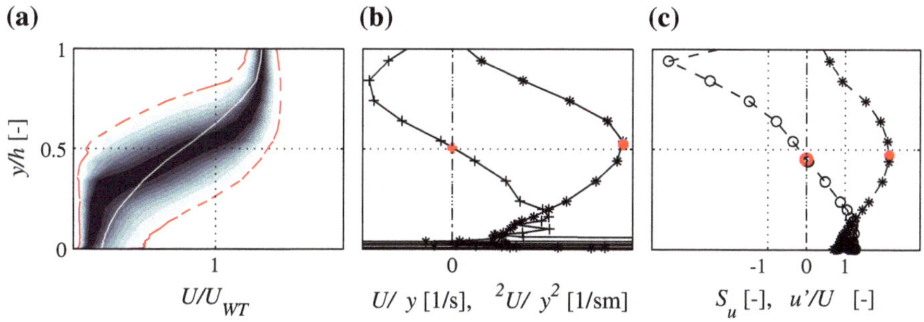

Fig. 2 **a** Contour map of the measured velocity pdf (normalized by its local maximum) at $x/L = 2$ (including a flow separated region) for $U_0 = 16$ m/s, with *red dashed lines* representing 1 % of the maximum value, and the *white solid* one depicting the mean velocity. **b** First (∗) and second (+) derivative of the mean streamwise velocity. **c** Skewness factor (∘) and rms (∗) of the streamwise velocity. *Red symbols* indicate extrema (zero-crossing or maximum) of respective quantities

Fig. 2, have been inspected. As apparent from the probability density function (pdf) map, the instantaneous velocity does not reach zero velocity, due to wall interference and buoyancy effects, besides the aforementioned directional insensitivity. Signal rectification causes the measured distribution to be more skewed towards the positive side than the actual velocity distribution and limits thereby the growth of the rms. Hence, the region of reverse flow can be discerned through the extrema of the skewness factor and rms which are related to the inflection point in the measured mean velocity profile as apparent from the 1st and 2nd derivative. This finding will henceforth be used as a diagnostic tool to discern measurement points that are within the separated region.

The mean velocity profiles upstream ($x/L = -1.3$ and -0.19), on ($x/L = 0.5$ and 0.85) and downstream ($x/L = 2.0$ and 3.5) the cylinder at both $U_0 = 8$ and 16 m/s are shown in Fig. 3a. The inflow profiles (i.e. $x/L = -2.5$) correspond to low Re TBLs at momentum-loss thickness Reynolds numbers of $Re_\theta \approx 500$ and 1000. Positions determined to have instantaneous back flow (although not fully separated) are marked in red. As evident for both Re, there is a back-flow region upstream the cylinder, the flow reattaches on the front part of the cylinder and separation occurs again beyond the top of the cylinder. Further downstream there is again reattachment on the flat plate but with a distinct difference for the two different free-stream velocities, with the higher velocity having a significantly smaller separated region and earlier reattachment. Although differences in effective blockage can have an effect on the separation, such differences in effective blockage are found to be small, since the velocity at $y/h = 3$ (normalised with the inlet velocity) did not differ with more than 1 % for the cases upstream and on top of the cylinder.

As also apparent from the rms profiles, depicted in Fig. 3b, the two Re cases are nearly identical upstream the cylinder, but differ slightly at the 90° position; a difference that is furthermore pronounced on the downstream side of the cylinder.

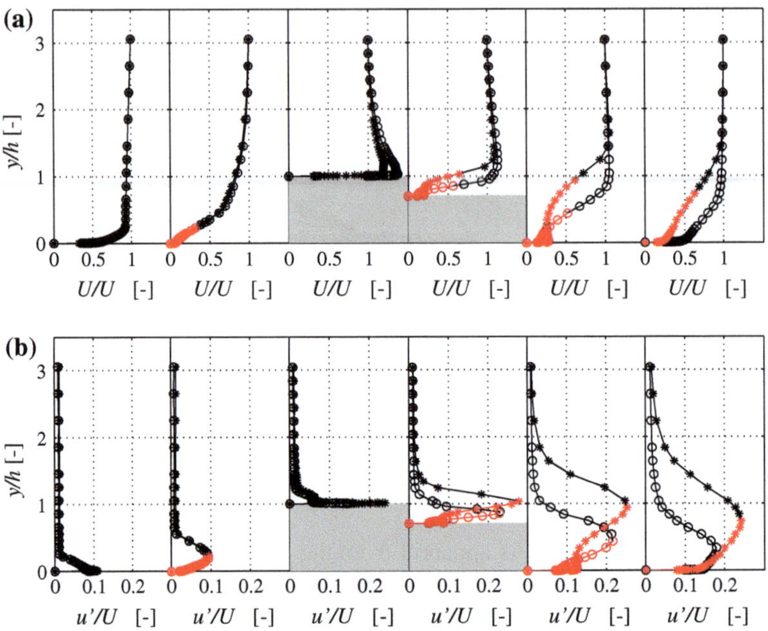

Fig. 3 Overview of the streamwise evolution of the measured streamwise **a** mean and **b** rms velocities at $U_0 = 8$ (*) and (**b**) 16 m/s (o) in the vicinity of the bump, i.e. $x/L = -1.3, -0.19, 0.5, 0.85, 2.0$ and 3.5. *Red* measurement points indicate positions detected to have back flow. Profiles are normalised by the mean velocity at $y/h = 3$

The agreement upstream the cylinder in both mean and rms values in outer scaling—despite doubling of Re—indicates that the TBL can be considered both fully developed and free of changed boundary conditions (e.g. blockage effects). The influence of the surface curvature and the induced pressure gradient is apparent from the peak location of the rms of the streamwise velocity component: while the peak location is related to the turbulence production for the most upstream station, the peak location is further away from the wall and coincides with the location of the inflection point (partially due to the directional insensitivity of the hot wire) as apparent from the color-marked (back flow) measurement points. As apparent, the internal boundary layer (with the associated knee point in the rms profile [5]) is thinner for the high speed case. It is also interesting to note that the thickness of the internal boundary layer is close to the momentum loss thickness of the external boundary layer in accordance with previous observations [4, 5].

Acknowledgments FRANCE project leader is Prof. Gunilla Efraimsson, KTH (also task leader for the numerical part of the project). The project is carried out in collaboration with Scania CV AB (Dr. Per Elofsson) and is partially financed by the Swedish Energy Agency.

References

1. Cavar, D., Meyer, K.E.: Investigation of turbulent boundary layer flow over 2D bump using highly resolved large eddy simulation. J. Fluids Eng. **133**, 111204 (2011)
2. Patel, V.C., Sotiropoulos, F.: Longitudinal curvature effects in turbulent boundary layers. Prog. Aerosp. Sci. **33**, 1–70 (1997)
3. Vernet, J., Örlü, R., Alfredsson, P.H.: Separation control by means of plasma actuation on a half cylinder approached by a turbulent boundary layer. J. Wind Eng. Ind. Aerodyn. **145**, 318–326 (2015)
4. Webster, D.R., DeGraaf, D.B., Eaton, J.K.: Turbulence characteristics of a boundary layer over a two-dimensional bump. J. Fluid Mech. **320**, 53–69 (1996)
5. Wu, X., Squires, K.D.: Numerical investigation of the turbulent boundary layer over a bump. J. Fluid Mech. **362**, 229–271 (1998)

Effects of a Wall on the Dynamics of Turbulence *Teardrops* and *Fingerprints*

Patrick Bechlars and Richard D. Sandberg

Abstract A direct numerical simulation of a compressible turbulent boundary layer has been carried out to investigate the characteristic distribution of turbulence as well as its dynamics. The focal point of this work is how the dynamics vary with the distance to the wall and how this affects the characteristic distribution. It is shown that the pressure terms are the main culprit for changes in turbulence dynamics.

1 Introduction

Turbulence, in many regards, is still a poorly understood phenomenon. Various methods exist to investigate its seemingly chaotic behavior. One of these methods is a characteristic analysis of the velocity gradient which was first introduced as *critical point analysis* [1]. This method uses the information about the eigenvalues of the velocity gradient tensor to draw conclusions about the local flow topology. With the distinction between different flow topologies one is able to investigate their frequency of occurrence in turbulent flows. This *characteristic distribution* is often assumed as *universal* [2] feature across many turbulent flows and the respective probability density function (pdf) has iso-contours that form a well-known *teardrop* shape in the QR-phase-space. The mechanism that causes such a characteristic distribution is the interplay of boundary conditions, dissipation and the cascading process. Studying this interplay in the same QR-phase-space as the distribution allows to understand alterations of this mechanism and their consequences to the respective distribution. For the first time the Lagrangian representation of the first three invariants of the velocity gradient P, Q, R are presented in a compressible form. The average dynamics are studied conditioned with the respective states in the QR-phase-space. Such an analysis is carried out and results are compared for different wall-normal locations

P. Bechlars (✉) · R.D. Sandberg
University of Southampton, Southampton, UK
e-mail: patrick.bechlars@gmail.com
URL: http://www.bechlars.org

R.D. Sandberg
e-mail: richard.sandberg@unimelb.edu.au

J. Peinke et al. (eds.), *Progress in Turbulence VI*,
Springer Proceedings in Physics 165, DOI 10.1007/978-3-319-29130-7_49

in a compressible zero pressure-gradient flat plate turbulent boundary layer (TBL). This reveals how the wall affects the dynamics of turbulence in a characteristic sense in this flow.

2 Characteristic Dynamics of Turbulence and Flow-Case

Based on the characteristic decomposition described in [1, 3] four local flow topologies can be distinguished. The frequency of occurrence of these topologies and its governing mechanisms will be investigated. The development and interaction of fluid particles forming the characteristic topologies can be described by the dynamics of the invariants of the velocity gradient tensor $A := \nabla \mathbf{u}$

$$\begin{aligned} \frac{dP}{dt} &= P^2 - 2\,Q - \mathrm{tr}(H) \\ \frac{dQ}{dt} &= QP - \frac{2}{3} P\,\mathrm{tr}(H) - 3\,R - \mathrm{tr}(AH^*) \\ \frac{dR}{dt} &= -\frac{1}{3} Q\,\mathrm{tr}(H) + PR + P\,\mathrm{tr}(AH^*) - \mathrm{tr}(A^2 H^*), \end{aligned} \tag{1}$$

where $P := -\mathrm{tr}(A)$, $Q := \frac{1}{2}\left(\mathrm{tr}(A)^2 - \mathrm{tr}(A^2)\right)$ and $R := -\det(A)$ are the invariants, $H := \nabla \frac{\nabla \cdot \tau - \nabla p}{\rho}$, $H^* := H - \frac{1}{3}\mathrm{tr}(H) I$, and τ is the viscous stress tensor, p is pressure and ρ is density.

To investigate the effect of a wall on the dynamics of turbulence we choose the canonical case of a TBL at a Mach number of $M = 0.5$. The simulated domain has a streamwise extent of $Re_\theta \approx 670 - 2300$. In terms of inflow boundary layer thickness δ_{99} the domain spans $L_x \approx 164.6$, $L_y \approx 16.7$ and $L_z = 7$ and is discretized on a grid with $7200 \times 260 \times 386$ points. The grid is stretched in the x-direction to keep a spacing of $\Delta x^+ \approx 5$. The first point off the wall ($y_{\mathrm{wall}} = 0$) is at $y^+|_{j=2} \approx 0.5 - 0.7$ and it is ensured to have at least 13 points within $y^+ < 10$. The wall has no-slip and an isothermal condition set to the adiabatic temperature of the freestream. Details for the discretization scheme can be found in [4].

3 Results

To reveal the effect of the wall on turbulence in a characteristic sense, three wall-normal locations (L1, L2, L3) are compared for the same streamwise location $Re_\theta \approx 1230$ in the TBL. L1 is in the logarithmic layer at $y^+ \approx 72$, L2 in the buffer layer at $y^+ \approx 17$ and L3 in the viscous sublayer at $y^+ \approx 6.7$. Data for the outer layer (not shown here) agrees well with the experimental dataset of a turbulent boundary layer at the same order of Reynolds number [5]. The characteristic distributions of

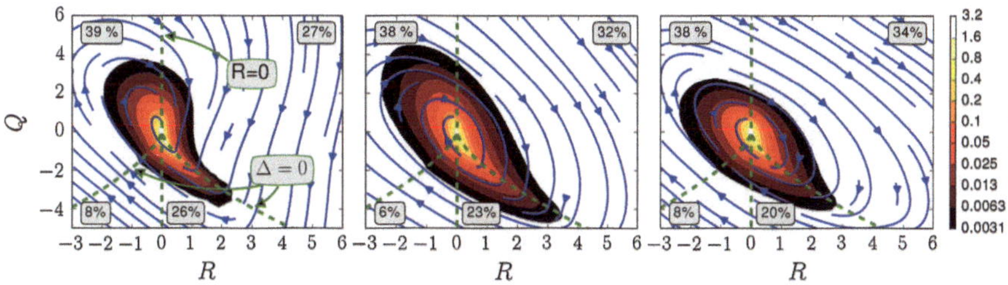

Fig. 1 The *dashed green lines* divide the QR-phase-space into the four sectors, covering a characteristic local flow topology each. The contours show the joint-pdf of Q and R and the %-boxes state the probability of a fluid particle having a state within the respective sector. The mean phase development of fluid particles is shown by *blue arrows*. *Left* to *right* L1, L2 and L3

local flow topologies are shown in Fig. 1 in form of joint-pdfs and agree well with [6]. The shape is changing from the well-known *teardrop* shape towards an oval shape when approaching the wall. Further, the probabilities for the occurrence of the respective topologies are changing with distance to the wall. To understand the mechanism behind these distributions, the mean dynamics of the invariant dynamics, see Eq. 1, are obtained by averaging $\frac{dQ}{dt}$ and $\frac{dR}{dt}$ conditioned with different states of the QR-phase-space and are shown via the trajectories in Fig. 1. Similar to the pdfs, the trajectories tend towards an oval shape while approaching the wall. To understand this variation the contribution of the single terms in Eq. 1 are investigated separately.

By neglecting the compressible effects (2–4 orders of magnitude lower than the remaining physics) the strength of this method becomes obvious: The contribution of the terms in Eq. 1, that are arising from the non-linear terms in the Navier–Stokes equations is invariant in the QR-phase-space, see Fig. 2 (left). I.e. the trajectories caused by non-linear effects have the same shape for all locations in the flow. The contribution of the diffusion to the overall QR-development, see Fig. 2 (right), shows minor differences for the respective locations. For all location the origin of the QR-phase-space is an attractor.

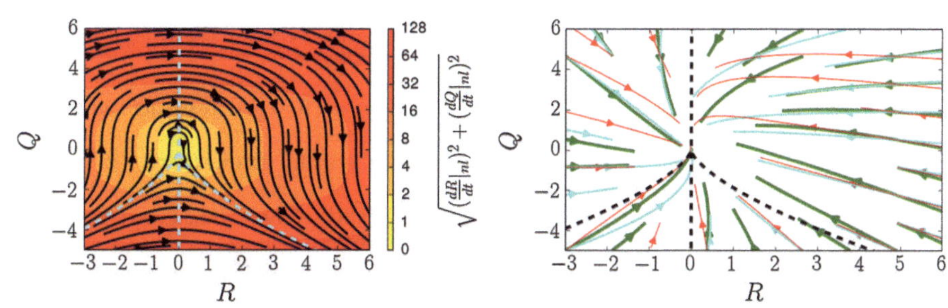

Fig. 2 *Left* QR-phase development caused by non-linear terms of NS only. *Right* QR-phase development caused by diffusion terms of NS only. L1: *green/thick*, L2: *cyan/normal*; L3: *red/thin*

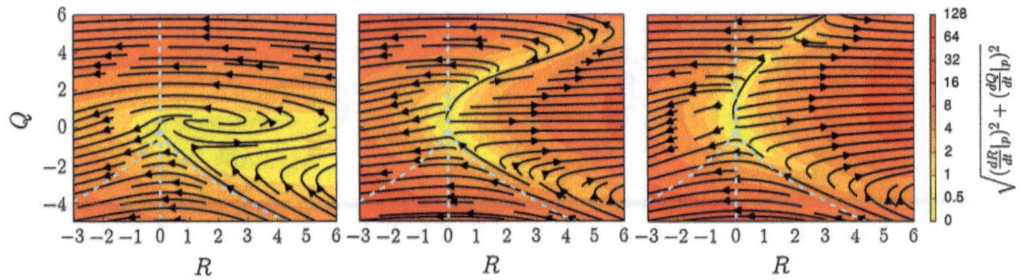

Fig. 3 *QR*-phase development caused by pressure terms of NS only. From *left* to *right* the plots are shown for L1, L2 and L3

The contribution of the pressure term, see Fig. 3, is varying strongly with the wall-normal location. For L1 the trajectories indicate that the pressure terms counteract the non-linear terms with the exception that they develop a swirl for positive R and Q that let these plots appear as *fingerprints* of pressure. Approaching the wall, this swirl moves away from the origin and unwraps. For L3 two distinct mechanisms, divided by a bow-shaped repellor, are present. On the left side of this repellor pressure still counteracts the non-linear terms whereas it supports them on the right.

4 Conclusions

The variation with wall-distance of the characteristic distribution suggests that the *teardrop* shape is not a universal feature of turbulence in a global sense. However, universality might hold if restricted to relatively small scales of motions.

The significant contribution to the variation of the invariant dynamics in the *QR*-phase-space is caused by the pressure terms in the Navier–Stokes equations. Universal models for the invariant dynamics in this phase-space have to capture the presented effects of pressure to work accurately.

References

1. Perry, A.E., Chong, M.S.: A description of eddying motions and flow patterns using critical-point concepts. Annu. Rev. Fluid Mech. **19**, 125–155 (1987)
2. Meneveau, C.: Lagrangian dynamics and models of the velocity gradient tensor in turbulent flows. Annu. Rev. Fluid Mech. **43**(1), 219–245 (2011)
3. Chong, M.S., Perry, E., Cantwell, B.J.: A general classification of three-dimensional flow fields. Phys. Fluids A: Fluid Dyn. **2**(5), 765 (1990)
4. Sandberg, R.D., Sandham, N.D., Suponitsky, V.: DNS of compressible pipe flow exiting into a coflow. Int. J. Heat Fluid Flow **35**, 33–44 (2012)
5. Elsinga, G.E., Marusic, I.: Evolution and lifetimes of flow topology in a turbulent boundary layer. Phys. Fluids **22**(1), 015102 (2010)
6. Chong, M.S., Soria, J., Perry, A.E., Chacin, J., Cantwell, B.J., Na, Y.: Turbulence structures of wall-bounded shear flows found using DNS data. J. Fluid Mech. **357**, 225–247 (1998)

Numerical Study of the Intermittency Region in Two-Fluid Turbulent Flow

S.V. Kraheberger, T. Wacławczyk and M. Wacławczyk

Abstract In this work we describe interactions between turbulence and water-air surface in the ensemble-averaged picture where, instead of a sharp interface between the phases we deal with a "surface layer" where the probability of the surface position is nonzero. Changes of the turbulent kinetic energy and the characteristic size eddies influence the width of the "surface layer". We present a numerical solution and convergence tests for the equation for the intermittency function α which describes the probability of finding the water phase at a given point and time.

1 Introduction

In the present work we investigate stratified turbulent air-water flows where the phases are separated by a deformable, but non-broken interface. We consider the statistical approach where the ensemble averaging of physical quantities, including these connected with the fluctuating surface, is performed. After the averaging, instead of a sharp boundary between the two phases, we receive a layer where the probability of the surface position is non-zero. We consider an intermittency function α, defined as an ensemble-average of the sharp-step indicator function $\alpha(\mathbf{x}, t) = \langle\chi(\mathbf{x}, t)\rangle$, where $\chi(\mathbf{x}, t) = 1$ if the water phase is present at $(\mathbf{x}, t)$ and 0 otherwise. The function α denotes the probability of finding the water phase at point $\mathbf{x}$ and time t.

S.V. Kraheberger (✉) · M. Wacławczyk
Chair of Fluid Dynamics, Department of Mechanical Engineering,
TU Darmstadt, Otto-Berndt Str. 2, 64287 Darmstadt, Germany
e-mail: kraheberger@fdy.tu-darmstadt.de

M. Wacławczyk
e-mail: martaw@fdy.tu-darmstadt.de

T. Wacławczyk
Institute for Numerical Methods in Mechanical Engineering,
TU Darmstadt, Dolivostr. 15, 64293 Darmstadt, Germany
e-mail: waclawczyk@fnb.tu-darmstadt.de

J. Peinke et al. (eds.), *Progress in Turbulence VI*,
Springer Proceedings in Physics 165, DOI 10.1007/978-3-319-29130-7_50

The region were $0 < \alpha(\mathbf{x}, t) < 1$ is called the *intermittency region* or the *surface layer*. The width of the surface layer is determined by the disrupting action of turbulent eddies on one hand and the stabilizing role of gravity and surface tension on the other hand, cf. Refs. [1, 2]. In our approach these two competing mechanisms are modelled, respectively, by the diffusion and contraction terms in the evolution equation for the function α, cf. [6, 7]. In this work we present results of simulations of a surface layer changing in time due to the changing kinetic energy and the characteristic eddy size. The width of the layer changing in time is compared with results of a priori tests.

2 Model for the Turbulence-Interface Interactions

If the ensemble averaging is applied to the advection equation of the Heaviside function χ, an unclosed correlation between the surface and velocity fluctuation $\langle \mathbf{u}' \cdot \nabla \chi \rangle$ appears, cf. [5, 7]. As it was proposed in Refs. [5–7], this term can be modelled by two counter-acting terms: diffusion and contraction leading to the following equation for the evolution of the surface layer α

$$\frac{\partial \alpha}{\partial t} + \langle \mathbf{u} \rangle \cdot \nabla \alpha = \nabla \cdot (D_t \nabla \alpha) + \nabla \cdot \left[C_t \alpha \, (1 - \alpha) \, \hat{\mathbf{n}} \right], \tag{1}$$

where coefficients D_t and C_t depend on turbulence statistics in the water phase and $\hat{\mathbf{n}} = -\nabla\alpha / |\nabla\alpha|$ denotes a vector normal to the isosurfaces of α. In the 1D stationary case and for constant coefficients D_t and C_t, Eq. (1) has an analytical solution

$$\alpha = \left[1 + \exp\left(\frac{C_t}{D_t} \left(y - \hat{y}_s \right) \right) \right]^{-1}, \tag{2}$$

where $\hat{y}_s$ denotes a mean position of the interface. In such a case the difference between the top and bottom limits of the intermittency layer, defined as the region where the probability of finding the surface equals 99.6 %, reads

$$t - b = \frac{24}{\sqrt{2\pi}} \frac{D_t}{C_t}. \tag{3}$$

It was proposed in Refs. [7, 9] to model the coefficient D_t as the product of the velocity scale q and the characteristic length scale of the eddy L, $D_t = c_D q L$. The coefficient $C_t = c_C q$ and the ratio c_D / c_C can be estimated from formula (3) such, that the proper width of the intermittency region is obtained.

3 Numerical Methods and Results

The present work aims to validate the above described model for the α-equation. For this purpose, input data for diffusion and antidiffusion coefficients D_t and C_t and reference data to compare results are provided by the a priori tests described in Refs. [7, 9]. Therein, the velocity field of two counter-rotating vortices was initialized in the water phase. Due to this initialized velocity field, the whole structure moves to the surface and, as soon as it reaches the surface, it splits into two eddies moving in opposite directions along the surface. The initial flat surface is deformed due to the eddies' motions, while the mean flow velocity is zero, $\langle \mathbf{u} \rangle = 0$.

In the a priori test described in Ref. [9], α-profiles were calculated at every 0.1 s from the instantaneous, exact surface position, where the simulation was performed from $t = 0$ to $t = 25.5$ s. Hence, in the present work we solve the α-equation providing input data every 0.1 s for diffusion and antidiffusion coefficients D_t and C_t. To define D_t and C_t, we use turbulent statistics which were calculated from the a priori test. Most satisfactory results were obtained when we chose $D_t = c_D\sqrt{2k/3}\,L$ where the kinetic energy k was averaged within the intermittency zone, while C_t was considered to be proportional to the mean turbulent intensity in the water side, $C_t = c_C \langle q \rangle$. The term $\langle q \rangle$ was found to be approximately constant during simulation time, $\langle q \rangle \approx 0.1\,\mathrm{ms}^{-1}$, see Ref. [7]. As stated above, the eddy's length scale $L = L(t)$ is assumed to change in time because the eddy-structure splits when it reaches the surface, i.e. $L(t < 12.5\,\mathrm{s}) = 2\,\mathrm{m}$ as long as both eddies rise to the surface and $L(t > 19.9\,\mathrm{s}) = 1\,\mathrm{m}$ as soon as the eddies start moving into opposite directions parallel to the surface. Between $12.5\,\mathrm{s} \leq t \leq 19.9\,\mathrm{s}$, $L(t)$ is defined by a linear interpolation. It was found that higher-order interpolation does not profoundly change results and is not necessary therefore.

For solving Eq. (1) numerically, we use similar methods as were described in a different context in Ref. [4]. For the discretization of α in time, we use a third-order total variation diminishing Runge-Kutta scheme cf. Ref. [3]. Results of such simulations show that the evolution of the intermittency region's width $|t - b|$ in time is reproduced very well, see Fig. 1. Since D_t and C_t are constant at each timestep, and are prescribed from the a priori data, it is possible to calculate the analytical α-profile from Eq. (2) at each timestep. The corresponding analytical width of the intermittency layer, $(t - b)_{\mathrm{analyt}}$, is then obtained by substituting α for α_{top}, α_{bottom} in (2) and solving it for y. This analytical solution $(t - b)_{\mathrm{analyt}}$ is also plotted in Fig. 1, and we observe that it is reproduced well. Although it is not part of the validation tests, we take advantage of getting quantitative results to test the numerical performance of our algorithm. Hence, the root-mean-square deviation of the numerical results from the analytical solution $(t - b)_{\mathrm{analyt}}$ is calculated

$$L_2^{t-b} = \sqrt{\frac{1}{n_t}\sum_{l=1}^{n_t}\left[(t-b)_{\mathrm{analyt},i} - (t-b)_{\mathrm{num},i}\right]^2}, \tag{4}$$

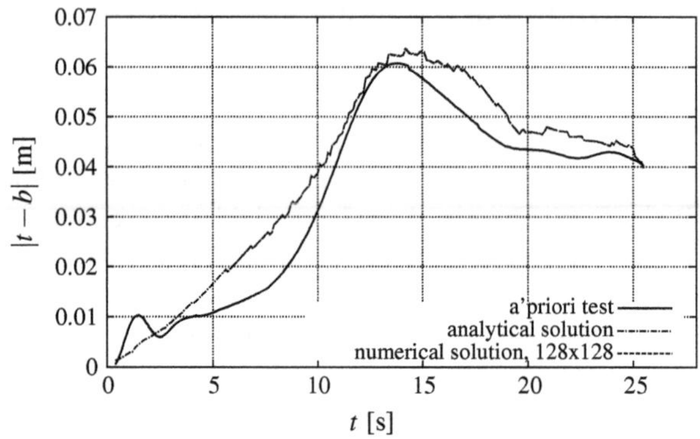

Fig. 1 Evolution of the intermittency layer's width $|t - b|$ in time. The diffusion coefficient $D_t(t) = c_D q(t) L(t)$ is defined using y-averaged kinetic energy $k(t)$ calculated from one-fluid velocity fluctuations

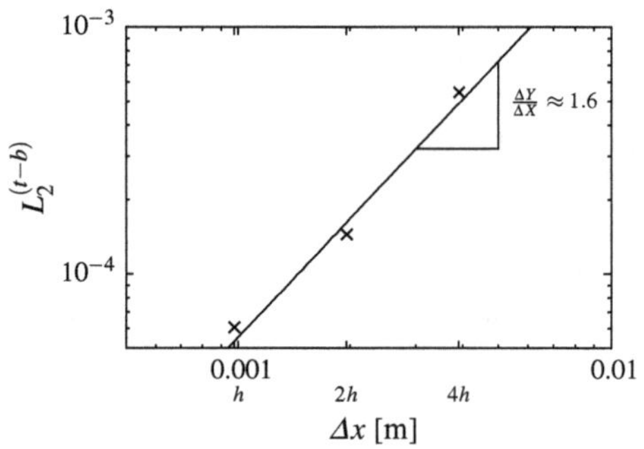

Fig. 2 Error compared to analytically computed width of the intermittency region using data from the a priori test, cf. Eq. (4). The grid width is parametrized by $h = 0.125/128$ m

where n_t denotes the number of timesteps. As can be seen in Fig. 2, L_2^{t-b} decreases with increasing grid resolution.

Although the above described definition of D_t and C_t is very simple, the results obtained for the width of the intermittency layer were satisfactory.

Acknowledgments This work was supported by Deutscher Forschungsgemeinschaft (DFG) funding grant No. WA 3097/3-1, WA 3098/2-1 "Modeling of turbulence-interface interaction in two-fluid systems".

References

1. Brocchini, M., Peregrine, D.H.: The dynamics of strong turbulence at free surfaces. Part 1. Description. J. Fluid Mech. **449**, 225–254 (2001)
2. Brocchini, M., Peregrine, D.H.: The dynamics of strong turbulence at free surfaces. Part 2. Free-surface boundary conditions. J. Fluid Mech. **449**, 255–290 (2001)
3. Gottlieb, S., Shu, C.-W.: Total variation diminishing Runge–Kutta schemes. Math. Comput. **67**, 73–85 (1998)
4. Olsson, E., Kreiss, G.: A conservative level set method for two phase flow. J. Comput. Phys. **210**, 225–246 (2005)

5. Wacławczyk, M., Oberlack, M.: Invariant closure proposals for the interface tracking in two-phase turbulent flows. J. Phys.: Conf. Ser. **318**, 042025 (2011)
6. Wacławczyk, M., Oberlack, M.: Closure proposals for the tracking of turbulence-agitated gas-liquid interfaces in stratified flows. Int. J. Multiph. Flow **37**, 967–976 (2011)
7. Wacławczyk, M., Wacławczyk, T.: A priori study for the modelling of velocity-interface correlations in the stratified air-water flows. Int. J. Heat Fluid Flow **52**, 40–49 (2015)
8. Wacławczyk, T., Gemici Ö. C., Schäfer, M.: Novel high-resolution scheme for interface capturing in multi-phase flow. In: Proceedings of the 6th International Conference on Multiphase Flow (2007)
9. Wacławczyk, T., Wacławczyk, M., Kraheberger, S.V.: Modelling of turbulence-interface interactions in stratified two-phase flows. J. Phys.: Conf. Ser. **530**, 012050 (2014)

Multi-scale Analysis of Turbulent Rayleigh-Bénard Convection

Riccardo Togni, Andrea Cimarelli and Elisabetta De Angelis

Abstract We report the results from a direct numerical simulation of turbulent Rayleigh-Bénard convection for Rayleigh number of 10^5 and Prandtl number of 0.7. The flow topology is characterized by the presence of coherent structures, the so-called thermal plumes, consisting of localized portions of fluid having a temperature contrast with the background. Two distinct events are identified close to the walls by using the wall-parallel divergence div_π of the velocity field: the impingement ($div_\pi > 0$) and the ejection of thermal plumes ($div_\pi < 0$). The impingement leads to the formation of larger velocity and temperature structures in the wall-parallel planes. Contrary to the classical picture of turbulence consisting of a direct transfer of energy from large toward smaller turbulent fluctuations, the impingement is conjectured to be probably responsible for a reverse transfer from small towards large scales in the near-wall region.

1 Introduction and Methodology

Thermally driven turbulence plays a major role in several natural and industrial processes, which range from atmospheric convection to cooling systems in nuclear power plant [1]. An idealized system for approaching all these cases is the Rayleigh-Bénard convection (RBC), which consists of a fluid layer heated from below and cooled from above in a vertically bounded domain. It is well known that the transport of thermal and kinetic energy across the fluid layer in turbulent RBC is carried out mainly by the so-called thermal plumes [3], which are commonly defined as localized portions of fluid having a temperature difference with the background [2]. Hot and cold plumes detach respectively from the bottom and the upper plate, moving toward the opposite wall driven by buoyancy forces. As nicely sketched by Kadanoff [4], thermal plumes carry out a self-sustained life cycle during which their morphology changes. In particular, the characteristic diameter of each structure enlarges as the distance from the impinging wall decreases due to the vertical constraint. The latter

R. Togni · A. Cimarelli · E. De Angelis (✉)
DIN, Università di Bologna, Via Fontanelle 40, 47121 Forlì, Italy
e-mail: e.deangelis@unibo.it

J. Peinke et al. (eds.), *Progress in Turbulence VI*,
Springer Proceedings in Physics 165, DOI 10.1007/978-3-319-29130-7_51

phenomenon, hereafter called impingement, is supposed to be related with a strong exchange of thermal and kinetic energy between turbulent scales, in particular, a transfer from small towards large scales. This conjecture can be verified by analyzing the budgets for the second-order structure functions, which are capable to capture both the local and the non-local phenomena of turbulence, i.e. the processes occurring simultaneously in physical space and in the space of turbulent scales, see Marati et al. [5] for a multi-scale analysis of a turbulent channel flow. An extensive analysis of the multi-scale energy budgets lies outside the purpose of the present work. Here, we want to investigate the life-cycle of thermal plumes in turbulent RBC in order to identify some fundamental events which could impress a clear footprint on the energy budgets. To that end, a data set coming from a direct numerical simulation (DNS) is employed. The Boussinesq equations are solved using a pseudospectral method which discretizes space with Fourier modes in the wall-parallel directions x and y, and with Chebychev polynomials in the wall-normal direction z. Time advancement is carried out with a fourth-order Runge–Kutta scheme for the nonlinear terms and a second-order accurate Crank-Nicholson scheme for the linear ones. The computational domain is a rectangular box of size $8 \times 8 \times 1$ along x, y, z respectively, where the Cartesian coordinate system is cell-centered, with the z-axis pointing in the direction opposite to that of gravity acceleration. The temperature of the lower and the upper walls are fixed at 0.5 and -0.5, respectively; no-slip boundary conditions are used on both horizontal plates, whereas periodic boundary conditions are imposed at the lateral sidewalls. The DNS is performed at $Pr = 0.7$ and $Ra = 1.7 \cdot 10^5$ using 128^2(horizontal) $\times$ 129 (vertical) modes and polynomials.

2 Analysis of the Flow Topology

According to the definition made in Sect. 1, thermal plumes are here displayed by isosurfaces of temperature. Further ways to extract these structures from the turbulent background consider thresholds of vertical velocity, thermal dissipation or vertical vorticity-temperature correlation. With respect to the latter, as an example, Zhou et al. [6] showed that the plumes are associated with a strong vertical vorticity component, hence the vorticity-temperature correlation is a suitable quantity for the identification. As can be seen in Fig. 1a, hot and cold plumes detach respectively from the lower and the upper wall, stretch across the domain and, finally, enlarge close to the opposite plate. Furthermore these structures have a sheet-like form at the beginning, whereas they take the appearance of a mushroom sufficiently away from the starting point. It seems quite evident from Fig. 1b that the sheet-like roots of thermal plumes create a fine network across the plates and that portions of fluid are emitted from the intersection spots. By considering the two-dimensional divergence of the velocity field in the xy-plane, it is possible to measure both the horizontal enlargement and narrowing of thermal plumes. It seems reasonable to apply this kind of analysis to the near-wall region, where both mushroom-like caps and sheet-like roots coexist, rather than the bulk, where the stalks have almost a constant section.

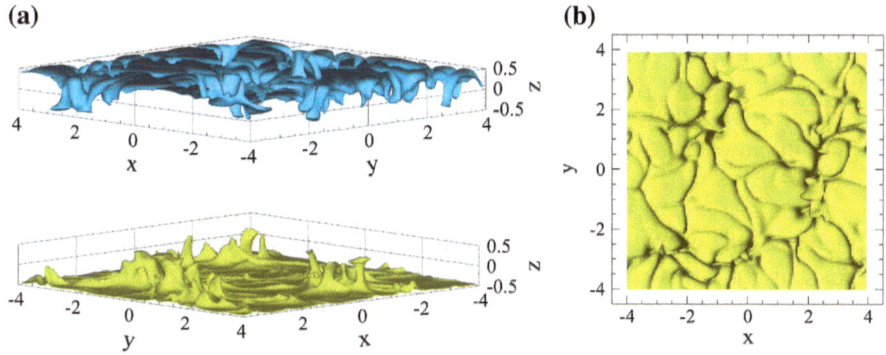

Fig. 1 **a** Isosurfaces of temperature ($\theta = 0.25$ *down* and $\theta = -0.25$ *up*) and **b** *top* view of the isosurface at $\theta = 0.25$

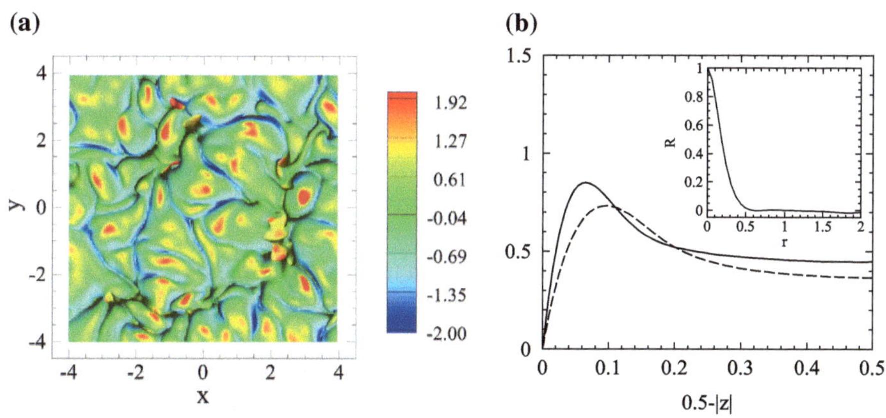

Fig. 2 **a** Top view of the isosurface at $\theta = 0.25$ coloured by div_π. **b** Main plot: $\langle div_\pi \rangle_+$ (*solid line*) and $-\langle div_\pi \rangle_-$ (*dashed line*) versus $z^* = 0.5 - |z|$. Inset: R versus r, evaluated at $z^* = 0.07$

Figure 2a shows an hot isosurface of temperature ($\theta = 0.25$) coloured with the horizontal divergence of the velocity field, $div_\pi = \partial u/\partial x + \partial v/\partial y$. Two distinct events can be identified in terms of the horizontal divergence: the emission ($div_\pi < 0$) and the impingement of thermal plumes ($div_\pi > 0$). As can be seen, concurrent regions of positive divergence are separated by thin filaments having a negative divergence, hence it is reasonable that sheet-like roots protrude from the thermal boundary layer mainly as a consequence of the mechanical action of impinging plumes. Furthermore, the intersection of different sheet-like roots leads to a large concentration of momentum, which is in turn responsible, together with the buoyancy forces, for the ejection of new structures.

Clearly, the impingement and the ejection are fundamental events in the self-sustained life cycle of thermal plumes. In order to locate the maximum of both phenomena, the contributions of impingement are separated from the ones of ejection. In the main plot of Fig. 2b the quantities $\langle div_\pi \rangle_+$ and $-\langle div_\pi \rangle_-$, where $\langle div_\pi \rangle_+ = \langle div_\pi \rangle$ for $div_\pi > 0$ and $\langle div_\pi \rangle_- = \langle div_\pi \rangle$ for $div_\pi < 0$, are displayed. These conditional statistics allow us to display the magnitude of both the impingement ($\langle div_\pi \rangle_+$) and the ejection ($\langle div_\pi \rangle_-$) as a function of the distance from the wall. As apparent, the impingement, $\langle div_\pi \rangle_+$, reaches its maximum at a distance from the wall $z^* = 0.5 - |z| = 0.07$, while the ejection, $-\langle div_\pi \rangle_-$, is peaked slightly further away from the wall, at $z^* = 0.10$, in accordance to the instantaneous field shown in Fig. 1a. Furthermore, the characteristic diameter of the impinging plume can be estimated by analyzing the two-points autocorrelation function in the wall-parallel plane at $z^* = 0.07$, where the impingement dominates the ejection. The two-points autocorrelation function is $R = \langle div_\pi(\boldsymbol{\pi}, z) div_\pi(\boldsymbol{\pi} + \mathbf{r}, z)\rangle \,/\, \langle div_\pi^2(\boldsymbol{\pi}, z)\rangle$, where $\boldsymbol{\pi} = (x, y)$ and $\mathbf{r} = (r_x, r_y)$ are respectively the position and the separation vector in the wall-parallel planes. Taking into account the statistical homogeneity and isotropy in the $x - y$ planes, the dependence of R on both the position vector $\boldsymbol{\pi}$ and the direction of $\mathbf{r}$ vanishes, hence $R = R(r, z)$, where $r = |\mathbf{r}|$. The inset of Fig. 2b displays R as a function of r evaluated at $z^* = 0.07$. As can be seen, R drops to zero at $r = 0.5$, hence this separation can be considered as the characteristic diameter of the impinging plumes.

3 Final Remarks

In conclusion, the rich dynamics observed in the near-wall region can be separated into impingement and emission events by using the planar divergence of the velocity field. From this topological analysis emerges that these events are related to each other and control the life cycle of turbulence. Since the impingement is stronger than the emission, as can be seen in Fig. 2b, a reverse transfer of thermal and kinetic energy from small towards large fluctuations is conjectured to take place in the near-wall region. However, given the multi-scale nature of these phenomena, a scale-by-scale budget is needed (and planned) in order to quantify and possibly verify these conjectures on the cascade processes occurring near the wall of RBC.

References

1. Ahlers, G.: Trend: turbulent convection. Physics **2**, 74 (2009)
2. Chillà, F., Schumacher, J.: New perspectives in turbulent Rayleigh-Bénard convection. Eur. Phys. J. E: Soft Matter Biol. Phys. **35**(7), 1–25 (2012)
3. Grossmann, S., Lohse, D.: Fluctuations in turbulent Rayleigh-Bénard convection: the role of plumes. Phys. Fluids **16**(12), 4462–4472 (2004). (1994-present)
4. Kadanoff, L.P.: Turbulent heat flow: structures and scaling. Phys. Today **54**(8), 34–39 (2001)
5. Marati, N., Casciola, C.M., Piva, R.: Energy cascade and spatial fluxes in wall turbulence. J. Fluid Mech. **521**, 191–215 (2004)
6. Zhou, Q., Sun, C., Xia, K.Q.: Morphological evolution of thermal plumes in turbulent turbulent Rayleigh-Bènard convection. Phys. Rev. Lett. **98**(7), 074501 (2007)

Numerical Analysis of Frazil Ice Formation in Turbulent Convection

A. Abbà, P. Olla and L. Valdettaro

Abstract We study the first steps of ice formation in fresh water turbulent convection (frazil ice regime). We explore the sensitivity of the ice formation process to the set of non-dimensional parameters governing the system. We model the mixture of ice crystals and water as a two-phase medium composed of water and ice particles of fixed diameter. We use the Boussinesq approximation and we integrate numerically the set of equations making use of a numerical code based on second order finite difference. A dynamic LES model for the subgrid scales is used.

1 Physical and Numerical Model

The first stage of ice formation at the supercooled free surface of oceans, rivers and lakes is called frazil ice, a suspension of small randomly-oriented crystals [1, 2]. We model the convective motion in the upper ocean mixed-layer as convection between two parallel plates, with the upper boundary corresponding to the air-ocean interface and the lower one corresponding to the mixed layer bottom. Assuming constant salinity, the system of water and frazil ice particles can be modelled as a two-component mixture, which is treated as a locally homogeneous fluid with averaged properties. In the dilute limit that we are considering, the effective viscosity of the ice-water mixture differs from that of pure fluid by a value of the same order of the concentration itself [3] and this viscosity increment can be neglected. The ice-water mixture density ρ is given by:

A. Abbà (✉)
Dip. di Scienze e Tecnologie Aerospaziali, Politecnico di Milano, Milan, Italy
e-mail: antonella.abba@polimi.it

P. Olla
ISAC-CNR and INFN, Sez. Cagliari, Monserrato, Italy
e-mail: olla@dsf.unica.it

L. Valdettaro
MOX, Dip. di Matematica, Politecnico di Milano, Milan, Italy
e-mail: lorenzo.valdettaro@polimi.it

J. Peinke et al. (eds.), *Progress in Turbulence VI*,
Springer Proceedings in Physics 165, DOI 10.1007/978-3-319-29130-7_52

$$\rho = \left(\frac{C_i}{\rho_i} + \frac{1 - C_i}{\rho_w}\right)^{-1} \tag{1}$$

in terms of ice mass concentration C_i, ice density ρ_i, assumed constant, and water density ρ_w related to the temperature by:

$$\rho_w = \rho_{w0}[1 - \alpha(T - T_0)], \tag{2}$$

ρ_{w0} is the density corresponding to the lower boundary temperature T_0 and α is the thermal expansion coefficient.

We use the Boussinesq approximation: we introduce a hydrostatic equilibrium without ice and we write the equations for the perturbations to this state. Assuming small ice concentration we can neglect density variations everywhere except in the buoyancy term. The resulting equations in nondimensional form are:

$$\frac{\partial \mathbf{u}'}{\partial t} + (\mathbf{u}' \cdot \nabla)\mathbf{u}' = -\nabla p' + \frac{Ra}{Pr}\left[\frac{1 - \rho_i}{\rho_i} C_i' + T'\right] e_z + \nabla^2 \mathbf{u}', \quad \nabla \cdot \mathbf{u}' = 0, \tag{3}$$

$$\frac{\partial T'}{\partial t} + \mathbf{u}' \cdot \nabla T' = -w' + \frac{1}{Pr}\nabla^2 T' + \beta C_i' f(z', T'), \tag{4}$$

$$\frac{\partial C_i'}{\partial t} + \mathbf{u}' \cdot \nabla C_i' = \frac{1}{Sc}\nabla^2 C_i' + w_r \frac{\partial C_i'}{\partial z} + \gamma C_i' f(z', T'). \tag{5}$$

where:

$$z' = \frac{z}{L}, \quad \mathbf{u}' = \frac{\mathbf{u}L}{\nu}, \quad T' = \frac{T - T_0}{\Delta T} - z', \quad C_i' = C_i \alpha \Delta T,$$

and z is the vertical coordinate, L is the distance between horizontal plates, ΔT is the temperature difference maintained between them. *Ra*, *Pr* and *Sc* are respectively Rayleigh, Prandtl and Schmidt numbers. Moreover

$$f(z', T') = \frac{T_i - T}{\Delta T} = \frac{T_i - T_0}{\Delta T} - z' - T'$$

Here T_i is the water freezing temperature. These equations depend on the nondimensional parameters w_r (the rising velocity of ice in water at rest), β and γ.

2 Numerical Results

Equations (3)–(5) were solved numerically using a fractional step approach [4] with a third order, three steps, explicit Runge-Kutta scheme. For the spatial discretization, we have employed a second order accurate finite difference scheme on a staggered grid. The effects of the small unresolved scales on the resolved ones are modelled by using the Subgrid Scale Model proposed by [5]. The turbulent viscosity, Prandtl and Schmidt number coefficients are computed using the dynamic procedure [6].

The domain has an extension of 2π in the horizontal directions and height equal to one. Results for Rayleigh number 10^6 and a resolution with $51 \times 51 \times 40$ collocation points are presented. The grid is stretched in the vertical direction in order to resolve the boundary layers. As usual in LES the resolution should capture the most energetic structures. For safety we have used a resolution larger than the one used for pure Rayleigh Bénard convection [7]. We use periodic boundary conditions in the lateral directions. On the bottom and top surfaces we fix the temperature and impose free-slip conditions for the velocity. In this series of simulations, we want to focus on the ice produced in the regions far from the upper layer. To this end we have used the boundary conditions $C_i' = 0$ at top and bottom. We assume the freezing temperature T_i to be the temperature in the middle of the hydrostatic layer (that is $\frac{T_i - T_0}{\Delta T} = \frac{1}{2}$). Due to the uncertainty in the determination of the ice particle rise velocity w_r and the ice concentration source term coefficient γ, we have conducted a sensibility analysis varying these two parameters. The value of the parameter β is not important for the linear phase, as it multiplies a term which becomes relevant only in the nonlinear stage. Typical curves of total ice concentration versus time are shown in Figs. 1 and 2. In the first figure we clearly distinguish a short transient followed by an exponential growth, ended by the nonlinear saturation. The production of ice is fast compared to the large eddies turnover time. As a consequence, during the linear phase, the velocity and temperature fields are almost constant and we get an almost pure exponential growth of ice concentration. In the second figure, instead, the exponential growth phase does not appear very clearly. This is because the time scale for the ice formation is large compared to the convection turnover time: the flow changes significantly during ice formation. We observe that the maximum rate of ice production is obtained when the rise velocity is of the same order of magnitude of the characteristic velocity of the thermal downwelling plumes (see Fig. 3). This can be easily understood: if the rise velocity is much larger than the velocity of

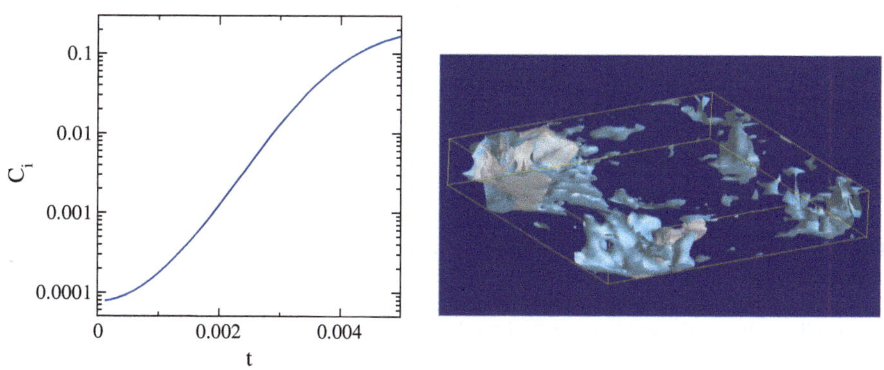

Fig. 1 Ice formation shorter than large convection plumes time scales: exponential growth, ice (*light blue*) is localized in cold (*grey*) downwelling plumes. $Ra = 10^6$, $\gamma = 10^4$, $w_r = 150$

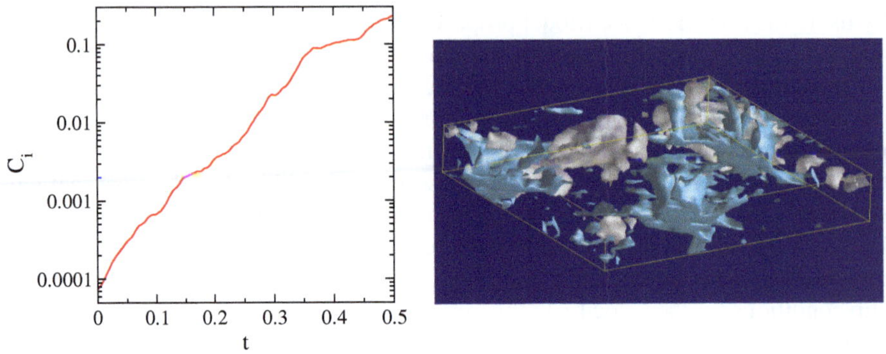

Fig. 2 Ice formation longer than large convection plumes time scales: irregular growth, ice distributed everywhere. $Ra = 10^6$, $\gamma = 10^3$, $w_r = 10$

downwelling plumes, then ice forms almost exclusively at the surface. If on the other hand the rise velocity is too small, then ice is entrained towards the bottom by the downwelling plumes and it melts due to the ambient large temperature. If w in the plumes and w_r are comparable, then ice can develop inside the whole extent of the cold downwelling plume.

We have devised a simple model that is able to capture the trend of the growth rate as a function of the relevant parameters in the fast regime: we consider the equation for ice concentration (5) inside the intense vertical plumes; we assume that velocity and temperature profiles are given and constant and we neglect horizontal velocities and horizontal derivatives. Eq. (5) reads

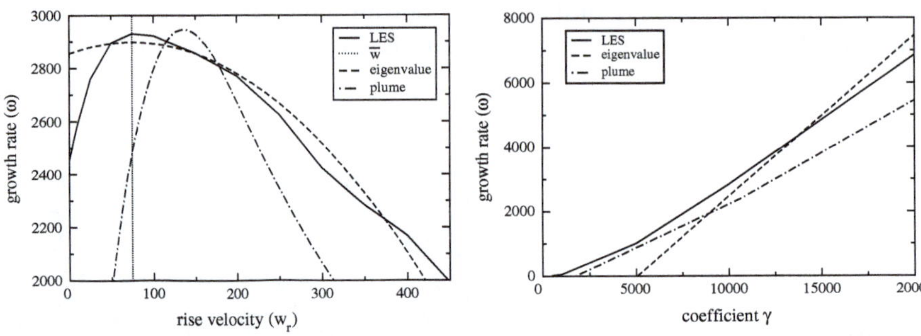

Fig. 3 *Left* growth rate of ice concentration versus w_r for $\gamma = 10^4$; *right* growth rate of ice concentration versus γ, $w_r = 150$. $Ra = 10^6$ in both cases. *Full line* growth rate from LES simulations. *Broken line* simplified 1-d model. *Dash-dotted line* simplified 1-d model applied to the most intense downwelling plume. Vertical line marks the velocity of the most intense downwelling plume. Definition of growth rate ω is $\omega = \log(< C'(t_2) > / < C'(t_1) >)/(t_2 - t_1)$ where $< C'(t_1) >$ and $< C'(t_2) >$ are total ice concentration at the beginning (t_1) and end (t_2) of the linear phase for ice concentration

$$\frac{\partial C_i}{\partial t} = -W\frac{\partial C_i}{\partial z} + \frac{1}{S}\frac{\partial^2 C_i}{\partial z^2} + \Gamma C_i, \tag{6}$$

where:

$$\frac{1}{S} = \frac{1}{Sc} + \frac{1}{Sc_{\text{turb}}}, \quad W = w + w_r, \quad \Gamma = \gamma(T_i - T).$$

This equation is decomposed into normal modes in time, where the eigenvalue with largest real part is the growth rate. The eigenvalues to this linear differential equation (supplemented by the boundary conditions at $z = \pm 1/2$) provide the growth rate. We plot it in Fig. 3. Broken line corresponds to the over simplified case in which $\mathbf{u}'$ and f in (5) are constant. In such case the eigenvalue Eq. 6 can be solved analytically and we get:

$$\omega = \Gamma - \frac{W^2 S}{4} - \frac{n^2\pi^2}{S} \qquad |n| \in \mathbb{N} \neq 0$$

The resulting growth rate is quadratic in w_r and linear in γ. A refined version (dash-dotted line) takes into account the actual vertical profiles for w' and T' by averaging the values obtained by the LES simulation inside the most intense downwelling plumes. Theoretical models of turbulent convection could provide appropriate profiles for temperature and vertical velocity distributions inside most intense plumes. Taking advantage of these models there would be no need to determine the profiles from the LES simulation. Such approach will be considered in a future work.

Acknowledgments This activity is in the frame of the PNRA project "PANACEA The role of frazil and PANcake ice in the mass and energy budgets of the AntarctiC sEA ice cover".

References

1. Kivisild, H.: Int. Assoc. Hydraul. Res. paper 1.0, 14 pp (1970)
2. Martin, S.: Ann. Rev. Fluid Mech. **13**, 379 (1981)
3. de Carolis, G., Olla, P., Pignagnoli, L.: J. Fluid Mech. **86**, 369–381 (2005)
4. Le, H., Moin, P.: J. Comp. Phys. **92**, 369 (1991)
5. Wong, V., Lilly, D.: Phys. Fluid. **6**(2), 1016 (1994)
6. Germano, M., Piomelli, U., Moin, P., Cabot, H.: Phys. Fluid A **3**(7), 1760 (1991)
7. Abbà, A., Cercignani, C., Valdettaro, L.: In: Proceedings from the Fourth ECCOMAS Computational Fluid Dynamics Conference. Greece, Athens (1998)

Large Eddy Simulation of Turbulent Flows: Benchmarking on a Rectangular Prism

L. Patruno, M. Ricci, A. Cimarelli, S. de Miranda, A. Talamelli and F. Ubertini

Abstract Preliminary results of a Large Eddy Simulation (LES) of rectangular cylinder performed with OpenFoam are presented. This is the preliminary part of a longer research project aimed at systematically study the ability of Computational Fluid Dynamics (CFD) techniques in reproducing the flow around slender bodies with sharp edges at high Reynolds numbers. In spite of the simple geometry, the problem is influenced by a number of parameters which makes its correct solution difficult to be achieved. The LES approach presented here appears to be a good candidate for this purpose but further analysis must be performed. Indeed, we highlight the need to adopt a finer resolution in the spanwise direction in order to capture the very anisotropic turbulent dynamics. Furthermore, it emerges the need of Direct Numerical Simulation (DNS) data in order to shed light on the compound role played by the turbulence model, the grid resolution and the inlet conditions.

1 Introduction and Numerical Approach

An accurate description of the unsteady wind loads on fixed structures (like large roofs, towers and bridge decks) is essential in their design process. Unfortunately, wind tunnel tests are highly expensive, require specific apparatus and long set-up operations. Thanks to the increase in computers power, the numerical simulation has become a valid complementary activity for the assessment of wind loads on structures and they promise to be an attractive alternative for the future. Numerical simulations of flows around fixed and vibrating bluff bodies have been extensively

L. Patruno (✉) · M. Ricci · S. de Miranda · F. Ubertini
DICAM, University of Bologna, Viale Risorgimento 2, 40134 Bologna, Italy
e-mail: luca.patruno@unibo.it

A. Cimarelli · A. Talamelli
Department of Industrial Engineering, Alma Mater Studiorum,
Università di Bologna, Via Fontanelle 40, 47121 Forlì, Italy
e-mail: andrea.cimarelli2@unibo.it

A. Talamelli
e-mail: alessandro.talamelli@unibo.it

J. Peinke et al. (eds.), *Progress in Turbulence VI*,
Springer Proceedings in Physics 165, DOI 10.1007/978-3-319-29130-7_53

proposed in the literature (see e.g. [1]). However, if analyzed in detail, their accuracy is often questionable so that their use in practical applications is not as widely spread as it could be expected. The main reason is that turbulence plays a fundamental role in determining the flow dynamic around the body and consequently the wind loads and its modeling still presents overwhelming difficulties. On the one hand, Unsteady Reynolds-Averaged Navier Stokes equations (URANS) based simulations are often not accurate and provide only a very approximated representation of the flow field. On the other hand, many of the proposed studies based on LES used not adequate grid resolution for the considered Reynolds numbers, leading to strong dependence of the results on the adopted turbulence model and discretization scheme. The present research, framed in the BARC (Benchmark on the Aerodynamics of a Rectangular 5:1 Cylinder) project [2], is aimed at testing LES simulations at high Reynolds numbers ($Re = 10^4 - 10^5$) around a rectangular prism with aspect ratio 1:5.

Here we report the first simulation of the present research project aiming at reproducing the LES data obtained in [2]. This first step aims at shedding lights on the strategy to follow in order to study the compromise between turbulence model and mesh resolution in order to keep the computational costs reasonable and at the same time increase significantly the accuracy of the results. In spite of the simple geometry, the problem is of high interest for both fundamental research (see e.g. fluid-structure interactions and turbulent wakes) and applications, providing useful informations on the aerodynamic behavior of bluff bodies widely employed in civil engineering (e.g. long span bridges decks or tall buildings).

The LES simulation here reported has been performed using the open-source code OpenFoam. The subgrid model is a standard Smagorinsky eddy viscosity model. PISO (pressure implicit with splitting of operator) pressure-velocity coupling and LUST (Linear Upwind Stabilized Transport) scheme for convected quantities are employed. The Reynolds number based on the vertical dimension of the rectangle is $Re = 2 \times 10^5$. The dimensions of the rectangular cylinder are $1 \times 0.25 \times 2$ while the computational domain is $40 \times 30 \times 2$ in the streamwise, vertical and spanwise directions respectively. In spite of the simple geometry the resolution requirements leads to a number of finite volumes of the order of 13 millions. The resolution is such that close to the wall we have $\langle \Delta x \rangle = 0.002$, $\langle \Delta y \rangle = 0.0004$ and $\langle \Delta z \rangle = 0.01$ respectively in the streamwise, vertical and spanwise direction.

2 Results and Final Comments

In principle, in LES, the large scale turbulent fluctuations are directly resolved allowing a better prediction of the flow properties with respect RANS. In the case of flows around sharp-edged bodies, this quality becomes very important. Indeed, the sensitivity of the problem on the properties of turbulence makes the prediction of relevant quantities very difficult to be achieved within an acceptable range of uncertainty. One of the most important quantities to analyse for the applications in civil engineering, is the pressure distribution. Indeed, such a quantity represents the main contribu-

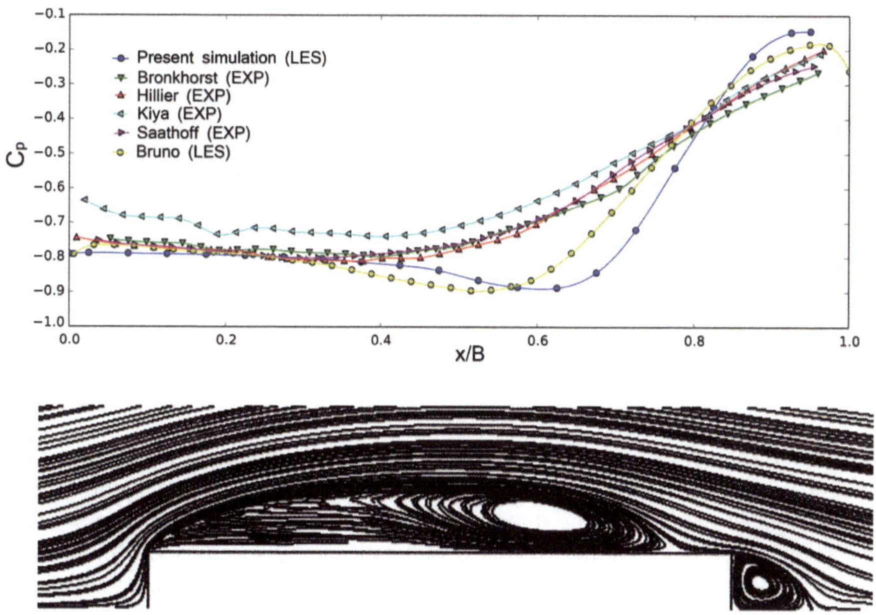

Fig. 1 *Top plot* time averaged distribution of the pressure coefficient, C_p, along the *top side* of the *rectangle*. The present LES data are compared with other simulations and experiments by various authors. *Bottom plot* streamlines of the time averaged velocity field

tion to the aerodynamic force and determines the aeroelastic properties of the body [4–6]. In the top plot of Fig. 1, the time averaged distribution of the pressure coefficient C_p is reported along the top side of the rectangle, together with experimental and LES results by different authors. Acceptable agreement is observed with the LES simulation performed in [3]. However, Fig. 1 also highlights that the numerical results significantly deviate from the experimental ones. This is due to the fact that the behavior of the pressure coefficient is mainly determined by the dimension and topology of the recirculating bubble taking place along the rectangle. As shown in the bottom plot of Fig. 1 by means of streamlines of the time averaged velocity field, the recirculating bubble starts from the separation of the flow at the sharp edge of the rectangle and ends far away downstream. This reattachment point determines the pressure recovery shown by C_p in the top plot of Fig. 1 and, hence, it is a very important flow feature that must be predicted. Since the reattachment point strongly depends on the turbulent properties of the flow, this means that the results of the numerical simulations are strongly correlated to the turbulence model, grid resolution, inlet conditions and numerical scheme used, thus explaining the differences between numerical and experimental results shown in Fig. 1 and more in general within the BARC project [2]. Let us consider here the problems related to the numerical mesh adopted. In LES, the anisotropy of the mesh and the resolution adopted should allow to compute the most energetic turbulent structures. In Fig. 2,

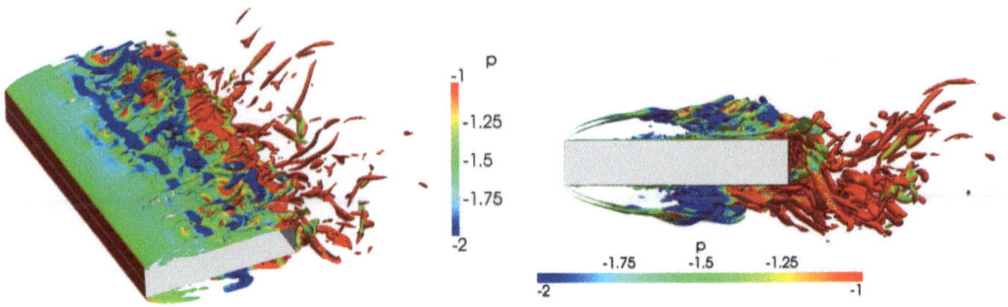

Fig. 2 Large Eddy Simulation of a rectangular cylinder at a Reynolds number, $Re = 2 \times 10^5$. *Top* and *bottom plots* show the flow topology by means of the instantaneous behaviour of isosurfaces of vorticity colored by pressure

the main turbulent structures populating the flow are shown by means of iso-surface of vorticity colored by pressure. As expected, turbulent structures are mainly aligned and elongated along the flow direction. As a consequence, the anisotropy of the mesh should be such that the spanwise and wall-normal spacings are smaller than the streamwise one. However, accordingly to the mesh adopted in [3] we use a mean streamwise grid spacing, $\langle \Delta x \rangle = 0.002$, five times smaller than the spanwise one $\Delta z = 0.01$. This choice is in contrast with the topology of the flow field shown in Fig. 2 and could explain, at least partially, the difference between numerical and experimental results shown in Fig. 1.

In conclusion, we have performed an LES simulation with OpenFoam trying to reproduce the reference LES data reported in [3]. The preliminary analysis shows an acceptable agreement with the data set reported within the BARC project but in the same time highlights a different strategy to follow in order to improve the accuracy of the data. Accordingly to the present results we plan to adopt meshes with a lower grid resolution in the streamwise direction with respect the spanwise one in order to correctly resolve the turbulent field. It also emerges the need for a DNS simulation at a lower Reynolds number in order to shed light on the compound role played by the turbulence model, the grid resolution and the inlet conditions.

References

1. Brusiani, F., de Miranda, S., Patruno, L., Ubertini, F., Vaona, P.: On the evaluation of bridge deck flutter derivatives using RANS turbulence models. J. Wind. Eng. Ind. Aerodyn. **119**, 39–47 (2013)
2. Bruno, L., Salvetti, M., Ricciardelli, F.: Benchmark on the aerodynamics of a rectangular 5:1 cylinder: an overview after the first four years of activity. J. Wind. Eng. Ind. Aerodyn. **126**, 87–106 (2014)
3. Bruno, L., Fransos, D., Coste, N., Bosco, A.: 3D flow around a rectangular cylinder: a computational study. J. Wind Eng. Ind. Aerodyn. **98**, 263–276 (2010)
4. de Miranda, S., Patruno, L., Ubertini, F., Vairo, G.: Indicial functions and flutter derivatives: a generalized approach to the motion-related wind loads. J. Fluids Struct. **42**, 466–87 (2013)

5. de Miranda, S., Patruno, L., Ubertini, F., Vairo, G.: On the identification of futter derivatives of bridge decks via RANS-based numerical models: benchmarking on rectangular prisms. Eng. Struct. **76**, 359–370 (2014)
6. Patruno L.: Accuracy of numerically evaluated flutter derivatives of bridge deck sections using RANS: effects on the flutter onset velocity. Engineering Structures, Under Review

Turbulent Flow of a Suspension of Rigid Spherical Particles in Plane Channels

Luca Brandt, Francesco Picano and Wim-Paul Breugem

Abstract Suspensions of solid particles are frequently found in applications and environmental flows. Several studies concern the rheological properties of suspensions in laminar flows, but much less is known of turbulent suspensions. The present work fills this gap providing DNS data on dense suspensions of neutrally-buoyant rigid sphere in a turbulent channel flow at the bulk Reynolds number of $Re = U_0 h/\nu = 2800$. We show that considering volume fractions $\Phi \leq 0.1$ the turbulent flow is similar to the unladen case with higher turbulence intensities. On the contrary, the flow behavior strongly changes at $\Phi = 0.2$ where the turbulence appears to be attenuated.

1 Introduction and Methodology

Fluids with a suspended solid phase frequently occur in environmental processes and in industrial applications. Sometimes laminar flow conditions may occur as in magmatic flows, but most often the flow is turbulent as in sediment transport in rivers or in industrial processes. When the suspended phase is not dilute, its effect on the fluid property is not negligible. Many studies dealt with the rheological properties of laminar suspensions from the dilute to the dense case [1]. It is well known that the

L. Brandt (✉)
SeRC (Swedish e-Science Research Centre) and Linné FLOW Centre, KTH Mechanics, 10044 Stockholm, Sweden
e-mail: luca@mech.kth.se

F. Picano
Department of Industrial Engineering, University of Padova, Via Venezia 1, 35131 Padua, Italy
e-mail: francesco.picano@unipd.it

W.-P. Breugem
Aero and Hydrodynamics Laboratory, Delft University of Technology, Leeghwaterstraat 21, 2628 Delft, The Netherlands
e-mail: w.p.breugem@tudelft.nl

J. Peinke et al. (eds.), *Progress in Turbulence VI*,
Springer Proceedings in Physics 165, DOI 10.1007/978-3-319-29130-7_54

presence of particles induces an increase of the effective viscosity of the suspension [2]. It has been observed that, at high shear rates and fixed volume fraction, the viscosity may increase also with the shear rate, so-called shcar-thickening [2, 3].

Turbulent and transitional flows are usually characterized by a shear rate that intermittently fluctuates in space and time. This feature, in combination with the peculiar rheological features of semi-dilute/dense suspensions, leads to unexplored physics in these chaotic flow regimes. As an example, the experiments in [4] show that in pipe flow and for relatively large particles, the critical Reynolds number at which transition to turbulence occurs cannot be simply rescaled considering the increase of the effective viscosity due to the presence of the solid phase. Recently, it has been shown by Lashgari et al. [5] that the flow of a suspension in the volume-fraction/Reynolds-number parameter space, $(\Phi - Re)$, can be divided into three regimes: viscous, turbulent and inertial shear-thickening. The present study aims to investigate the property of wall turbulence and related particle behavior in the dense case and at a Reynolds number at which the flow can be considered turbulent. The previous DNS study of turbulent channel flow laden with finite-size particles in [6] investigates volume fractions up to $\Phi = 0.07$. The authors show that particles interact with the near wall turbulent structures inducing an increase of the turbulent drag especially at the highest volume fraction they considered, i.e. $\phi = 0.07$. In the present work we extend the analysis to higher volume fractions, up to $\Phi = 0.2$. We find an increase of the turbulence activity up to $\Phi = 0.1$, where the flow shows high level of fluctuations, but a decrease at $\Phi = 0.2$. The attenuation of the turbulent transverse velocity fluctuations reduces the particle mixing across the channel.

Direct numerical simulations have been performed by using an algorithm that fully describes the coupling between the solid and fluid phases. The fluid phase is evolved following the incompressible Navier-Stokes equations by second order finite differences on a staggered mesh. The finite-size particles are evolved by a Lagrangian algorithm that solves the linear and angular momentum equations. The coupling between the two phases is directly achieved by using an Immersed Boundary Method. When the particle gap distance is smaller than the mesh size, lubrication correction are used to correctly reproduce the close interaction between particles. A soft-collision model with restitution coefficient of 0.97 is adopted when the particle collide. The code has been fully validated as shown in [5, 7] for more details.

Simulations of turbulent channel flow are performed in a domain of length $6h$, $2h$ and $3h$ in the (x) streamwise, (y) wall-normal and (z) spanwise directions, respectively. Here h denotes the channel half width. Periodic boundary conditions are enforced in the x and z directions, while no slip is prescribed at the wall. The bulk velocity is kept constant by adjusting the mean pressure gradient fixing the bulk Reynolds number, $Re_h = U_0 h/\nu = 2800$ (ν the fluid kinematic viscosity). The domain is discretized by a cubic mesh of $864 \times 288 \times 432$ points in the x, y and z directions. Rigid neutrally-buoyant spheres are considered; ratio between the particle radius and the channel half-width fixed at $a/h = 1/18$. Three different volume fractions, $\Phi = 0.05; 0.1; 0.2$, are performed changing the particle number up to 10000 at $\Phi = 0.2$.

2 Results and Discussion

Figure 1 reports snapshots of the particle configurations on the streamwise-wallnormal plane for different volume fractions Φ. The (a) panel reports the particle positions at the beginning of the statistical stationary state $t = 0$, (b) the flow at $t = 20\,h/U_0$ with particles color-coded by their initial wall-normal position at $t = 0$. The increase of the particle volume fraction is visible for the case $\Phi = 0.2$ that does not exhibit clear void regions in the spanwise extension. At this volume fraction, the particle configuration seems quite ordered near the wall exhibiting an evident geometrical layering. The color map used in the panel (b) of Fig. 1 show how the particle mix in 20 large-eddy turnover times h/U_0. While the particles appear similarly mixed at $\Phi = 0.05$ and $\Phi = 0.1$, the dense case of $\Phi = 0.2$ shows a much slower mixing rate. This behavior suggests a strong change in the turbulent dynamics that is at the basis of the dispersion properties of the fluid system.

To quantify how turbulence is modulated by the increase of the particle volume fraction, we present the fluid phase statistics in Fig. 2. Panel (a) shows the mean velocity profile normalized by the bulk velocity U/U_0 as a function of the wall normal distance y/h. The mean fluid velocity is quite similar to the unladen case up to $\Phi \leq 0.1$, while at $\Phi = 0.2$ the profile appears higher in the bulk flow region and smaller around $y = h/5$. It seems that this dense case exhibit a mean velocity profile between the laminar parabola and the turbulent unladen mean profile, though still more similar to turbulent case. It should be noted however that the wall shear of the mean velocity profile is similar for all the cases and actually slightly increases with the volume fraction, i.e. higher friction coefficient.

Panels (b, c and d) display the root-mean-square (rms) of the fluid velocity fluctuations in the streamwise u_{rms}, wall-normal v_{rms}, and spanwise w_{rms} directions versus

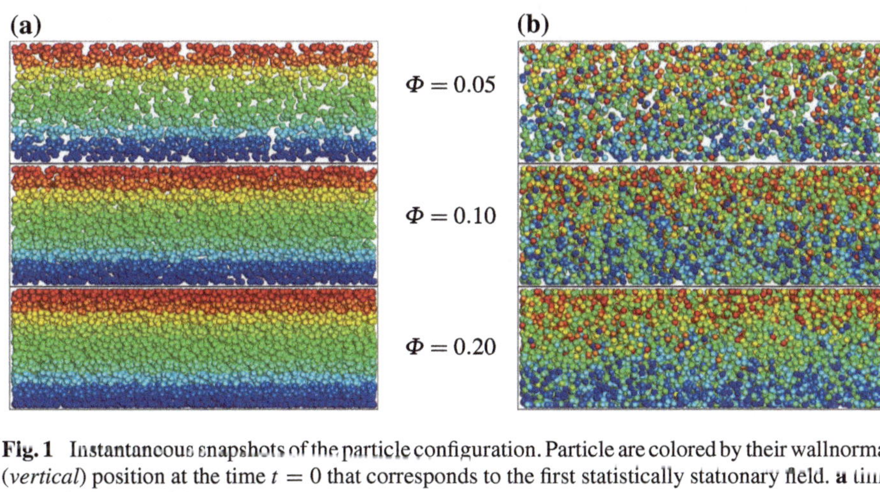

Fig. 1 Instantaneous snapshots of the particle configuration. Particle are colored by their wallnormal (*vertical*) position at the time $t = 0$ that corresponds to the first statistically stationary field. **a** time $t = 0$; **b** time $t \simeq 20\,h/U_0$

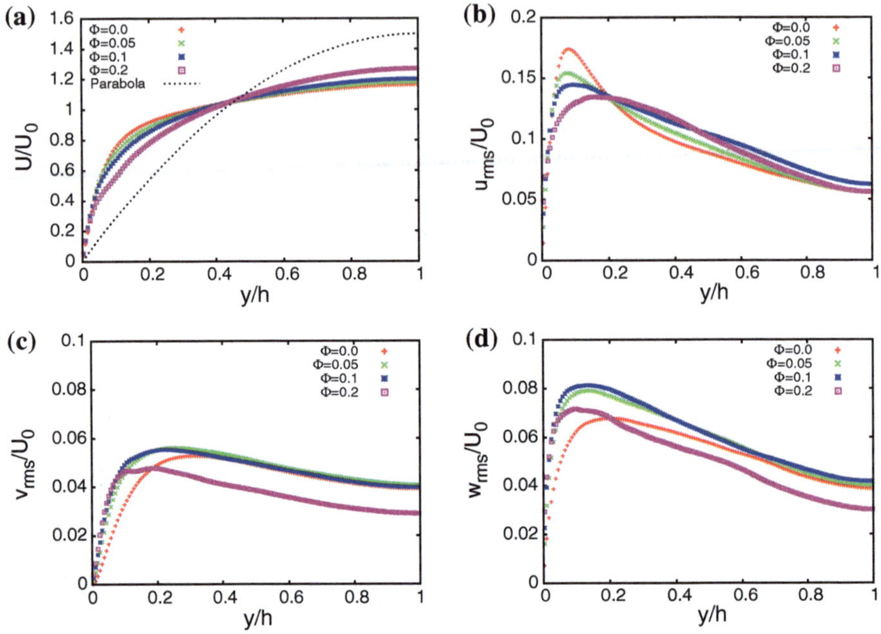

Fig. 2 Velocity single-point statistics versus channel height y/h in outer units. Panel **a** mean fluid velocity U/U_0; Root-mean-square of the fluid velocity fluctuations: panel **b** u_{rms} streamwise, panel **c** v_{rms} wall-normal, panel **d** w_{rms} spanwise component

y/h. The peak of the streamwise velocity fluctuations monotonically decreases with the volume fraction Φ, while it shows similar levels in the bulk. Conversely, the peaks of the other two components are not smaller than the unladen cases. For the dense case, $\Phi = 0.2$, we found smaller fluctuation levels ($20 \div 30\,\%$ less) for the spanwise and wall-normal components than the unladen case in the bulk of the flow. Hence we find a strong attenuation of the turbulence in the dense case, while cases with $\Phi = 0.05 \div 0.1$ show similar turbulence levels as in the unladen case. The reduced turbulent activity at $\Phi = 0.2$, especially in the wall-normal direction, explains the reduced particle mixing.

Acknowledgments This work was supported by the European Research Council Grant No. ERC-2013-CoG-616186, TRITOS and the COST Action MP1315: *Flowing matter*. The authors acknowledge computer time provided by SNIC (Swedish National Infrastructure for Computing).

References

1. Guazzelli, É., Morris, J.: A Physical Introduction to Suspension Dynamics. Cambridge University Press, Cambridge (2011)
2. Stickel, J., Powell, R.: Annu. Rev. Fluid Mech. **37**, 129 (2005)

3. Picano, F., Breugem, W.P., Mitra, D., Brandt, L.: Phys. Rev. Lett. **111**(9), 098302 (2013)
4. Matas, J.P., Morris, J.F., Guazzelli, E.: Phys. Rev. Lett. **90**(1), 014501 (2003)
5. Lashgari, I., Picano, F., Breugem, W., Brandt, L.: Phys. Rev. Lett. **113**, 254502 (2014)
6. Shao, X., Wu, T., Yu, Z.: J. Fluid Mech. **693**, 319 (2012)
7. Picano, F., Breugem, W.P., Brandt, L.: J. Fluid Mech. **764**, 463 (2015)

MIX
Papier aus verantwortungsvollen Quellen
Paper from responsible sources
FSC® C105338

If you have any concerns about our products,
you can contact us on
ProductSafety@springernature.com

In case Publisher is established outside the EU,
the EU authorized representative is:
Springer Nature Customer Service Center GmbH
Europaplatz 3, 69115 Heidelberg, Germany

Printed by Libri Plureos GmbH
in Hamburg, Germany